HIGH-PERFORMANCE MATERIALS AND ENGINEERED CHEMISTRY

Innovations in Physical Chemistry: Monograph Series

HIGH-PERFORMANCE MATERIALS AND ENGINEERED CHEMISTRY

Edited by

Francisco Torrens, PhD
Devrim Balköse, PhD
Sabu Thomas, PhD

APPLE
ACADEMIC
PRESS

Apple Academic Press Inc.
3333 Mistwell Crescent
Oakville, ON L6L 0A2
Canada

Apple Academic Press Inc.
9 Spinnaker Way
Waretown, NJ 08758
USA

© 2018 by Apple Academic Press, Inc.

First issued in paperback 2021

No claim to original U.S. Government works

ISBN-13: 978-1-77463-640-4 (pbk)
ISBN-13: 978-1-77188-598-0 (hbk)

Library of Congress Control Number: 2017953051

Trademark Notice: Registered trademark of products or corporate names are used only for explanation and identification without intent to infringe.

Library and Archives Canada Cataloguing in Publication

High-performance materials and engineered chemistry / edited by Francisco Torrens, PhD, Devrim Balköse, PhD, Sabu Thomas, PhD.

(Innovations in physical chemistry : monograph series)
Includes bibliographical references and index.
Issued in print and electronic formats.

ISBN 978-1-77188-598-0 (hardcover).--ISBN 978-1-315-18786-0 (PDF)
1. Chemical engineering. 2. Materials. I. Haghi, A. K., editor
II. Series: Innovations in physical chemistry

TP155.H54 2017	660	C2017-905898-3	C2017-905899-1

CIP data on file with US Library of Congress

Apple Academic Press also publishes its books in a variety of electronic formats. Some content that appears in print may not be available in electronic format. For information about Apple Academic Press products, visit our website at **www.appleacademicpress.com** and the CRC Press website at **www.crcpress.com**

ABOUT THE EDITORS

Francisco Torrens, PhD

Francisco Torrens, PhD, is Lecturer in physical chemistry at the Universitat de València, Spain. His scientific accomplishments include the first implementation at a Spanish university of a program for the elucidation of crystallographic structures and the construction of the first computational-chemistry program adapted to a vector-facility supercomputer. He has written many articles published in professional journals and has acted as a reviewer as well. He has handled 26 research projects, has published two books and over 350 articles, and has made numerous presentations.

Devrim Balköse, PhD

Devrim Balköse, PhD, is currently a retired faculty member in the Chemical Engineering Department at Izmir Institute of Technology, Izmir, Turkey. She graduated from the Middle East Technical University in Ankara, Turkey, with a degree in chemical engineering. She received her MS and PhD degrees from Ege University, Izmir, Turkey, in 1974 and 1977 respectively. She became Associate Professor in macromolecular chemistry in 1983 and Professor in process and reactor engineering in 1990. She worked as a research assistant, assistant professor, associate professor, and professor between 1970 and 2000 at Ege University, Turkey. She was Head of the Chemical Engineering Department at Izmir Institute of Technology, Izmir, Turkey, between 2000 and 2009. She is now a retired faculty member in the same department. Her research interests are in polymer reaction engineering, polymer foams and films, adsorbent development, and moisture sorption. Her research projects focused on nanosized zinc borate production, ZnO polymer composites, zinc borate lubricants, antistatic additives, and metal soaps.

Sabu Thomas, PhD

Sabu Thomas, PhD, is Professor of Polymer Science and Engineering at the School of Chemical Sciences and Director of the International and Inter University Centre for Nanoscience and Nanotechnology at Mahatma Gandhi University, Kottayam, Kerala, India. The research activities of Professor

Thomas include surfaces and interfaces in multiphase polymer blend and composite systems; phase separation in polymer blends; compatibilization of immiscible polymer blends; thermoplastic elastomers; phase transitions in polymers; nanostructured polymer blends; macro-, micro- and nano-composites; polymer rheology; recycling; reactive extrusion; processing–morphology–property relationships in multiphase polymer systems; double networking of elastomers; natural fibers and green composites; rubber vulcanization; interpenetrating polymer networks; diffusion and transport; and polymer scaffolds for tissue engineering. He has supervised 68 PhD theses, 40 MPhil theses, and 45 Masters thesis. He has three patents to his credit. He also received the coveted Sukumar Maithy Award for the best polymer researcher in the country for the year 2008. Very recently, Professor Thomas received the MRSI and CRSI medals for his excellent work. With over 600 publications to his credit and over 23,683 citations, with an h-index of 75, Dr. Thomas has been ranked fifth in India as one of the most productive scientists. He received his BSc degree (1980) in Chemistry from the University of Kerala, B.Tech. (1983) in Polymer Science and Rubber Technology from the Cochin University of Science and Technology, and PhD (1987) in Polymer Engineering from the Indian Institute of Technology, Kharagpur.

CONTENTS

LIST OF CONTRIBUTORS

Cristóbal N. Aguilar
Food Research Department, School of Chemistry, Universidad Autónoma de Coahuila, Saltillo, Coahuila 25280, México

Sharma Amit L
Optical Devices and Systems, CSIR-Central Scientific Instruments Organisation, Chandigarh, India

Devrim Balköse
Department of Chemical Engineering, İzmir Institute of Technology, Gulbahce, Urla İzmir, Turkey

Ondrej Bošák
Institute of Materials Science, Faculty of Materials Science and Technology in Trnava, Slovak University of Technology, Bratislava, Slovak Republic

Gloria Castellano
Departamento de Ciencias Experimentales y Matemáticas, Facultad de Veterinaria y Ciencias Experimentales, Universidad Católica de Valencia San Vicente Mártir, Guillem de Castro-94, València E-46001, Spain

Samiha F. Deriase
Egyptian Petroleum Research Institute, Nasr City Cairo 11727, Egypt

Miguel A. Esteso
U.D. Química Física, Facultad de Farmacia, Universidad de Alcalá, 28871 Alcalá de Henares, Madrid, Spain

Muhammad Faisal
Department of Physics, PES Institute of Technology, Bangalore South Campus, Bangalore 560100, Karnataka, India

Nour Sh. El-Gendy
Egyptian Petroleum Research Institute, Nasr City Cairo 11727, Egypt

Dipak K. Goswami
Department of Physics, Indian Institute of Technology Kharagpur, Kharagpur 721302, West Bengal, India

Barış Gümüş
Department of Chemical Engineering, İzmir Institute of Technology, Gulbahce, Urla İzmir, Turkey

Palle Kiran
Department of Mathematics, Rayalaseema University, Kurnool 518002, Andhra Pradesh, India

Marián Kubliha
Institute of Materials Science, Faculty of Materials Science and Technology in Trnava, Slovak University of Technology, Bratislava, Slovak Republic

Vemuri SRS Praveen Kumar
Optical Devices and Systems, CSIR-Central Scientific Instruments Organisation, Chandigarh, India

Vladinír Labaš
Institute of Materials Science, Faculty of Materials Science and Technology in Trnava, Slovak University of Technology, Bratislava, Slovak Republic

Ramón Larios-Cruz
Food Research Department, School of Chemistry, Universidad Autónoma de Coahuila, Saltillo, Coahuila 25280, México

Victor M. M. Lobo
Department of Chemistry and Coimbra Chemistry Centre, University of Coimbra, Coimbra 3004535, Portugal

Stanislav Minárik
Research Centre of Progressive Technologies, Slovak University of Technology, Bratislava, Slovak Republic

Julio C. Montañez
Food Research Department, School of Chemistry, Universidad Autónoma de Coahuila, Saltillo, Coahuila 25280, México

Kumar Mukesh
Optical Devices and Systems, CSIR- Central Scientific Instruments Organisation, Chandigarh, India

Kumari Neelam
Optical Devices and Systems, CSIR- Central Scientific Instruments Organisation, Chandigarh, India

Ahmed M. Omer
Polymer Research Department, Advanced Technologies and New Materials Research Institute (ATNMRI), City of Scientific Research and Technological Applications (SRTA-City), New Borg El-Arab, Alexandria 21934, Egypt

Sukanchan Palit
Department of Chemical Engineering, University of Petroleum and Energy Studies, Energy Acres, Post-Office-Bidholi via Premnagar, Dehradun 248007, Uttarakhand, India

Arely Prado-Barragán
Department of Biotechnology, Universidad Autónoma Metropolitana Unidad Iztapalapa, Delegación Iztapalapa, Distrito Federal 09340, México

Parinam Krishna Rao
Optical Devices and Systems, CSIR- Central Scientific Instruments Organisation, Chandigarh, India

Ana C. F. Ribeiro
Department of Chemistry and Coimbra Chemistry Centre, University of Coimbra, Coimbra 3004535, Portugal

Rosa M. Rodríguez-Jasso
Food Research Department, School of Chemistry, Universidad Autónoma de Coahuila, Saltillo, Coahuila 25280, México

Héctor A. Ruiz
Food Research Department, School of Chemistry, Universidad Autónoma de Coahuila, Saltillo, Coahuila 25280, México

Maysa M. Sabet
Laboratory of Bioorganic Chemistry of Drugs, Institute of Experimental Pharmacology and Toxicology, Bratislava 84104, Slovakia

Cecilia I. A. V. Santos
Department of Chemistry and Coimbra Chemistry Centre, University of Coimbra, Coimbra 3004535, Portugal

Diana C. Silva
Department of Chemistry and Coimbra Chemistry Centre, University of Coimbra, Coimbra 3004535, Portugal

Pedro S. P. Silva
Department of Physics, CFisUC, Universidade de Coimbra, Rua Larga, P-3004516 Coimbra, Portugal

Ladislav Šoltés
Laboratory of Bioorganic Chemistry of Drugs, Institute of Experimental Pharmacology and Toxicology, Bratislava 84104, Slovakia

Nimmakayala V. V. Subbarao
Center for Nanotechnology, Indian Institute of Technology Guwahati, Guwahati 781039, Assam, India

Parinam Sunita
Optical Devices and Systems, CSIR- Central Scientific Instruments Organisation, Chandigarh, India

Heru Susanto
Department of Information Management, College of Management, Tunghai University, Taichung, Taiwan

Tamer M. Tamer
Laboratory of Bioorganic Chemistry of Drugs, Institute of Experimental Pharmacology and Toxicology, Bratislava 84104, Slovakia

Carmen Teijeiro
U.D. Química Física, Facultad de Farmacia, Universidad de Alcalá, 28871 Alcalá de Henares, Madrid, Spain

Francisco Torrens
Institut Universitari de Ciència Molecular, Universitat de València, Edifici d'Instituts de Paterna, P. O. Box 22085, València E-46071, Spain

Merve Türk
Department of Chemical Engineering, İzmir Institute of Technology, Gulbahce, Urla İzmir, Turkey

R. R. Usmanova
Ufa State Technical University of Aviation, Ufa 450000, Bashkortostan, Russia

Fatma Ustun
Department of Chemical Engineering, İzmir Institute of Technology, Gulbahce, Urla İzmir, Turkey

Katarína Valachová
Laboratory of Bioorganic Chemistry of Drugs, Institute of Experimental Pharmacology and Toxicology, Bratislava 84104, Slovakia

Karar Vinod
Optical Devices and Systems, CSIR-Central Scientific Instruments Organisation, Chandigarh, India

G. E. Zaikov
N. M. Emanuel Institute of Biochemical Physics, Russian Academy of Sciences, Moscow 119991, Russia

LIST OF ABBREVIATIONS

AFM	atomic force microscopy
ANOVA	analysis of variance
APCVD	atmospheric pressure CVD
AR	antireflection
ASTM	American Society for Testing Materials
BCB	benzocyclobutene
CCD	central composite design
CCFCD	central composite face centered design
CCI	coherence correlation interferometer
CdS	cadmium sulfide
CDSS	clinical decision support system
COM	completion of melt onset temperature
CP	cloud point
CPCs	conducting polymer composites
CVD	chemical vapor deposition
dc	direct current
DOD	D-optimal design
DOS	density of state
DPA	differential pulse voltammetry
DST	Department of Science and Technology
DWNT	double wall
EHR	electronic health record
EMR	electronic medical record
ER	electrorheological
FFAs	free fatty acids
FP	flash point
FrFD	fractional factorial design
FTIR	Fourier transform infrared spectroscopy
HIM	health information management
HIT	health information technology
HMDE	hanging mercury drop electrode
HR	high-reflection
HUD	head up display
HWOT	half wave optical thickness

ICOP	International Conference on Optics and Photonics
EMC	electromagnetic compatibility
EMI	electromagnetic interference
ICPs	intrinsically conducting polymers
IP	induction period
JCB	Jatropha curcas-biodiesel
LPCVD	low pressure CVD
MBE	molecular beam epitaxy
MBR	membrane bioreactor
MF	microfiltration
MIM	metal–insulator–metal
MnO_2	manganese dioxide
MNu	mean value of Nusselt number
MOCVD	metal organic CVD
MOLB	modulation of only the lower boundary
NF	nanofiltration
OA	orthogonal arrays
OFETs	organic field-effect transistors
OLEDs	organic light emitting diodes
OSCs	organic solar cells
PAni	polyaniline
PECVD	plasma enhanced CVD
PM	particulate matter
PMMA	hydrophobic poly-(methylmethacrylate)
PP	pour point
PS	polystyrene
PVA	poly(vinylalcohol)
PVD	physical vapor deposition
PVP	poly(vinyl phenol)
RCCD	rotatable central composite design
RFID	radio frequency identification
RMSE	root mean square error
RO	reverse osmosis
SBR	styrene-butadiene rubber
SEM	scanning electron microscopy
SEM	scanning electron microscope
SMP	soluble microbial products
SMUs	source measure units
SS	subthreshold swing
SSE	sum of squares error

SSF	solid-state fermentation
SST	treatment sum of squares
SWNT	single wall
TAN	total acid number
TEM	transmission electron microscopy
UF	ultrafiltration
UHVCVD	ultra high vacuum
US-DOE	US-Department of Energy
VNA	vector network analyzer
WBOS	Wei biogenic oxidative system
WCO	waste cooking oils
WFSFO	waste frying sunflower oil
XRD	X-ray diffraction analysis
XRF	X-ray fluorescence

PREFACE

An integrated approach to materials and chemistry is necessitated by the complexity of the engineering problems that need to be addressed, coupled with the interdisciplinary approach that needs to be adopted to solve them.

This is an advanced book for all those looking for a deeper insight into the theories, concepts and applications of modern materials and engineered chemistry. The book comprises innovative applications and research in varied spheres of materials and chemical engineering. The purpose of this book is to explain and discuss the new theories and case studies concerning materials and chemical engineering and to help readers increase their understanding and knowledge of the discipline.

This book covers many important aspects of applied research and evaluation methods in chemical engineering and materials science that are important in chemical technology and in the design of chemical and polymeric products.

The integrated approach used in this fosters an appreciation of the interrelations and interdependencies among the various aspects of the research endeavors. The goal is to help readers become proficient in these aspects of research and their interrelationships, and to use that information in a more integrated manner.

This book:

- highlights some important areas of current interest in polymer products and chemical processes
- focuses on topics with more advanced methods
- emphasizes precise mathematical development and actual experimental details
- analyzes theories to formulate and prove the physicochemical principles
- provides an up-to-date and thorough exposition of the present state of the art of complex materials.

In this book you can also find an integrated, problem-independent method for multi-response process optimization. In contrast to traditional approaches, the idea of this method is to provide a unique model for the optimization of engineering processes, without imposition of assumptions relating to the type of process.

PART I
High-Performance Materials

CHAPTER 1

OPTICAL THIN FILM FILTERS: DESIGN, FABRICATION, AND CHARACTERIZATION

PARINAM SUNITA, VEMURI SRS PRAVEEN KUMAR, KUMAR MUKESH, KUMARI NEELAM, PARINAM KRISHNA RAO, KARAR VINOD, and SHARMA AMIT L*

Optical Devices and Systems, CSIR-Central Scientific Instruments Organisation, Chandigarh, India

Corresponding author. E-mail: amitsharma_csio@yahoo.co.in

CONTENTS

ABSTRACT

This chapter outlines the introduction and importance of optical filters in precision optics applications and its development process. First section gives a brief overview of the thin film optical filters, materials used, their types, and applications. Second section discusses the development process of the optical filters. Third section shows some examples design, fabrication, and characterization aspects of optical filter.

1.1 INTRODUCTION

Thin film optical filters are one of the important components which are used in precision optics applications like optics, avionics, sensors, fiber optics, and space applications.[1–3] It consists of one or more thin layers of material deposited on an optical component such as a lens or mirror, which alters the way in which the optic component reflects and transmits light due to light wave interference (see Fig. 1.1) and the differences in refractive indices of layers and substrate.[4] The thickness of the layers of coating material must be in the order of the desired wavelength. Metals, metal oxides, dielectrics, or composites having desired optical constants in the wavelength range of interest are used for optical coating. Commonly used metals are gold (Au), silver (Ag), aluminum (Al), titanium (Ti), chromium (Cr), copper (Cu), iron (Fe), platinum (Pt), etc., which can be deposited as either single layer or multi-layers. Dielectric materials are magnesium fluoride (MgF_2), titanium dioxide (TiO_2), aluminum oxide (Al_2O_3), silicon dioxide (SiO_2), hafnium dioxide (HfO_2), lithium fluoride (LiF), zinc sulfide (ZnS), indium tin oxide, etc., which are commonly used for coating.

These materials can be deposited on a variety of substrate materials such as optical grade glass (BK7, SF6, and SK2 used in precision optics), plastics, metals, semiconductors, and ceramics. Various theoretical and experimental investigations have been carried out on properties of metal oxide thin films.[5–9] TiO_2 is one of the semi-conductor metal oxides, which has gained a lot of attention in past few years due to its excellent physical, chemical, and optical properties.[10–14] It has a high band energy and exhibits high transparency in the visible region. TiO_2 has tunable refractive index, low thermal stress, and high stability, which make it suitable for various applications like sensors, optical coatings, self-cleaning, anti-fogging,[15] anti-reflective coatings,[16] band pass, and band stop filters, where it is used in combination with other dielectric materials.[17,18]

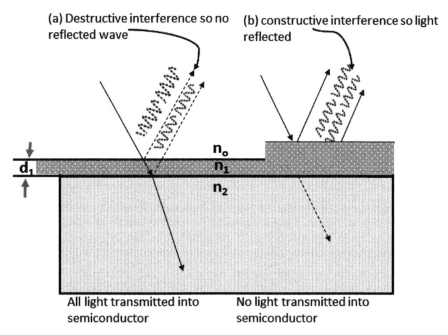

(a) Destructive interference so no reflected wave

(b) constructive interference so light reflected

n_o

d_1 n_1

n_2

All light transmitted into semiconductor

No light transmitted into semiconductor

FIGURE 1.1   Principle of interference phenomenon.

1.1.1 TYPES OF OPTICAL COATINGS

Antireflection (AR) coating is a special type of optical coating, which is applied to the surface of lenses to reduce unwanted reflections from surfaces. It is commonly used on spectacle and photographic lenses. This coating improves the efficiency of the system as it reduces the loss of light. Whenever, a ray of light moves from one medium to another (from air to glass), some portion of the light is reflected from the surface (known as the interface) between the two media[19–21] (Fig. 1.1). A number of different effects are used to reduce the reflection. The simplest is to use a thin layer of material at the interface, with an index of refraction between those of the two media. The reflection is minimized when:[22]

$$n_1 = \sqrt{n_0 n_s} \qquad\qquad (1.1)$$

where n_1 is the index of the thin layer, and n_0 and n_s are the indices of the two media.

Another type of coating is high-reflection (HR) coating, which work the opposite way to AR coatings. These coatings are used to produce mirrors of high reflectance value. They reflect more than 99.99% of the light falling on their surface. The general idea is usually based on the periodic layer system composed from two materials, one with a high index, such as zinc sulfide ($n = 2.32$) or titanium dioxide ($n = 2.4$) and low index material, such as magnesium fluoride ($n = 1.38$) or silicon dioxide ($n = 1.49$). This periodic system significantly enhances the reflectivity of the surface in the certain wavelength range called band-stop, whose width is determined by the ratio of the two used indices only (for quarter-wave system). More complex optical coatings exhibit high reflection over some range of wavelengths, and anti-reflection over another range, allowing the production of dichroic thin-film optical filters. Further, the narrow band pass optical filters have been extensively utilized for fiber optic communication technologies. Various other examples of optical coatings[23-32] (see Fig. 1.2) are metallic and front surface mirrors with protective coatings, hot and cold mirrors, dichroic mirrors and beam splitter coatings, band pass, short pass, long pass, wide band pass, notch filters, neutral density filters, transparent conducting films, antiglare films for night driving, coatings for beam combiners, and head up display (HUD) system.

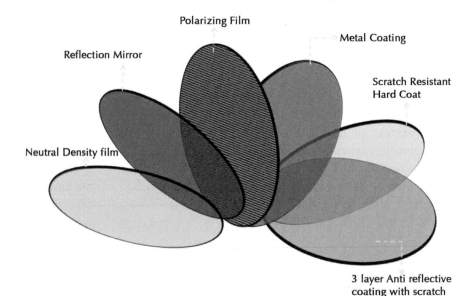

FIGURE 1.2   Different types of optical coatings.

Optical filters have applications in other areas like medical application (UV absorbing coatings), electrical applications (shielding and display coatings), night vision (IR coatings and night driving filter), coatings for wind screens, humidity sensor, analytic instrumentation, laser applications, ophthalmic applications, lens coatings, military applications, surveillance and targeting systems, binocular coatings, and astronomy (space telescopes).

1.2 DEVELOPMENT PROCESS

Optical filter development process mainly consists of three steps—design, fabrication, and characterization (see Fig. 1.3). Each step plays a pivotal role in the development of the optical filters.

1.2.1 DESIGN

The design of optical thin film coatings has always remained as a key research area in the field of optical filters as it has outpaced the manufacturing capabilities of most of the advanced deposition processes with the incorporation of graded index, graded thickness, and apodization concepts. Advances in computation techniques have resulted in better synthesis and optimization methods to match the required filter performance. An efficient design of an optical thin film filter largely depends on the choice of suitable material combinations, a good starting design coupled with a powerful optimization algorithm. Starting design includes the selection of various design parameters like substrate and coating materials, thickness, refractive index, number of layers, angle of incidence, etc. Algorithms like genetic algorithm,[33,34] particle swarm optimization,[35] etc. are used for the design and optimization of the filters.

Design procedure includes:

1. Filter specifications are finalized based on the application requirements.
2. Materials are selected depending upon their dispersion behavior.
3. Targets are set and optimization algorithm is chosen.
4. An initial design with suitable number of layers and thickness values is specified.

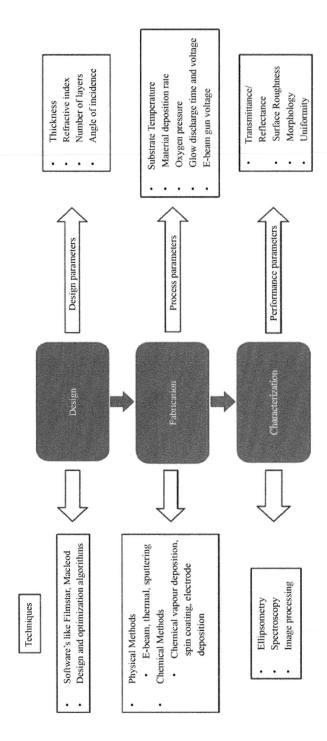

FIGURE 1.3   Optical filter development process.

1.2.2 DEPOSITION PROCESS

A thin film is deposited onto a substrate or onto previously deposited layers. Deposition process mainly involves following steps:

a. coating plant,
b. substrate cleaning and preparation,
c. material preparation,
4. coating process.

1.2.2.1 COATING PLANT

Chamber is cleaned before the coating process is started. Calibration test runs are often required to determine tooling factors for particular machines and materials. The geometrical distance of the substrate from the source, chamber vacuum, and temperature as well as evaporation rate may all affect the final coating. Conditions during the process must be monitored and reproduced as closely as possible from run to run. All the jigs and fixtures that are required to hold the substrates, especially unusual shapes and sizes, can be produced as per the requirement.

1.2.2.2 SUBSTRATE CLEANING AND PREPARATION

Cleanliness and preparation of substrates is essential for success in thin film work. Condensation rate and adhesion of the deposit are critically dependent of conditions on the surface. Even a thin layer of grease can have such a gross effect on a molecular scale as to alter completely the characteristics of the layer. Substrates and optical components are cleaned in an ultrasonic bath with an alkali based cleaner. Then they are rinsed in hot water and submerged in isopropyl alcohol to prevent watermarks. Drying and dust removal finally makes them ready for the coating process. Any slight marks found on the substrates mean that the whole process must be repeated. After cleaning, the items are secured in special jigs; this stops any movement of the piece and vibration from the coating plant affecting the coating. Any jigs that are not held in house are manufactured when required.

1.2.2.3   MATERIAL PREPARATION

The preparation of materials for deposition or coating depends upon the type of deposition to be performed (which is discussed in Section 1.2.2.4). The materials can be either in solid or in chemical (gaseous) form. This chapter mainly considers the solid materials which are usually in tablet, palettes, or powdered form. Tablets or palettes are powdered using mortar and pestle before they can be used for deposition. The selected materials during design process are filled into the boats/crucibles and are kept in the crucible holder inside the chamber of the deposition plant. The material of the boat depends upon the type of material being used. For materials with high melting point usually molybdenum or tungsten boats are used.

1.2.2.4   COATING PROCESS

There are a number of methods in which thin films can be grown or deposited[36,37] on a substrate. The type of deposition technique chosen for a particular filter depends on the application of that filter as well as mechanical and environmental stress stability requirements.

Deposition techniques of thin films are broadly divided into two categories:

- physical vapor deposition (PVD),
- chemical vapor deposition (CVD).

1.2.2.4.1   Physical Vapor Deposition

PVD refers to "physical" movement of the bulk material in a high vacuum environment toward the substrate in its vapor phase which condenses on the substrate to form the thin film. Physical deposition uses mechanical, electromechanical, or thermodynamic means to produce a thin film of solid. Since engineering materials require high energies to be held together and chemical reactions are not used to store these energies, physical deposition systems tend to require a low-pressure vapor environment to function properly. In order to make the particles of material escape its surface, the source is placed in an energetic and entropic environment. A cooler surface which draws energy from the particle source as they arrive is kept facing the source, allowing them to form a solid layer. The whole system is kept in a vacuum deposition chamber, so that the particles can travel as freely as possible. Films deposited by physical means are commonly *directional*, rather than *conformal*, since particles tend to follow a straight path. Various PVD methods are discussed below:

- *Electron beam evaporation technique*

    The process of electron beam evaporation technique involves evaporation of the bulk material by a high energetic electron beam in which the material's phase changes from solid to liquid phase to the vapor phase which condenses on to the substrate.

The E-beam evaporator (Fig. 1.4) consists of a gun source, in which the electrons are thermionically emitted from heated filaments that are shielded from direct line of sight of both the evaporant charge and substrate. The cathode potential is biased negatively with respect to a nearby grounded anode by anywhere from 4 to 20 kv and this serves to accelerate the electrons. In addition, a transverse magnetic field is applied that serves to deflect the electron beam in 270° circular arc and focus it on the hearth and evaporant charge at ground potential.

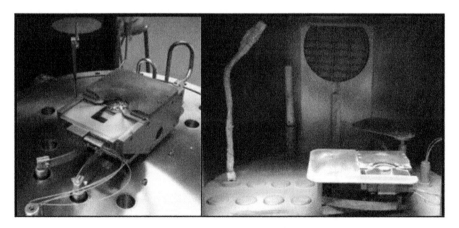

FIGURE 1.4   E-beam gun inside the chamber of electron beam evaporation plant.

- *Thermal evaporation*

    In the thermal evaporation technique (Fig. 1.5), the vapor is produced by heating the material to be deposited till it evaporates through two electrodes. The material finally condenses in the form of thin film on to the substrate surface and on the vacuum chamber walls.

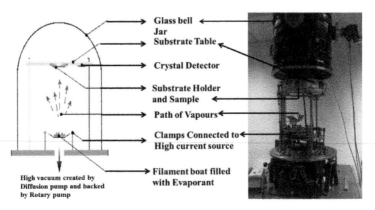

FIGURE 1.5 Thermal evaporation plant.

- *Sputtering*

    It is a process whereby atoms are ejected from a solid target material and gets deposited on the substrate by bombarding the target by high-energy particles like atoms or ions. Momentum transfer from the bombarding atoms or ions is used to eject target material to be deposited on the substrate. In this, the target is bombarded with argon ions to cause the target materials to come into the vapor phase (Fig. 1.6). Other PVD methods are molecular beam epitaxy (MBE), pulsed laser deposition, cathodic arc deposition (arc-PVD), electro hydrodynamic deposition (electrospray deposition).

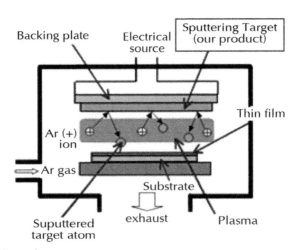

FIGURE 1.6 Sputtering process.

1.2.2.4.2   *Chemical Vapor Deposition (CVD)*

CVD[36,38] generally uses a gas-phase precursor, often a halide, hydride or oxide of the materials to be deposited. In this process, the chemical precursors are transported in the vapor phase to decompose, combine, or react with other precursors on a heated substrate to form a film. The films may be epitaxial, polycrystalline, or amorphous depending on the materials and reactor conditions. CVD has become the major method of film deposition for the semiconductor industry due to its high throughput, high purity, and low cost of operation. Depending upon the vacuum conditions maintained inside the CVD reactor, the activation process of the precursors as well as the type of thin films formed on the substrate, these are broadly classified as low pressure CVD (LPCVD), atmospheric pressure CVD (APCVD), ultra high vacuum CVD (UHVCVD), plasma enhanced CVD (PECVD), metal organic CVD (MOCVD), etc. Detailed description of CVD techniques is beyond the scope of this document and the basic principle behind CVD techniques are presented here for the sake of completeness.

Layer thickness monitoring is essential for thin film work. Even small percentage errors can cause unwanted and detrimental effects in the final product. The thickness of the growing layer can be monitored by an oscillating quartz crystal arrangement which is exposed to the evaporant in the chamber. As the crystal becomes coated the oscillation frequency of the crystal falls and this is translated into a thickness value by the controller.

1.2.3   *MEASUREMENT AND ANALYSIS*

The characterization of the coated thin films is an important task in the process of optical coatings. The estimation of the optical constants and there surface properties have a significant effect on the performance of the optical filters.[39,40] The design of the optical filters is dependent on the above mentioned properties of the thin film which in turn is dependent on the selection of material for coating. Various techniques are there for characterization of the coated film, for example, spectroscopic, morphological, image processing, film quality test, etc.

1.2.3.1   *SPECTROSCOPIC TECHNIQUES*

Spectroscopy is the study of the reflection or transmission properties of a substance as a function of wavelength. A spectrophotometer is the combination of two devices, a spectrometer, and a photometer. Spectrometer is used for producing light of any selected wavelength or color while a photometer is used for measuring the intensity of light. Spectrophotometers are mainly classified on the basis of different measurement techniques, wavelengths they work with, how they acquire a spectrum, sources of intensity variation. Different types of spectrophotometers found in the market are UV–VIS-NIR and Gonio spectrophotometers, reflectometer, ellipsometer, and Fourier transform infrared spectroscopy (FTIR).

1.2.3.2   MORPHOLOGICAL TECHNIQUES

These techniques are used to study the topography, structure, shape of a surface, variation of roughness, and size of the crystallites with thickness and growth rate. Some of the techniques are atomic force microscopy (AFM),[41,42] scanning electron microscopy (SEM),[43,44] transmission electron microscopy (TEM),[36] contact mechanical profiler,[36] coherence correlation interferometer (CCI),[45] and image processing.[46,47]

- AFM (see Fig. 1.7) provides a 3D profile of the surface on a nanoscale, by measuring *forces* between a sharp probe (<10 nm) and surface at very short distance (0.2–10 nm probe-sample separation). It usually operates in three modes—contact mode, non-contact mode and tapping mode. Contact mode AFM is one of the more widely used scanning probe modes, and operates by rastering a sharp tip (made either of silicon or Si_3N4 attached to a low spring constant cantilever) across the sample.
- SEM (see Fig. 1.8) is used for inspecting topographies of specimens at very high magnifications using a piece of equipment called the scanning electron microscope. The magnifications can go to more than 300,000×. During SEM inspection, a beam of electron is focused on a spot volume of the specimen, resulting in the transfer of energy to the spot. These bombarding electrons, also referred to as primary electrons, dislodge electrons from the specimen itself. The dislodged electrons, also known as secondary electrons, are attracted and collected by a positively biased grid or detector, and then translated into a signal. To produce the SEM image, the electron beam is swept across the area being inspected, producing many such signals. These signals are then amplified, analyzed, and translated into images of the topography being inspected. Finally, the image is shown on a CRT.

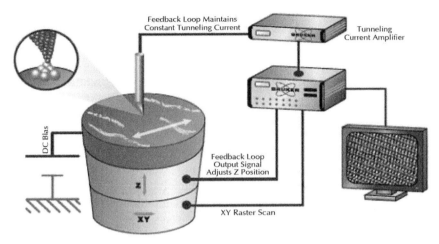

FIGURE 1.7 Principle of AFM.

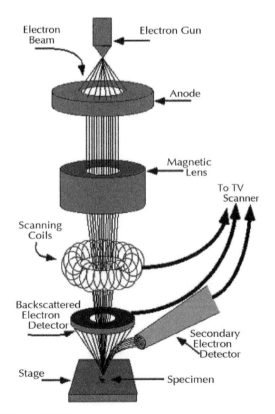

FIGURE 1.8 Working of SEM.

- TEM (Fig. 1. 9) uses a highly energetic electron beam (100 keV to 1 MeV) to image and obtain structural information from thin film samples. The electron microscope consists of an electron gun, or source, and an assembly of magnetic lenses for focusing the electron beam. Apertures are used to select among imaging modes and to select features of interest for electron diffraction work. The sample is illuminated with an almost parallel electron beam, which is scattered by the sample. Features in the sample that cause scattering have darker contrast in a bright-field image than those that cause little or no scattering. An electron diffraction pattern can be generated from a particular area in a bright-field image (such as a particle or grain) by using a selected area aperture. Dark-field images are formed from a single diffracted beam and are used to identify all the areas of a particular phase having the same crystalline orientation. Magnifications from about 100× up to several hundred thousand times can be achieved in the TEM.

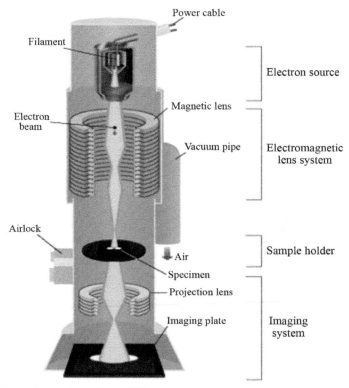

FIGURE 1.9   Schematic outline of TEM.

- Contact mechanical profiler is a contact type mechanical profiler that uses conical stylus diamond tip which record height variation of surface along a straight line at a time being in contact with surface. The profile meter is mounted on epoxy granite construction on anti-vibration mounts and provides a firm support for the column and work piece. The stylus moves over the surface at a constant speed, and an electrical signal is produced by the transformer. These electrical signals are amplified and undergo analog-to-digital conversion. The resulting digital profile is stored in the computer and can be analyzed subsequently for roughness or waviness parameter. This instrument offers dimensional, form and texture measurements simultaneously with high accuracies and gauge repeatability. Talysurf contact mechanical profiler is shown in Figure 1.10.

FIGURE 1.10   Contact mechanical profiler (Talysurf PGI 120).

- The CCI methods are based on the cross-coherence analysis of two low-coherence light beams, the object beam being reflected from the object, whilst the reference beams is reflected from a reference mirror. The high-contrast interference pattern arises if the optical path length in the object arm is equal to the optical path length in the reference arm. For each object point a correlogram is recorded during the movement of the object. The position of the corresponding object point along the x axis can be measured by another measuring system for the maximum of the correlogram. An interference signal is helpful for accurately determining this position. CCI 6000 non-contact profiler is shown in Figure 1.11. It is a non-contact optical profiler. It makes use of coherence correlation algorithm and a high-resolution digital camera array to measure the surface. A three dimensional image of the surface is generated by scanning the surface in a 'Z' direction by measuring the fringes. The information obtained by fringe measurement is processed by a dedicated software, which transforms this fringe data into a quantitatively three dimensional image.

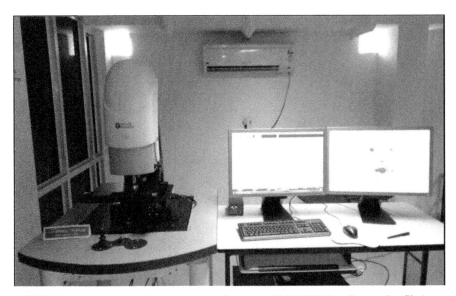

FIGURE 1.11   Coherence Correlation Interferometer (CCI 6000 Non-Contact Profiler).

1.2.3.3 IMAGE PROCESSING TECHNIQUES

Image processing is a vast area which has applications in many fields like image sharpening and restoration, medical field, remote sensing, transmission and encoding, machine/robot vision, color processing, pattern recognition, video processing, microscopic imaging, etc. It is subcategory of digital signal processing which uses computer algorithms to extract beneficial information from the digital images by modifying or enhancing them.[46] These days researchers are using image processing in the field of optical thin films for analyzing various surface characteristics.[47-50] Parameters like surface roughness measurement are normally done through the use of stylus type instruments. The major disadvantage of using stylus profilometry for such measurements is that it requires direct physical contact, which limits the measuring speed. Alternate techniques like fast Fourier transform and digital filtering process can be used.[51]

1.3   CASE STUDIES

1.3.1   DESIGN OF MULTILAYER INTERFERENCE FILTER

Alternately stacked ultrathin multilayer dielectric optical interference filters were designed using combinations of zinc sulfide (ZnS), cadmium sulfide (CdS), and magnesium fluoride (MgF_2) with refractive indices 2.35, 2.529, and 1.38, respectively. All filters were designed with quarter wave thickness for the design wavelength of 550 nm for normal incidence angle. Refractive index of BK7 glass was taken wherever substrate refractive index was required in design. All the materials and substrates were assumed to be absorption free. Further designs of multilayer broadband high reflective interference and narrowband filters are discussed.

Design equations, multilayer broad band reflective interference filter

$$\text{Glass}/(\text{HL})^3\text{H}/\text{air}$$
$$\text{Glass}/(\text{HL})^6\text{H}/\text{air}$$

where H represents high refractive index material and L represents low refractive index material.

The 7-layer design of ZnS/MgF_2 and CdS/MgF_2 is shown in Figures 1.12 and 1.13. Here, "a" and "b" (of Figs. 1.12 and 1.13), respectively, shows refractive index and physical thickness with respect to layer number. Figures 1.12c and 1.13c show their respective reflectance versus wavelength performance calculated using Filmstar™ Software in the visible wavelength region (400–800 nm). Reflectance values of ~95.52% and ~95.74 were obtained for ZnS and CdS designs with a broadband full width half maximum bandwidth (FWHM) of 345 and 362 nm, respectively.

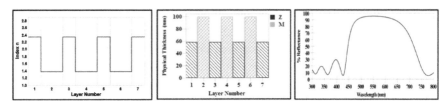

FIGURE 1.12 (a) Refractive index versus layer number of 7 layer ZnS/MgF$_2$ design, (b) Physical thickness vs. layer number of 7 layer ZnS/MgF$_2$ design, and (c) Reflectance versus wavelength performance of 7 layer ZnS/MgF$_2$ design.

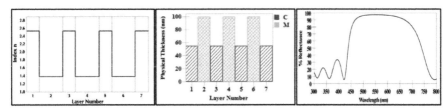

FIGURE 1.13 (a) Refractive index versus layer number of 7 layer CdS/MgF$_2$ design, (b) Physical thickness vs. layer number of 7 layer CdS/MgF$_2$ design, and (c) Reflectance versus wavelength performance of 7 layer CdS/MgF$_2$ design.

Further, alternatively stacked 13 layers of ZnS/MgF$_2$ and CdS/MgF$_2$ were designed (Figs 1.14 and 1.15, respectively). A peak reflectance of ~99% and ~97.52 were obtained for ZnS and CdS designs with a broadband FWHM bandwidth of 229 and 256 nm, respectively.

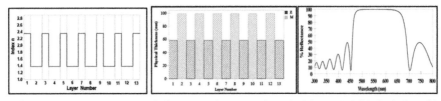

FIGURE 1.14 (a) Refractive index versus layer number of 13 layer ZnS/MgF$_2$ design, (b) Physical thickness vs. layer number of 13 layer ZnS/MgF$_2$ design, and (c) Reflectance versus wavelength performance of 13 layer ZnS/MgF$_2$ design.

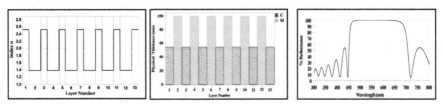

FIGURE 1.15 (a) Refractive index versus layer number of 13 layer CdS/MgF$_2$ design, (b) Physical thickness vs layer number of 13 layer CdS/MgF$_2$ design, and (c) Reflectance versus wavelength performance of 13 layer CdS/MgF$_2$ design.

1.3.2   DESIGN OF NARROW BAND FILTER

A narrow band filter was designed by introducing a spacer layer of half wave optical thickness (HWOT) thickness of L-refractive index, between quarter wave thicknesses of HL pairs. The thicknesses of the layers were taken as quarter waves to the respective wavelength.

Design Equation

$$\text{Glass}/(HL)^3 2L(LH)^3/\text{air}$$

where H represents high refractive index material and L represents low refractive index material.

The refractive index versus layer number and physical thickness versus layer number of 13 layer micro cavity filter with ZnS/MgF$_2$ combination and the corresponding transmittance spectra at normal angle of incidence in visible region are shown in Figure 1.16 a–c, respectively. It can be observed from the figure that a transmittance of 95.72% is achieved within the pass band region with a FWHM of around 8 nm.

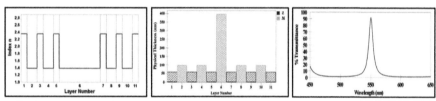

FIGURE 1.16   (a) Refractive index versus layer number of 13 layer ZnS/MgF$_2$ micro-cavity design, (b) Physical thickness vs. layer number of 13 layer ZnS/MgF$_2$ micro-cavity design and (c) Reflectance versus wavelength performance of 13 layer ZnS/MgF$_2$ micro-cavity design.

Further, the same optical interference (reflective) filter designed using CdS/MgF$_2$ and its transmission plot is shown in Figure 1.17a–c, respectively. It is observed that the transmittance within the pass band region is around 95.742% and the FWHM is of ~5 nm.

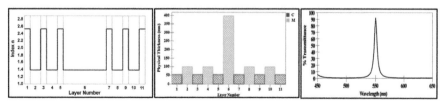

FIGURE 1.17   (a) Refractive index versus layer number of 13 layer CdS/MgF$_2$ micro-cavity design, (b) Physical thickness vs. layer number of 13 layer CdS/MgF$_2$ micro-cavity design, (c) Reflectance versus wavelength performance of 13 layer CdS/MgF$_2$ micro-cavity design.

1.3.3   FABRICATION AND CHARACTERIZATION OF TiO₂ THIN FILMS

Four TiO_2 thin films with different thicknesses have been deposited on the optical grade glass substrates in different batches. Transmittance and optical properties like absorption coefficient (α), extinction coefficient (K), and optical band gap energy were determined in the wavelength range of 350–800 nm. Surface morphology of the coated films was studied and roughness was estimated. Further, adhesion, and hardness tests for all the films were performed using MIL standards.

1.3.3.1   MATERIALS AND METHODS

Four TiO_2 films with thickness values 50, 130, 300, and 450 nm were deposited on optical grade glass substrate (Schott make) with refractive index of 1.52 in different batches under identical conditions using E-beam evaporation technique (Fig. 1.18). The coating system (PLS 570, Pfeiffer Vacuum) consists of one e-beam source, four liners made of molybdenum (for evaporation of materials), substrate holder and heater, glow discharge, quartz crystal monitor (Intellemetrics IL 820), and optical thickness monitor (Intellemetrics IL 552) for monitoring thickness. TiO_2 material with 99.99% purity (Umicore, Liechtenstein) was taken in granulated form in molybdenum liners. Ex situ and in situ cleaning of the substrates was performed. For ex situ cleaning, the substrates were ultrasonicated in acetone medium prior to mounting for deposition. The substrates were subjected to glow discharge cleaning in presence of argon gas. The chamber was evacuated to base pressure of 2×10^{-5} mbar prior to deposition and argon gas was flushed in a controlled manner to ensure further cleaning. A plasma discharge with a discharge current of around 100–200 mA was generated for 4–5 min. After glow discharge cleaning, thin film was deposited on the substrate with chamber pressure maintained at 2×10^{-4} mbar and substrate temperature 250°C. A stable flow of high purity oxygen was regulated and maintained by e-beam gun controlled system. The thickness of the films during deposition was controlled by optical thickness monitor and rate was controlled by quartz crystal monitor. This cleaning and deposition procedure was repeated three times with different thicknesses for depositing the remaining TiO_2 films on different substrates. Table 1.1 shows the deposition parameters maintained during e-beam deposition.

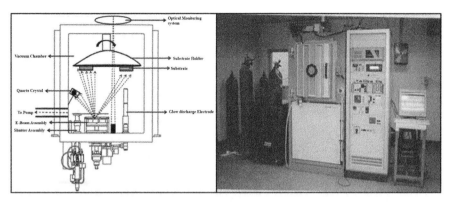

FIGURE 1.18   Electron beam evaporation plant.

TABLE 1.1   Deposition Parameters Used for E-Beam Deposited TiO_2 Thin Films.

S. No.	Deposition Parameters	Corresponding Values
1	Substrate temperature	250°C
2	Initial base pressure	2.432×10^{-5} mbar
3	Substrate used	BK7, glass
4	Oxygen gas pressure	2×10^{-4} mbar
5	Glow discharge gas	Argon (Ar)

Optical and morphological studies on the coated films were performed using different characterization tools which were discussed in next section.

1.3.3.2   ANALYSIS

Transmittance and optical constants like absorption coefficient, extinction coefficient, and energy band gap were calculated. For surface properties studies, roughness was estimated. Adhesion and hardness tests were performed to check the durability of the coating and its scratch resistance.

Transmittance (T)

Measurement of transmittance of coated films was carried out using Perkin Elmer, Lambda 9 UV–VIS-NIR spectrophotometer in the wavelength range of 350–800 nm. Transmission spectra of TiO_2 films were plotted in Figure 1.19. A high transparency of the films can be noticed in a large wavelength domain from 400 to 800 nm. Also, it can be observed that with increase in thickness of the film, numbers of interference fringes were increased. The curve shows the interference fringes on the high transmittance (low absorption) region of wavelength.

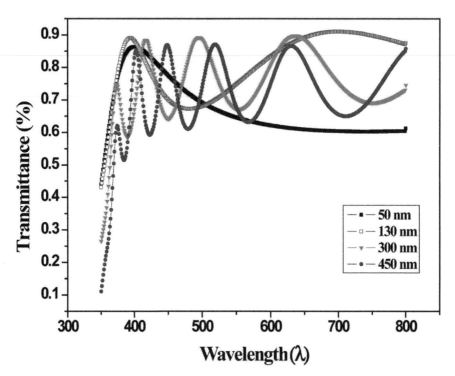

FIGURE 1.19   Transmittance profiles of TiO₂ thin films with thicknesses 50, 130, 300, and 450 nm.

Absorption Coefficient (α)

The absorption coefficients of the prepared thin films were calculated from eq 1.1:[11]

$$\alpha = \frac{1}{d}\ln\left(\frac{1}{T}\right) \qquad (1.1)$$

where d is the thickness of the film and T is the transmittance.

Figure 1.20 shows the relation of absorption coefficient as a function of wavelength for TiO₂ thin films. It can be seen that TiO₂ thin films have high value of absorption coefficient which increases the probability of occurrence of direct transitions. Also, it can be noted that there is a decrease in absorption coefficient with increase in thickness.

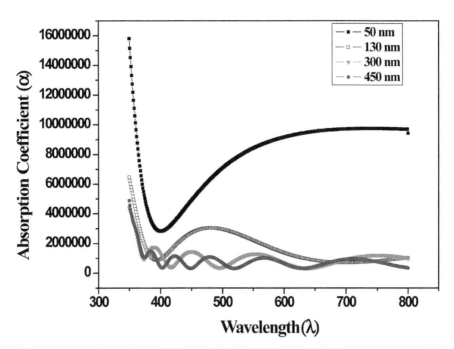

FIGURE 1.20 Absorption coefficients of TiO$_2$ thin films with thicknesses 50, 130, 300, and 450 nm.

Extinction Coefficient (K)

Extinction coefficients of the thin films were deduced using eq 1.2:[3]

$$K = \frac{\alpha\lambda}{4\pi} \tag{1.2}$$

Figure 1.21 shows the relation of extinction coefficient as a function of wavelength. It can be seen that the behavior is similar to that of absorption coefficient. The extinction coefficient of the thin films is varying approximately between 0.1×10^8 and 7×10^8 and decreasing with increase in thickness.

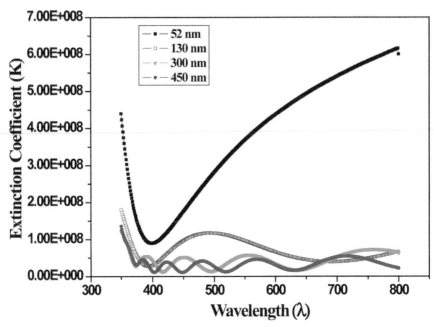

FIGURE 1.21   Extinction coefficients of TiO_2 thin films with thicknesses 50, 130, 300, and 450 nm.

Optical Band Gap

Direct band gap energy of the TiO_2 films was evaluated from $(\alpha h\nu)^2$ versus photon energy $(h\nu)$, plots (Fig. 1.22a–d) using Tauc's shown in eq 1.3:[4]

$$\alpha = \frac{A\left(h\nu - E_g\right)^{1/2}}{h\nu} \tag{1.3}$$

where A is constant and E_g is optical band gap energy. The plotted data was fitted with a straight line and E_g was estimated from the intercept of these linear fits with photon energy axis. It is observed that the band gap energy for the TiO_2 films is approximately 3.25 ± 0.5 eV for all the four thin film samples. This slight variation in the band gap with thickness may be due to the improvement in the crystallinity of the phase.

Surface Morphology

For surface morphology, Coherence Correlation Interferometer (CCI 6000) non-contact profiler was used and surface roughness was measured. It makes use of coherence correlation algorithm and a high-resolution digital camera array to measure the surface (measures of fringes). Figure 1.23a–d shows CCI images of TiO_2 thin films with thicknesses of 50, 130, 300, and 450 nm, respectively. It is observed that the surface morphology varies with the thickness. Figure 1.24a–d shows the 2-d profiles of the four TiO_2 thin film samples with thicknesses 50, 130, 300, and 450 nm taken using CCI. It shows the height of peaks and troughs through which the roughness of the film can be estimated. It can be noticed that 450 nm thickness sample shows minimum peaks and troughs representing more uniform coating and least surface roughness. The estimated roughness from the samples is shown in Table 1.2. It can be clearly noticed from table that roughness of the TiO_2 thin films is decreasing with the increasing thickness. One of the reasons for the same may be improvement in the crystallinity of the film as reported earlier in case of optical band gap. Other dependent factors can be the uniformity of the substrate on which the films were deposited and environmental conditions during deposition as the thin films were deposited in different batches.

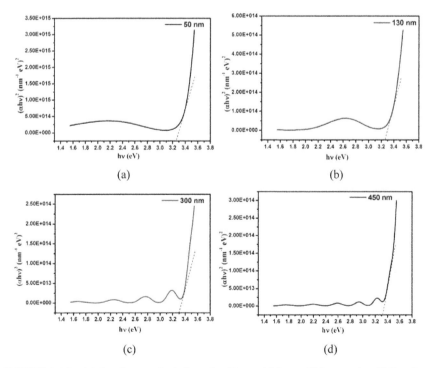

FIGURE 1.22 (a) Band gap calculations for 50 nm thickness TiO_2 sample, (b) Band gap calculations for 130 nm thickness TiO_2 sample, (c) Band gap calculations for 300 nm thickness TiO_2 sample, and (d) Band gap calculations for 450 nm thickness TiO_2 sample.

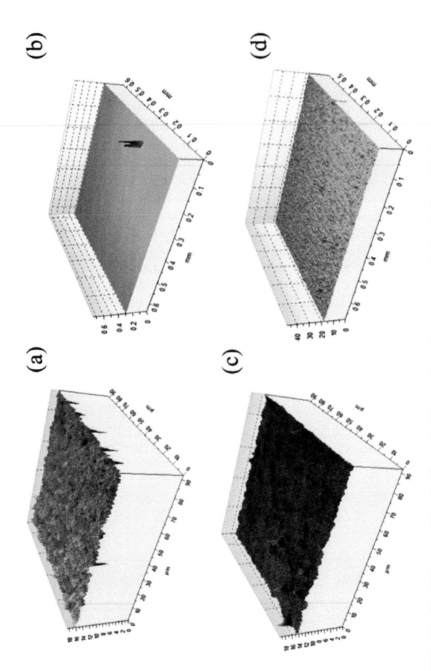

FIGURE 1.23 (a) CCI images of TiO$_2$ thin films with thickness 50 nm, (b) CCI images of TiO$_2$ thin films with thickness 130 nm, (c) CCI images of TiO$_2$ thin films with thickness 300 nm, and (d) CCI images of TiO$_2$ thin films with thickness 450 nm.

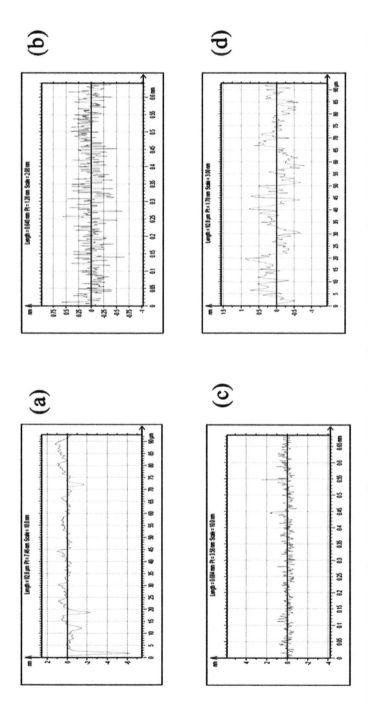

FIGURE 1.24 (a) 2-d profiles of TiO$_2$ thin films with thickness 50 nm taken using CCI, (b) 2-d profiles of TiO$_2$ thin films with thickness 130 nm taken using CCI, (c) 2-d profiles of TiO$_2$ thin films with thickness 300 nm taken using CCI, and (d) 2-d profiles of TiO$_2$ thin films with thickness 450 nm taken using CCI.

TABLE 1.2   Estimated Roughness of the TiO$_2$ Samples Using CCI.

S. No.	Thickness of Sample (nm)	Roughness (nm)
1	50	0.367
2	130	0.305
3	300	0.279
4	450	0.277

Adhesion and Hardness Test

To test the durability of the coating, the coated samples were subjected to MIL standard adhesion (no. MIL-F-48616) and hardness tests [52]. A ½″ wide adhesive tape (Fig. 1.25) conforming to MIL standard I of L-T-90 (MIL-C-48497A) was applied on the coated film samples from one end to other. Then it was removed suddenly to assess the adhesion of the coating by visually examining the surface conditions. It was found that the edges

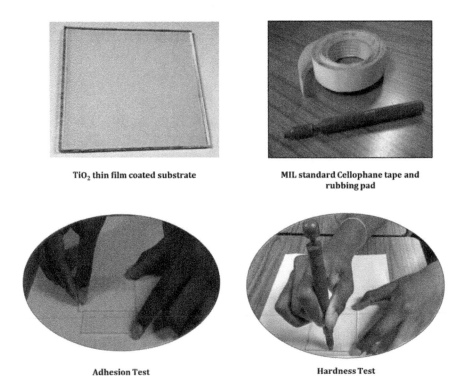

TiO$_2$ thin film coated substrate

MIL standard Cellophane tape and rubbing pad

Adhesion Test

Hardness Test

FIGURE 1.25   Adhesion and hardness test for TiO$_2$ films.

were smooth and coating was not detached from anywhere. For testing hardness (Fig. 1.25), the coated samples were subjected to a moderate abrasion by rubbing the coated surface with a MIL standard MIL–E-12397 rubbing pad across the sample from one point to another over the same path for 20 strokes with a force of 2.5 pounds. The abrasion tester was held approximately normal to the surface under test during rubbing operation. After the operation, the sample was cleaned and dried for visual examination for any physical damage to the coating. It was found that all the samples withstood 20 strokes without peeling off and also no visible scratch could be seen.

KEYWORDS

- **thin films**
- **optical coatings**
- **physical vapor deposition techniques**
- **computational techniques**
- **metal oxides**

REFERENCES

1. Lee, D.; Xie W.; Yang, M.; Zhang, Y.; Zhuang, Z. *Optical Fiber Relative-Humidity Sensor Using Fabry–Perot Cavity Formed by E-beam Evaporated Dielectric Films*, In Asia Pacific Optical Sensors Conference, International Society for Optics and Photons, 89240I–89240, 2013.

2. Mazur., M.; Wojcieszak D., Domaradzki, J.; Kaczmwerek, D.; Song, S.; Placido, F. TiO$_2$/SiO$_2$ Multilayer as an Antireflective and Protective Coating Deposited by Microwave Assisted Magnetron Sputtering. *Opto. Electro. Rev.* **2013,** *21* (2), 233–238.

3. Wang, X.; Masumoto, H.; Someno, Y.; Hirai, T.; Helicon Plasma Deposition of a TiO2/SiO2 Multilayer Optical Filter with Graded Refractive Index Profiles. *Appl. Phys. Lett.* **1998,** *72* (25), 3264–3266.

4. Habubi, N. F.; Mishjil, K. A.; Rashid, H. G.; Mansour H. L. Design and Fabrication of Edge Filter Using Absorbed ZnS Single Layer Prepared by Flash Evaporation Technique. *Modern Phys. Lett. B.* **2010,** *24* (28), 2821–2829.

5. Sta, M. Jlassi.; Hajji, M.; Boujmil, M. F.; Jerbi, R.; Kandyla, M.;. Kompitsas, M.; Ezzaouia, H. Structural and Optical Properties of TiO$_2$ Thin Films Prepared by Spin Coating. *J. Sol. Gel. Sci. Technol.* **2014,** *72,* 421.

6. El-Nahass, M. M.; Ali, M. H.; El-Denglawey, A. Structural and Optical Properties of Nano-Spin Coated Sol–Gel Porous TiO$_2$ Films. *Trans. Nonferrous Met. Soc. China.* **2012,** *22,* 3003–3011.

7. Zhao, B. X.; Zhou, J. C.; Rong, L. Y. Microstructure and Optical Properties of TiO_2 Thin Films Deposited at Different Oxygen Flow Rates. *Trans. Nonferrous Met. Soc. China.* **2010,** *20* (8), 1429–1433.

8. Ghrairi, N.; Bouaicha, M. Structural, Morphological, and Optical Properties of TiO_2 Thin Films Synthesized by the Electro Phoretic Deposition Technique. *Nanoscale Res. Lett.* **2012,** *7* (1), 1–7.

9. Parinam Sunita; Vemuri SRS Praveen Kumar; Mukesh Kumar; Parinam Krishna Rao; Neelam Kumari; Vinod Karar; Amit L. Sharma. In *Estimation of Optical Constants and Thicknesses of E-beam Deposited Metal Oxide Films by Envelope Method,* International Conference on Optics and Photonics (ICOP), Kolkata, February 20–22, 2015;Calcutta University.

10. Narasimha Rao, k.; Murthy, M. A.; Mohan, S. Optical Properties of Electron Beam Evaporated TiO_2 Films. *Thin Solid Films.* **1989,** *176* (2), 181–186.

11. Narasimha Rao, K.; Mohan, S. Optical Properties of Electron Beam Evaporated TiO_2 Films Deposited in Ionized Oxygen Medium. *J. Vac. Sci. Technol. A.* **1990,** 8(4), 3260–3264.

12. Selhofer Hubert; Elmar Ritter; Robert Linsbod. Properties of Titanium Dioxide Films Prepared by Reactive Electron-Beam Evaporation from Various Starting Materials. *Appl. Opt.* **2002,** *41* (4), 756–762.

13. Solovana, M. N.; Maryanchuka, P. D.; Brusb, V. V.; Parfenyuka, O. A. Electrical and Optical Properties of TiO_2 and TiO_2:Fe Thin Films. *Inorg. Mater.* **2012,** *48* (10), 1154–1160.

14. Daniyan, A. A.; Umoru, L. E.; Olunlade, B. Preparation of Nano-TiO_2 Thin Film Using Spin Coating Method. *J. Miner. Mater. Charact. Eng.* **2013,** *1,* 138–144.

15. Watanabe, T.; Nakajima, A.;Wang, R.; Minabe, M.; Koizumi, S.; Fujishima, A.; Hashimoto, K. Photocatalytic Activity and Photoinduced Hydrophilicity of Titanium Dioxide Coated Glass. *Thin Solid Films.* **1999,** *351,* 260–263.

16. Jayasinghe, R. C.; Perera, A. G. U.; Zhu, H.; Zhao, Y. Optical Properties of Nanostructured TiO_2 Thin Films and Their Application as Antireflection Coatings on Infrared Detectors. *Opt. Lett.* **2012,** *37* (20), 4302–4304.

17. Kumari, N.; Kumar, M.; Rao, P. K.; Sharma, A. L.; Karar, V. In *Design and Development of Multilayer Antireflection Coatings for Avionics Displays Devices,* National Conference on Emerging Trends in Aircraft Design and Manufacturing, ATHAL Press and Media, India, 71–73, 2014.

18. Kumar, M.; Kumari, N.; Rao, P. K.; Sharma, A. L., Karar, V., *Performance Comparison of Silica-Titania and Silica-Alumina Multilayer Reflective Filters for Avionics Displays,* National Conference on Emerging Trends in Aircraft Design & Manufacturing ATHAL Press & Media, India, 57–59, 2014.

19. Angus Macleod. *Thin Film Optical Filters, 3rd ed.;* Institute of Physics Publishing: Bristol and Philadelphia, 2001.

20. Sh. A. Furman; Tikhonravov, A. V. *Basics of Optics of Multilayer Systems;* Atlantica Séguier Frontières: Paris, 1992.

21. Baumeister, P. W. *Optical Coating Technology;* SPIE Press: Bellingham, 2004.

22. Ronald, R. Willey. *Field Guide to Optical Thin Films,* Vol. FG 07, SPIE Press: Bellingham, 2006.

23. Lee, C. C.; Chen, S. H.; Kuo, C. C.; Wei, C. Y. Achievement of Arbitrary Bandwidth of a Narrow Bandpass Filter. *Opt. Exp.* **2007,** *15* (23), 15228–15233.

24. Willey, R. R. Achieving Narrow Bandpass Filters Which Meet the Requirements for DWDM. *Thin Solid Films.* **2001**, *398*, 1–9.
25. Finkelstein, N. D.; Lempert, W. R.; Miles, R. B. Narrow-Linewidth Passband Filter for Ultraviolet Rotational Raman Imaging. *Opt. Lett.* **1997**, *22* (8), 537–539.
26. Piegari, A.; Bulir, J.; Variable Narrowband Transmission Filters with a Wide Rejection Band for Spectrometry. *Appl. Opt.* **2006**, *45* (16), 3768–3773.
27. Piegari, A.; Bulir, J.; Krasilnikova Sytchkova, A. Variable Narrow-Band Transmission Filters for Spectrometry from Space. 2. Fabrication Process. *Appl. Opt.* **2008**, *47* (13), C151–C156.
28. Tan, M.; Lin, Y.; Zha, D. Reflection Filter with High Reflectivity and Narrow Bandwidth. *Appl. Opt.* **1997**, *36* (4), 827–830.
29. Zheng, S. Y.; Lit, J. W. Design of a Narrow-Band Reflection IR Multilayer. *Canadian J. Phys.* **1983**, *61* (2), 361–368.
30. Augustsson, T. Proposal of a DMUX with a Fabry-Perot All-Reflection Filter-based MMIMI Configuration. *IEEE Photonic. Tech. L.* **2001**, *13* (3), 215–217.
31. Postava, K.; Pistora, J.; Kojima, M.; Kikuchi, K.; Endo, K.; Yamaguchi, T. Thickness Monitoring of Optical Filters for DWDM Applications. *Opt. Exp.* **2003**, *11* (6), 610–616.
32. Lee, C. C.; Wu K. In Situ Sensitive Optical Monitoring with Proper Error Compensation, *Opt. Lett.* **2007**, *32* (15), 2118–2120.
33. Vázquez, J. M.; Li, Z.; Liu, Y.; Tangdiongga, E.; Zhang, S.; Lenstra, D.; Dorren, H. J. S. Optimization of Optical Band-Pass Filters for All-Optical Wavelength Conversion Using Genetic Algorithms. *IEEE J. Quant. Electron.* **2007**, *43* (1), 57–64.
34. Li, D. G.; Watson, A. C. In *Computational Intelligence and Multimedia Applications, 1999. ICCIMA'99.* Proceedings of the Third International Conference, IEEE, ICCIMA, *1999.*
35. Rabady, R. I.; Ababneh, A. Global Optimal Design of Optical Multilayer Thin-film Filters Using Particle Swarm Optimization. *Optik.* **2014**, *125* (1), 548–553.
36. Milton Ohring. *Materials Science of Thin Films (Second Edition): Deposition and Structure;* Elsevier Inc.: New York, 2002.
37. Holland, L. *Vacuum Deposition of Thin Films*; Chapman & Hall: London, 1970.
38. Pierson, Hugh, O. *Handbook of Chemical Vapor Deposition: Principles, Technology and Applications*; William Andrew: New York, 1999.
39. Thaidun, M. B.; Venkata Rao, L.; Raja Mohan Reddy; Venkata Chalapathi, G. Structural, Dielectric and Optical properties of Sputtered TiO_2 Nano-films. *IOSR J. Appl. Physics.* **2013**, *4*, 49–53.
40. S. R. S. Praveen Kumar, V.; Sunita, P.; Saraf, M.; Kumar, M.; Rao, P.; K. Kumari, N.; Sharma, A. L. In *Estimation of Optical Constants and Thicknesses of E-beam Deposited TiO2 Thin Films by Envelope Method*, Elsevier Proceedings Exploring Basic and Applied Sciences Conference EBAS'14, Jalandhar, 277–280, 2014.
41. Blanchard, C. R. Atomic Force Microscopy. *Chem. Educ.* **1996**, *1* (5), 1–8.
42. Goddard III, W. A.; Brenner, D.; Lyshevski, S. E.; Iafrate, G. J., Eds.; *Handbook of Nanoscience, Engineering, and Technology.* CRC Press: Boca Raton, 2007.
43. Yao, N.; Wang, Z. L. Eds.; *Handbook of Microscopy for Nanotechnology*, Boston: Kluwer Academic Publishers: New York, 2005, pp 287–319.
44. Tseng, A. A. *Nanofabrication: Fundamentals and Applications*; World Scientific: Singapore, 2008.
45. Kaplonek, W.; Lukianowicz, C. *Coherence Correlation Interferometry in Surface Topography Measurements.* INTECH Open Access Publisher: Poland, 2012.

46. Gonzalez, R. C.; Woods, R. E. *Digital Image Processing*; Prentice-Hall: Upper Saddle River, NJ, 2002.

47. Re, G. L.; Lopresti, F.; Petrucci, G.; Scaffaro, R. A Facile Method to Determine Pore Size Distribution in Porous Scaffold by Using Image Processing. *Micron.* **2015,** *76,* 37–45.

48. Sun, W.; Romagnoli, J. A.; Tringe, J. W.; Létant, S. E.; Stroeve, P.; Palazoglu, A. (2009). Line Edge Detection and Characterization in SEM Images Using Wavelets. *IEEE Trans. Semicond. Manuf.* **2009,** *22* (1), 180–187.

49. Asha, V.; Bhajantri, N. U.; Nagabhushan, P. Similarity Measures for Automatic Defect Detection on Patterned Textures. International Journal of Information and Communication Technology. **2012,** *4* (2–4), 118–131.

50. Widiatmoko, E.; Abdullah, M. In *A Method to Measure Pore Size Distribution of Porous Materials Using Scanning Electron Microscopy Images*, The Third Nanoscience and Nanotechnology Symposium (NNSB-2010), Vol. 1284, No. 1, pp 23–26, AIP Publishing: Indonesia, 2010.

51. Tien, C. L.; Yang, H. M.; Liu, M. C. The Measurement of Surface Roughness of Optical Thin Films Based on Fast Fourier Transforms. *Thin Solid Films.* **2009,** *517* (17), 5110–5115.

52. Rancourt, J. D. *Optical Thin Films: User Handbook.* SPIE Press: Bellingham, 1996.

53. Image References

1. https://commons.wikimedia.org/wiki/File:Thermal_evaporation1.PNG

2. http://www.kobelcokaken.co.jp/target/english/index.html

3. http://blog.brukerafmprobes.com/category/guide-to-spm-and-afm-modes/

4. https://www.purdue.edu/ehps/rem/rs/sem.html

5. http://www.hk-phy.org/atomic_world/tem/tem02_e.html

CHAPTER 2

FABRICATION OF EXCEPTIONAL AMBIENT STABLE ORGANIC FIELD-EFFECT TRANSISTORS BY EXPLOITING THE POLARIZATION OF POLAR DIELECTRIC LAYERS

NIMMAKAYALA V. V. SUBBARAO[1] and DIPAK K. GOSWAMI[2,*]

[1]Center for Nanotechnology, Indian Institute of Technology Guwahati, Guwahati 781039, Assam, India

[2]Department of Physics, Indian Institute of Technology Kharagpur, Kharagpur 721302, West Bengal, India

[*]Corresponding author. E-mail: dipak@phy.iitkgp.ernet.in

CONTENTS

ABSTRACT

In this chapter, we have studied the systematic effects of different polymeric dielectric layers on the operational and environmental stability of CuPc based organic field-effect transistors (OFETs). Hydrophilic poly(vinylalcohol) (PVA) and hydrophobic poly-(methylmethacrylate) (PMMA) polymers were used as organic dielectric materials in addition to the anodized alumina layer. PVA, PMMA, and PVA/PMMA single and bilayers are used as dielectric materials. We have studied the performance and stability of the devices under vacuum and ambient air conditions. PVA based devices are highly sensitive to polar ambient gases and degrade very easily. In order to overcome that we have used PMMA layer together with the PVA layer as bilayer combination. The hysteresis and bias-stress stability were studied under both the conditions. The variation of drain current under bias-stress was explained by the modified stretched exponential function by introducing dipole polarization relaxation term. The possible mechanism for the stability of the device was explained by proposing a model for the polarization of hydroxyl groups present in PVA layer under vacuum and humidity conditions. We demonstrated how the systematic exploration of the engineered dielectrics could provide a meaningful step toward optimizing the organic semiconductor/dielectric interface, thereby implementing highly ambient stable and high-performance OFETs.

2.1 INTRODUCTION

Over the last three decades, organic semiconductors have been extensively studied to understand the physics of the molecules for scientific and technological reasons.[1,2] The main interesting feature of the molecules is that the tunability of chemical properties by functionalization with different moieties. The research has long been a subject of scientific interest for many chemists and physicists for the broad scope of research and applications.[3,4] These organic semiconducting molecules are the key components for organic field-effect transistors (OFETs), organic light emitting diodes (OLEDs), and organic solar cells (OSCs). The development of stable organic semiconductors is still behind the demand of the real applications because of their instability under ambient conditions. Thus, air-stable, high-performance, and low cost organic semiconductors are still needed for further advancement of the organic electronics technology. OFETs have recently attracted considerable interest due to their unique large-area and low-cost applications in a variety of fields like transparent flexible devices,[5,6] radio frequency identification (RFID) tags,[7,8] and sensors.[9-11]

The operational and environmental stabilities of OFETs are the main challenges and have been the subject of serious research interest in the recent years.[12] The device performance and stabilities are mainly depends on the nature of the molecules or polymers, the device engineering and the conditions used for the fabrication. The device stability is typically represented by stability under gate bias-stress and ambient air under long-term operation. The gate dielectrics materials are one of the important components play a crucial role to have a huge impact on the electrical performance of OFETs.[13] Different types of inorganic and organic dielectric materials were used in the literature for the fabrication of OFETs.[14-22] SiO_2, Al_2O_3, TiO_2, HfO_2, etc. metal oxide dielectric materials were used. The processing of metal oxides requires high temperatures and sophisticated equipments. On the other hand, the use of organic dielectric materials is increased due to their easy processing conditions and offer an alternative solution to the traditionally used inorganic dielectric materials, even though; they have low dielectric constant than non-polymeric materials.[23-27] Organic dielectric materials are mainly the polymers with high molecular weight and can be classified into hydrophobic and hydrophilic.[28] The growth of the organic semiconducting molecules on the surface of the polymer depends on its nature because of the favorable interaction between organic semiconductor and dielectric material.[3,29] The hydrophilic dielectric material contains hydroxyl groups which absorb water molecules easily and act as trap states.[30] This is quite cumbersome as the trapped or absorb moisture will dramatically alter the desired electrical properties of the OFET devices. The use of polymer dielectric materials increased recently due to their highly flexible, tractable processing, good chemical stability, and readily tunable properties. The main drawback is that they have lower thermal stability which limits their wider applications. There are numerous number of polymer dielectric materials, such as poly(vinylalcohol) (PVA), poly-(methylmethacrylate) (PMMA), poly(vinyl phenol) (PVP), polystyrene (PS), benzocyclobutene (BCB), etc. Polymers can be fabricated fairly easily into thin film by solution casting, spin coating, and dip coating methods. These films can be processed at lower temperature due to low glass transition and melting temperature.

The nature of the dielectric material greatly affects the performance of the device. The OFET devices typically exhibit hysteresis and bias-stress based on the charge trapping at the interface of organic semiconductor and dielectric materials. The hysteresis leads to the instability of the device. The hysteresis can be avoided for the stability application; however, this is exploited in the memory applications where the hysteresis used for the read/write operations. The bias-stress can also be used to estimate the stability of the device. Generally, under gate bias-stress I_{DS} will decay with time due to charge trapping at the interface of organic semiconductor and dielectric material.[31] Instead of decay, the I_{DS} increase under bias-stress was also observed in polarizable gate dielectrics, called anomalous bias-stress effect.[32,33] The increase/decrease of I_{DS} under bias-stress significantly affects the stability of the devices. The degradation of the devices under ambient conditions is also a major impediment for the practical applications.[34] In such conditions, active semiconducting channel of the OFETs absorbs water from ambient moisture, which are act like charge carrier traps at the semiconductor–dielectric interface, resulting poor carrier mobility and hysteresis in the transfer characteristics.[35,36] These molecules can also be trapped within the dielectric layers by penetrating through the semiconductor.[37] In presence of gate field, the trapped water molecules can be polarized and affects the device performance, results a huge hysteresis in transfer characteristics and decay of bias-stress curve.[38] The degradation further pronounced when the polymer dielectrics contain polar hydroxyl group such as PVA or PVP.[39] These hydroxyl groups can easily attract water molecules because of their hydrophilic nature.[40] Hence, the hydroxyl groups in the organic dielectrics play a crucial role in the stability of the OFETs.[35] This has been an issue in the fabrication of OFETs and attempts have been taken to reduce the concentration of hydroxyl groups by cross-linking the polymer dielectric materials.[41,42] When the polar dielectric layer is separated from a semiconductor channel by non-polar dielectric, the adsorbed water molecules do not come in contact directly with the hydroxyl groups of polar polymers. This non-polar polymer layer diminishes the number of charge traps at the interface and creates a compatible platform for the growth of organic molecules. In such conditions, polarized water molecules under gate field interact with the hydroxyl groups through the non-polar layer. As a result, hydroxyl groups will polarize much faster than before, resulting enhanced device stability by reducing hysteresis.[26,43]

In this chapter, we report high-performance, low operating voltage, and p-channel OFET devices based on the PVA, PMMA single layer, and PVA/PMMA bilayer dielectrics along with the Al_2O_3 layer. The effect of the dielectric material on the performance of the device was studied under vacuum and ambient conditions. The PMMA/PVA/Al_2O_3 based device showed enhanced mobility and stability with a high reproducibility under ambient conditions. The OFET device fabricated with PVA/Al_2O_3 showed less performance than PMMA/Al_2O_3 and PMMA/PVA/Al_2O_3 devices. The degradation of the device with PVA/Al$_2$O$_3$ dielectric was studied in both the vacuum and ambient conditions. FTIR spectroscopy reveals an existence of —OH functional groups in the PVA and PMMA/PVA films, which are responsible for the hysteresis and degradation of the device.

2.2 EXPERIMENTAL SECTION

OFETs with top-contact and bottom gate configurations were fabricated on glass substrates by thermally evaporated ~100 nm thick aluminum layer as a gate electrode. The top surface of the aluminum was anodized to convert the aluminum into thick barrier alumina, which can act as inorganic dielectric layer having high dielectric constant and low leakage current. Anodic oxidation is an easy and cost effective method to grow nanometer-controlled aluminum oxide (Al$_2$O$_3$) on the aluminum surface. Anodization process was carried out at room temperature using 0.001 M citric acid monohydrate (C$_6$H$_8$O$_7$H$_2$O) as electrolyte solution. Aluminum deposited glass slides were immersed into an aqueous electrolyte solution to form working electrode (anode) and platinum mesh served as counter electrode (cathode). Keithley 2400 source meter was connected to the cathode and anode terminals. To perform the anodic oxidation, a constant current density $j = 0.4$ mA/cm^2 was driven through the electrolyte. Constant current across the growing insulating Al$_2$O$_3$ layer was maintained by ramping up the voltage up to a limiting anodization voltage V_A, which was then maintained for several minutes until the current had completely decayed to below ~6 μA. The samples were rinsed in high-purity hot milli Q water and dried on a hot plate remove the acid molecules and water.[44] To study the effect of dielectric material, PVA, PMMA, and PMMA/PVA polymer films were spin coated to fabricate single and bilayer polymer dielectric materials. The PVA dielectric was prepared by dissolving PVA (Sigma-Aldrich, M_w = 76,500–81,000 kg/mol) 30 mg/mL in de-ionized water. The polymer solution was spin coated onto the anodized glass slides at 3000 rpm for 60 s, followed by annealing at 100°C for 1 h in vacuum oven. Similarly, 30 mg PMMA dissolved in 1 mL of anisole to get 70 nm thick layer. Bilayer dielectric device was fabricated by spin coating PMMA and PVA. After spin coating of PVA, PMMA (10 mg/mL dissolved in anisole) was spin coated at 3000 rpm for 60 s to get thickness about 30 nm, followed by vacuum annealing at 120°C for 1 h to remove residual solvent and to improve the film quality. These single and bilayer polymer dielectric films were characterized by Perkin Elmer FTIR spectrophotometer by spin coating onto the KBr pellets. CuPc films of about ~50 nm were

deposited on to the polymer dielectrics in a high vacuum thermal evaporator (<3 × 10⁻⁶ mbar) with a substrate temperature of 80°C at a rate 0.02
nm/s. The dielectric and semiconductor film thickness were measured by the
profilometer (Veeco, Dektak 150). The surface morphology of the polymer
dielectrics and organic semiconductor films were characterized in air by
the atomic force microscopy (AFM) (Agilent AFM/STM 5500) in tapping
mode. Finally, 50 nm thick copper was deposited as source and drain electrodes through a shadow mask with a deposition rate of about 0.5 nm/s. The
channel length (L) and channel width (W) are defined as 25 and 750 μm,
respectively. The schematic illustration of the single and bilayer dielectric
based OFET devices was shown in Figure 2.1

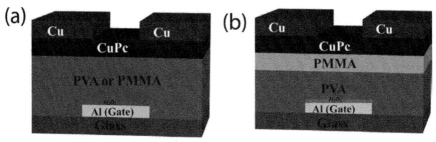

FIGURE 2.1   Schematic illustration of (a) single dielectric layer (PVA or PMMA) based
OFET (b) bilayer (PVA/PMMA) based OFET. Al used as gate electrode material and anodized
alumina used as inorganic dielectric material to reduce the leakage current. CuPc used as
organic dielectric material and copper used as source drain contact material.

Electrical characterization of the OFETs was carried out in a probe
station (Lake Shore, <3 × 10⁻⁴ mbar) with different environments under
dark and normal light conditions. All the devices were tested under vacuum
(~1 × 10⁻⁴ mbar) as well as in ambient (70% relative humidity) using Lake
Shore probe station connected with Keithley 4200 semiconductor characterization system (SCS). All capacitance–electric field (C–V) measurements, leakage current, current–voltage (I–V) characteristics, and bias-stress
measurement of OFETs were collected with a Keithley 4200-SCS semiconductor parameter analyzer with two source measure units (SMUs). The key
device parameters such as μ_{FE}, on/off current ratio (I_{off} / I_{on}), and threshold
voltage (V_{Th}) were extracted using equation of drain current (I_{DS}) in the saturation region as

$$I_{DS} = \frac{\mu_{FE} C_{diel} W}{2L} \left(V_{GS} - V_{Th} \right)^2$$

(2.1)

where C_{diel} is the capacitance of the dielectric layer per unit area, V_{GS} and V_{DS} are the gate-source voltage and drain-source voltage, respectively.[45] All the data listed in the chapter are average values of at least 10 devices on each of the samples, fabricated with common gate electrode and independent source drain electrodes.

2.3   RESULTS AND DISCUSSION

The performance of the OFET depends on the nature of the dielectric material and the concentration of hydroxyl groups (—OH groups) present. The —OH functional groups act like traps and also interact with the water molecules, leading to the degradation of the device. To understand the nature of the single and bilayer polymer dielectric materials, spectroscopic measurements were performed at room temperature. The polymer films were deposited on the KBr pellets by spin coating technique and characterized by using FTIR spectroscopy to find out the functional groups and their concentration. Prior to the measurements, the films were heated in vacuum oven to remove the solvent molecules and used immediately for characterization. Figure 2.2 shows the FTIR spectra in transmittance mode against wave number. The

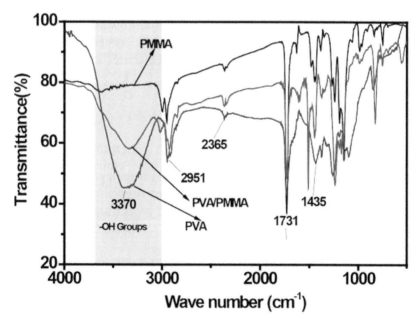

FIGURE 2.2   FTIR spectra of the polymer thin films coated on the KBr pellets and measured at room temperature.

region 3000–3700 cm^{-1} represents O—H broadening stretching peak indicates the existence of —OH functional groups. The PMMA film does not show any peak related to —OH groups. The films PVA and PVA/PMMA show the existence of hydroxyl groups. It can be concluded that the PVA has the hydrophilic nature and PMMA has the hydrophobic nature without any —OH groups. The concentration of hydroxyl groups decreases with the passivation of PMMA layer on the PVA layer as shown in Figure 2.2. The broad peak ranging from 1260 to 1000 cm^{-1} can be explained owing by the C—O (ester bond) stretching vibration in the PMMA.[46]

The same compositions of films were coated on the anodized alumina gate substrates to fabricate the OFET devices. The leakage current density and capacitance of the PVA/Al$_2$O$_3$, PMMA/Al$_2$O$_3$, and PMMA/PVA/Al$_2$O$_3$ layers were measured using a metal–insulator–metal (MIM) configuration by deposition of copper as the top electrode. The typical plots of the leakage current density versus voltage (J–V) and capacitance versus frequency (C–F) and are shown in Figure 2.3a, b. The values of leakage current density for the PVA/Al$_2$O$_3$, PMMA/Al$_2$O$_3$, and PMMA/PVA/Al$_2$O$_3$ dielectric systems were very low, ~5 × 10^{-10}, ~4 × 10^{-9}, and ~2 × 10^{-9} A/cm^2, respectively. In general, dense dielectric films with high k can effectively prevent leakage current through the films and reduces the capacitance because the capacitance is inversely proportional to the dielectric thickness. The capacitance was measured to be 80, 23, and 55 nF/cm^2 at 100 kHz for the PVA/Al$_2$O$_3$, PMMA/Al$_2$O$_3$, and PMMA/PVA/Al$_2$O$_3$ dielectric systems, respectively. The effective dielectric constant k for the dielectrics could be derived using a model of two or three capacitors in a series $\dfrac{1}{C_i} = \dfrac{1}{C_{Al_2O_3}} + \dfrac{1}{C_{PVA(or)PMMA}}$, $\dfrac{1}{C_i} = \dfrac{1}{C_{Al_2O_3}} + \dfrac{1}{C_{PVA}} + \dfrac{1}{C_{PMMA}}$. The $k_{A/PVA}$, $k_{A/PMMA}$, and $k_{A/PVA/PMMA}$ can be determined by the following equation, $C_i = \dfrac{k\varepsilon_0 A}{d}$, where ε_0 is the vacuum permittivity and d the thickness of the dielectric layer, the dielectric constant $k_{A/PVA}$, $k_{A/PMMA}$, and $k_{A/PVA/PMMA}$ were determined to be 9, 2.6, and 6.3, respectively. The observed good dielectric properties and low level of leakage currents of these layers were expected to enable OFETs which can be operate at low voltage levels.

The effect of a k on the mobility (μ_{FE}) of OFETs has been generally explained with two hypotheses being contrary to each other: a μ_{FE} increased or decreased as the k value of the dielectric increased. The former case suggested that a higher k value could induce more free carriers into the channel formed at the semiconductor/dielectric interface during application of a V_{GS} to the device, which allowed for the traps present near the interface to fill at a relatively lower V_{GS}. Hence, V_{Th} approached 0 V, and the channel conductance and μ_{FE} in the saturation regime increased. On the other hand, several studies have reported that a high polarizability in a dielectric increased with k is able to broaden the density of state (DOS) near the edge of the conduction/valence bands, leading to trap formation that enables to cause the decreased μ_{FE} of a device. Here we considered both the systems to investigate how k influenced the OFET performances.

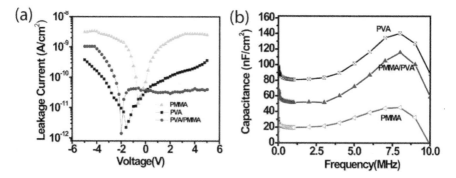

FIGURE 2.3   (a) Leakage current vs. applied voltage and (b) capacitance vs. frequency curves of PVA, PMMA single layers, and PVA/PMMA bilayer with 10 nm Al_2O_3 layer.

To gain a better understanding of the effect of morphology on the charge carrier mobility, AFM has been utilized to describe the distinctive features of the polymer and organic semiconductor layers. Tapping mode AFM images (Fig. 2.4) were recorded on the same devices used for OFET measurements. Figure 2.4(a), (b), (c), and (d) shows the AFM images of PVA, PMMA surfaces and CuPc molecules deposited on PVA and PMMA surfaces, respectively. Average surface roughness of PVA was 0.42 nm, whereas PMMA has 2.12 nm. Polymer coating made the surface extremely smooth which is highly essential for the semiconductor deposition. So, the interface traps can be reduced and enhances the device performance. AFM image of Figure 2.4(c) and (d) confirms the grain type structure of CuPc molecule on the insulator substrate deposited at the 80°C substrate temperature. This type of morphology is due to the strong propensity of planar CuPc to from one dimensional structure. The roughness was found to be 7–8 nm. The formation of grain boundaries may then act as carrier scattering sites and contribute to the reduction of the field-effect mobility.

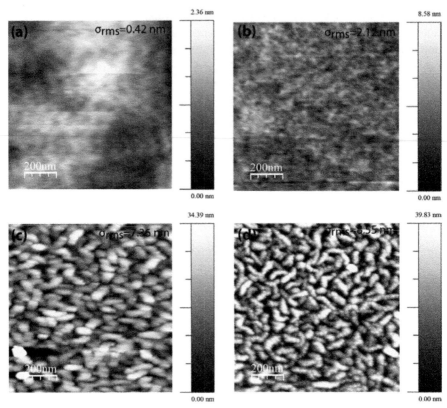

FIGURE 2.4 Surface morphology of (a) PVA, (b) PMMA, (c) CuPc deposited on PVA surface, and (d) CuPc deposited on PMMA surface.

A comparative analysis was performed to investigate the effect of PVA/Al_2O_3, PMMA/Al_2O_3, and PVA/PMMA/Al_2O_3 dielectric layers on the device performance. All the thin-film OFETs grown on the different dielectrics displayed reproducible characteristics in the low-voltage regime of $<|12|$ V, and the I–V characteristics were satisfactorily described by the standard field-effect transistor equations working in the accumulation mode. The devices showed excellent output and transfer characteristics measured under both vacuum and ambient conditions with relative humidity 65%. Figure 2.5 shows the output and transfer curves for a PVA/Al_2O_3 dielectric based device measured under vacuum and ambient conditions. CuPc used as the p-type organic semiconductor. The device showed excellent p-channel characteristics under vacuum with anti-clock wise hysteresis, where the on-to-off current higher than the off-to-on current in both the transfer and output curves (Fig. 2.5a,c). There are two possible mechanisms that can link the sweep change with the current (I_{ds}) increase: (1) slow polarization of the gate dielectric, or (I. Murtaza, #18503) charge trapping and detrapping at the semiconductor/dielectric interface. When the gate is swept from off-to-on, polar species in the dielectric medium are polarized at the interface between the semiconductor and the gate electrode. If the polarized species do not respond quickly to the changes in the gate electric field, they may remain polarized without being able to return to their original sites during the on-to-off gate sweep. As a result, the on-to-off transfer curve shifts to higher V_{GS} values compared to the off-to-on sweep.[47]

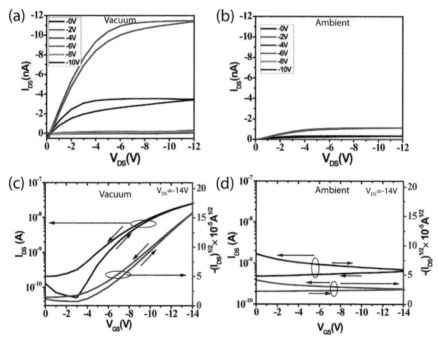

FIGURE 2.5 (a,b) Output and (c,d) transfer characteristics of a representative device with PVA as the single dielectric material under vacuum (a and c) and ambient air (b and d) conditions. The device exhibited hysteresis under vacuum. Under ambient, the device not showed any good field-effect behavior.

Under ambient conditions, we did not observe any output or transfer characteristics with reasonable quality to deduce the hole mobility. The device properties were degraded completely as shown in Figure 2.5b, d. The transfer curves did not show any linear and saturation regions. At this point, we can conclude that this dielectric material does not exhibit field-effect behavior under ambient, due to the presence of interface —OH groups. This is caused by the energy fluctuation induced by randomly oriented dipole moments of polymer gate dielectrics and water molecules which are absorbed from the ambient. The random dipole moments broaden density-of-state distribution of semiconductors and thus promote charge trapping. The broadening of distribution of DOS is more severe when the gate dielectric is more polar. The average field-effect hole mobility in a saturation regime under vacuum was derived from the linear fits of $(I_{DS})^{1/2}$ versus V_{GS}. The mobility and threshold voltage values derived under vacuum were largely different when estimated from the forward and reverse sweeps because of a large hysteresis. From the forward sweep the calculated mobility is 0.0012 cm²/(Vs) with a threshold voltage (V_{Th}) of −4.23 V. Conversely, from the reverse sweep, the mobility decreased to 0.0007 cm²/(Vs), with a V_{Th} of −3.15 V. The device parameters were calculated under vacuum to be $I_{on}/I_{off} < 10^3$ and a low subthreshold swing (SS) value of 2.13 and 3.92 V/dec for both the directions. Table 2.1 summarizes the device parameters including I_{on}/I_{off}, threshold voltages, the SS, and hysteresis gap. The hysteresis gap (ΔV_T) is defined as the difference in threshold voltage between the backward and forward V_{GS} sweeps, as determined by the linear extrapolation method.[47] The hysteresis gap found to be 1.08 V under vacuum conditions.

Figure 2.6 shows the output and transfer curves of the OFETs fabricated with PMMA/Al$_2$O$_3$ dielectric material and measured under vacuum and ambient conditions. The device showed improved performance under ambient condition contrary to the PVA/Al$_2$O$_3$ dielectric based OFET. The device exhibited higher mobilities and negligible hysteresis under both the vacuum and ambient conditions. An average hole mobility 0.0083 (cm²/Vs) and 0.0087 (cm²/Vs), threshold voltage of −2.53 and −3.25 V, and on/off ratio of ~10⁴ and ~10³ were calculated under vacuum and ambient conditions, respectively. The mobilities are found to be seven times higher than the previous device values. The improved mobility for device compared to PVA device, possibly due to the reduction of traps at the interfaces by the hydrophobic PMMA layer. There is no significant degradation in overall OFETs performance under ambient except an increase in the mobility by ~5%. The excellent device performance with the PMMA dielectric is attributed to the combined effects of the acquisition of high-quality films on hydrophobic interface layer and leads to the less number of interface traps. The extracted device parameters of the devices are shown in Table 2.1.

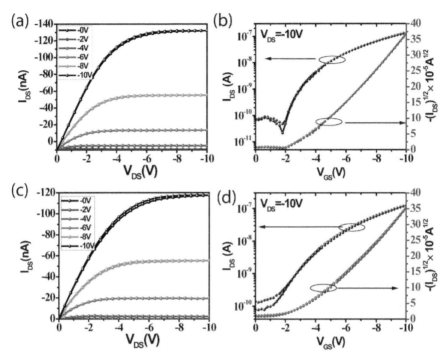

FIGURE 2.6 (a,b) Output and (c,d) transfer characteristics of a representative device with PMMA as the single dielectric material under vacuum (a and b) and ambient air (c and d) conditions.

TABLE 2.1 Summary of the Electrical Parameters for CuPc OFETs Fabricated Using Different Dielectric Materials and Measured under Vacuum and Humidity Conditions.

Device	Measurement		C_i (nF/ cm^2)	μ (cm^2/ Vs)	V_{Th} (V)	S (V/dec)	$I_{on/off}$	N_{trap}
PVA/Al$_2$O$_3$	Vacuum	FS		0.0012	−4.23	2.13	~0.71 ? 10^3	1.74 ? 10^{13}
		RS	80	0.0007	−3.15	3.92	~0.15 ? 10^3	3.24 ? 10^{13}
	Ambient air							
PMMA/Al$_2$O$_3$	Vacuum		23	0.0083	−2.53	2.34	~1 ? 10^4	0.55 ? 10^{13}
	Ambient air			0.0087	−3.25	1.98	~1 ? 10^3	0.46 ? 10^{13}
PMMA/PVA/ Al$_2$O$_3$	Vacuum	FS		0.0042	−4.23	1.63	~1 ? 10^3	0.91 ? 10^{13}
		RS	55	0.0027	−2.86	2.03	~1 ? 10^3	1.14 ? 10^{13}
	Ambient air	FS		0.012	−3.75	1.96	~1 ? 10^3	1.10 ? 10^{13}
		RS		0.012	−3.75	1.96	~1 ? 10^3	1.10 ? 10^{13}

μ, field-effect mobility; V_{Th}, threshold voltage; S, subthreshold slope; $I_{on/off}$, on/off current ratio, where FS is forward sweep and RS reverse sweep.

Figure 2.7 shows the typical dual sweep output and transfer characteristic curves of PMMA/PVA/Al$_2$O$_3$ dielectric based OFET device operated under both the vacuum and ambient humidity conditions. Under vacuum, the device exhibited anticlockwise hysteresis like PVA device as shown in Figure 2.7(a). The maximum drain current obtained at $V_{GS} = -10$ V is ~100 nA. Whereas under ambient conditions, the device showed nil hysteresis in both the output and transfer curves as shown in Figure 2.7(b and c). The extracted device parameters of the devices are shown in Table 2.1. The I_{DS} was increased from 100 to 230 nA, when switching the atmosphere from vacuum to ambient (~60 RH%). An average mobility of 0.0042 and 0.0027 cm^2/Vs, threshold voltage of −4.23 and −2.86 V, and on/off ratio of ~10^3 and ~10^3 were measured under vacuum for FS and RS of the transfer curve, respectively. The hysteresis gap found to be 1.37 V under vacuum conditions, which is higher than the gap calculated for the PVA/Al$_2$O$_3$ device. Under ambient conditions, the mobility of 0.012 cm^2/Vs, threshold voltage of −3.75 V, and on/off ratio of ~10^3 were calculated. The mobility enhanced by four times under ambient conditions and by two times than the PMMA/Al$_2$O$_3$ device.

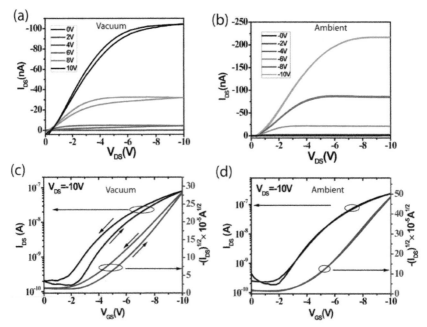

FIGURE 2.7 (a,b) Output and (c,d) transfer characteristics of a representative device with PVA/PMMA as bilayer dielectric material under vacuum and ambient air conditions. The device exhibited hysteresis under vacuum whereas it is completely eliminated in the ambient with enhanced properties.

The reason for the enhanced performance can be explained as follows: The hydroxyl groups (–OH) in the polymer dielectric material act like electron traps and degrade the performance of the device under long-term operation, when they are in the interface of dielectric/organic semiconductor. But, the effect is entirely different, when polar dielectric is capped by a non-polar dielectric. As a result, hydroxyl groups will try to orient slowly in the field direction generated by the applied gate voltage. There are several reports that demonstrated that the moisture is the main cause of bias-stress instability with polymer dielectrics.[48] Slow polarization also increases the capacitance of the gate dielectric, resulting in a large increase in I_{DS} and the overestimation of mobility. However, the complete absence of hysteresis in the case of PMMA dielectric based devices may be due to the more hydrophobic nature. Although the higher dielectric constant of the gate insulator provides a greater degree of charge injection in the channel using low permittivity polymers allows avoiding trap states by reducing energetic disorder. The number of traps at the interface mainly related to the nature of the interface. The trap density (N_{trap}) at the interfaces can be calculated using the following equation:

$$N_{trap} = \left[\frac{S.\log(e)}{kT/q} - 1 \right] . \frac{C_i}{q} \qquad (2.2)$$

where C_i is the gate dielectric capacitance per unit area and q is the elementary charge. As listed in Table 2.1, the N_{trap} values drastically decreased with the PMMA dielectric layer. Theoretically, the N_{trap} value can represent the extent of structural disorders and/or defects in the thin films and show strong correlation with the transistor performance. Such drastic reduction of N_{trap} value for the PMMA/Al$_2$O$_3$ device clearly represents the hydroxyl-free polymers induces the growth of organic active layers with fewer traps than the polar dielectrics. The observed hysteresis was also found to be closely related to both the trapping of charge carriers at the interface between organic semiconductors and dielectrics, and polarization of the gate dielectrics. The aforementioned N_{trap} reduction at the PMMA, PMMA/PVA interfaces indicates that a considerable fraction of the trap sites was removed, because of the hydroxyl-free PMMA interface. The PMMA devices exhibited low trap density of the order of ~10^{13} than compared to the other two devices. In general, most organic-based devices are known to lack the stability in air when exposed to oxygen and moisture. The improved operational stability resulted from the hydrophobic PMMA layer preventing the penetration of oxygen and moisture responsible for the charge trapping at the semiconductor/dielectric interfaces and capacitance of the inner PVA layer increased. Low trap density and enhanced mobility are achieved with the PMMA/PVA/Al_2O_3 device. Such enhancement suggests that the coating of interface layer with PMMA scheme can be suitable for the production of high-performance devices that can operate under ambient conditions.

In order to investigate the stability of the three devices, bias-stress measurements were performed under vacuum and ambient conditions. The devices were subjected to 30 min of dc bias-stress to measure the variation of I_{DS} with the time. Figure 2.8 shows the time dependent I_{DS} behavior of the devices under constant bias-stress ($V_{GS} = V_{DS} = -10$ V) in vacuum and humidity. In vacuum, PMMA device exhibited the decay in drain current. But, PVA and PMMA/PVA devices exhibited the anomolous behavior of I_{DS} with current increases initially and decays after some time. Under ambient conditions, all the devices exhibited the usual I_{DS} decay under 1 h of bias-stress. The decay in I_{DS} during electrical bias-stress is commenly related to the charge carrier trapping at the interface of dielctric and organic semiconductors channel, or in the bulk of the semiconductor. The normal I_{DS} decay can be expressed with stretched exponential function.[49,50] Generally in the polar polymer dielectrics, the permenent dipoles (i.e., −OH groups) are strongly attached to the backbone of the polymer with an orientation perpendicular or parallel to the main chain.[51] These dipoles will orient slowly under the applied electric field with a relaxation time in the range of seconds to days. Thus, it is expected that the polar molecules, that is, −OH groups in the PVA and PVP are slowly polarized under gate bias. As the polarization increases with time, the aligned dipoles induce additional charges in the accumulation region. Therefore, we observed an enhancement of drain current up to a certain time untill the polarization of the dipoles saturates. The time require to reach the maximum current depends on the polarization time (τ_p) for the particular dielectric material. Therefore, it is necessary to introduce the orientation of dipoles with respect to time under the continuous gate bias in the expression of I_{DS}. Hence, we modified the traditional model used for I_{DS} decay by introduicing the polarization term. The additional induced current due to the polarization of the dipoles can be considered as[52,53]

$$I_{DS}(t) = I_0(0) \, exp\left[-\left(\frac{t}{\tau_d}\right)^\alpha\right]$$

(2.3)

$$I_{DS}(t) = I_0(0) \left[\left\{ 1 - exp\left(-\left(\frac{t}{\tau_p} \right)^\alpha \right) \right\} + exp\left(-\left(\frac{t}{\tau_d} \right)^\beta \right) \right] \qquad (2.4)$$

where α and β represents the dispersion parameters describing the distribution of activation energy for charge inducing in the active channel and charge trapping in the interface and channel. τ_p and τ_d represent the characteristic time constants for charge inducion and charge trapping. $I_0(0)$ is the drain current at time $t = 0$, β is the dispersion parameter whose value is $0 < \beta < 1$, and τ_d is the characteristic decay time. The curves are fitted using the above two equations and the fitting parameters are summarized in Table 2.2. Under ambient conditions the calculated β and τ_d values are calculated to be 0.423 and 850s; 0.352 and 12,630 s for the PMMA and PMMA/PVA dielectric based devices, respectively. Under vacuum, the PMMA device only exhibited the decay with $\beta = 0.757$ and $\tau_d = 5374$ s. The other devices show the anomolous bias-stress effect with the fitting parameters $\alpha = 0.553$ and 0.732, $\beta = 0.725$ and 0.647, $\tau_p = 752$ and 485 s, and $\tau_d = 3210$ and 3309 s, respectively, for the PVA and PMMA/PVA devices. It is clearly indicating that the trilayer PMMA/PVA/Al$_2$O$_3$ dielectric can effectively reduce the interface traps and increases the decay time to about one order of magnitude higher than bilayer PMMA/Al$_2$O$_3$.

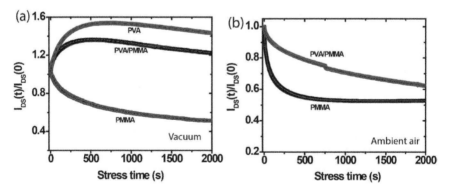

FIGURE 2.8 Bias-stress measurement for 30 min under (a) vacuum and (b) ambient conditions for the PVA, PMMA, and PMMA/PVA devices. Fitting to stretched exponentials are also shown as thin lines.

TABLE 2.2   Bias-Stress Fitting Parameters of the Three Devices under Vacuum and Ambient Conditions.

Device	Ambient			Vacuum		
	β	τ_d	α	β	τ_p	τ_d
PVA/Al$_2$O$_3$			0.553	0.725	752	3210
PMMA/Al$_2$O$_3$	0.423	850		0.757		5374
PMMA/PVA/Al$_2$O$_3$	0.352	12630	0.732	0.647	485	3309

2.4   CONCLUSION

In conclusion, we have demonstrated effect of polymer dielectric single and bilayers on the performance of the device in vacuum and humidity conditions. The anomalous bias-stress and hysteresis were observed under vacuum conditions for the dielectric layers contains PVA as at least one layer and the phenomenon was explained by the slow polarization of the bulk dipoles under the gate field effect. We have used stretched exponential functions to fit the variation of drain current under both the vacuum and ambient conditions. The device containing PMMA/PVA as bilayer dielectric is showed a high mobility and ambient stability. The passivation of PVA surface with PMMA reduced the trap density and enhanced the OFET performance by reducing the trapping and de-trapping. Our results show that the degradation of the device under ambient conditions can be significantly tuned by engineering the polymer dielectric structure and exploited to create useful high-performance and stable electronic devices for real applications.

ACKNOWLEDGMENTS

The financial support provided by the Department of Science and Technology (DST), Government of India under the projects DST/TSG/ME/2008/45, and DST/TSG/PT/2009/23 to carry out this research work is fully acknowledged. Authors would also like to acknowledge Prof. P. I. Iyer, Department of Chemistry, IIT Guwahati for his support to carry out experiments. Organic Electronics Laboratory, IIT Guwahati, where experiments were carried out, is acknowledged.

KEYWORDS

- **organic semiconductors**
- **organic field-effect transistors**
- **forward sweep**
- **mobility**
- **threshold voltage**

REFERENCES

1. Burroughes, J. H.; Bradley, D. D. C.; Brown, A. R.; Marks, R. N.; Mackay, K.; Friend, R. H.; Burn, P. L.; Holmes, A. B. Light-Emitting-Diodes Based on Conjugated Polymers. *Nature.* **1990,** *347,* 539–541.

2. Forrest, S. R. The Path to Ubiquitous and Low-Cost Organic Electronic Appliances on Plastic. *Nature.* **2004,** *428,* 911–918.

3. Coropceanu, V.; Cornil, J.; Silva, D. A.; Olivier, Y.; Silbey, R.; Bredas, J. L. Charge Transport in Organic Semiconductors. *Chem. Rev.* **2007,** *107,* 926–952.

4. Sirringhaus, H. Device Physics of Solution-Processed Organic Field-Effect Transistors. *Adv. Mater.* **2005,** *17,* 2411–2425.

5. Sekitani, T.; Zschieschang, U.; Klauk, H.; Someya, T. Flexible Organic Transistors and Circuits with Extreme Bending Stability. *Nat. Mater.* **2010,** *9,* 1015–1022.

6. Jeong, J. K. The Status and Perspectives of Metal Oxide Thin-Film Transistors for Active Matrix Flexible Displays. *Semicond. Sci. Technol.* **2011,** *26,* 034008.

7. Myny, K.; Steudel, S.; Smout, S.; Vicca, P.; Furthner, F.; van der Putten, B.; Tripathi, A. K.; Gelinck, G. H.; Genoe, J.; Dehaene, W.; Heremans, P. Organic RFID Transponder Chip with Data Rate Compatible with Electronic Product Coding. *Org. Electron.* **2010,** *11,* 1176–1179.

8. Yeh, K. H.; Lo, N. W.; Tsai, K. Y.; Li, Y. J.; Winata, E. A Novel RFID Tag Identification Protocol: Adaptive n-Resolution and k-Collision Arbitration. *Wireless Pers. Commun.* **2014,** *77,* 1775–1800.

9. Kergoat, L.; Piro, B.; Berggren, M.; Horowitz, G.; Pham, M. C. Advances in Organic Transistor-Based Biosensors: From Organic Electrochemical Transistors to Electrolyte-Gated Organic Field-Effect Transistors. *Anal. Bioanal. Chem.* **2012,** *402,* 1813–1826.

10. Wang, L.; Fine, D.; Sharma, D.; Torsi, L.; Dodabalapur, A. Nanoscale Organic and Polymeric Field-Effect Transistors as Chemical Sensors. *Anal. Bioanal. Chem.* **2006,** *384,* 310–321.

11. Minamiki, T.; Minami, T.; Kurita, R.; Niwa, O.; Wakida, S.; Fukuda, K.; Kumaki, D.; Tokito, S. A. Label-Free Immunosensor for IgG Based on an Extended-Gate Type Organic Field-Effect Transistor. *Materials.* **2014,** *7,* 6843–6852.

12. Bobbert, P. A.; Sharma, A.; Mathijssen, S. G. J.; Kemerink, M.; de Leeuw, D. M. Operational Stability of Organic Field-Effect Transistors. *Adv. Mater. (Weinheim, Ger.)* **2012,** *24,* 1146–1158.

13. Kim, S. H.; Lee, J.; Park, N.; Min, H.; Park, H. W.; Kim, D. H.; Lee, H. S. Impact of Energetically Engineered Dielectrics on Charge Transport in Vacuum-Deposited Bis(triisopropylsilylethynyl)pentacene. *J. Phys. Chem. C.* **2015,** *119,* 28819–28827.

14. Ortiz, R. P.; Facchetti, A.; Marks, T. J. High-k Organic, Inorganic, and Hybrid Dielectrics for Low-Voltage Organic Field-Effect Transistors. *Chem. Rev.* **2010,** *110,* 205–239.

15. Chang, J. W.; Wang, C. G.; Huang, C. Y.; Tsai, T. D.; Guo, T. F.; Wen, T. C. Chicken Albumen Dielectrics in Organic Field-Effect Transistors. *Adv. Mater.* **2011,** *23,* 4077–4081.

16. Tiwari, S. P.; Zhang, X. H.; Potscavage, W. J.; Kippelen, B. Low-Voltage Solution-Processed n-Channel Organic Field-Effect Transistors with High-k HfO2 Gate Dielectrics Grown by Atomic Layer Deposition. *Appl. Phys. Lett.* **2009,** *95,* 223303.

17. Lee, E. K.; Kim, J. Y.; Chung, J. W.; Lee, B. L.; Kang, Y. Photo-Crosslinkable Polymer Gate Dielectrics for Hysteresis-Free Organic Field-Effect Transistors with High Solvent Resistance. *Rsc Adv.* **2014,** *4,* 293–300.

18. Tetzner, K.; Schroder, K. A.; Bock, K. Photonic Curing of Sol-Gel Derived HfO2 Dielectrics for Organic Field-Effect Transistors. *Ceram. Int.* **2014,** *40,* 15753–15761.

19. Yaginuma, S.; Yamaguchi, J.; Itaka, K.; Koinuma, H. Pulsed Laser Deposition of Oxide Gate Dielectrics for Pentacene Organic Field-Effect Transistors. *Thin Solid Films.* **2005,** *486,* 218–221.

20. Park, Y. M.; Daniel, J.; Heeney, M.; Salleo, A. Room-Temperature Fabrication of Ultra-thin Oxide Gate Dielectrics for Low-Voltage Operation of Organic Field-Effect Transistors. *Adv. Mater.* **2011,** *23,* 971–974.

21. Park, Y. M.; Desai, A.; Salleo, A.; Jimison, L. Solution-Processable Zirconium Oxide Gate Dielectrics for Flexible Organic Field Effect Transistors Operated at Low Voltages. *Chem. Mater.* **2013,** *25,* 2571–2579.

22. Huang, C.; Katz, H. E.; West, J. E. Solution-Processed Organic Field-Effect Transistors and Unipolar Inverters Using Self-Assembled Interface Dipoles on Gate Dielectrics. *Langmuir.* **2007,** *23,* 13223–13231.

23. Ye, X.; Lin, H.; Yu, X. G.; Han, S. J.; Shang, M. X.; Zhang, L.; Jiang, Q.; Zhong, J. High Performance Low-Voltage Organic Field-Effect Transistors Enabled by Solution Processed Alumina and Polymer Bilayer Dielectrics. *Synth. Met.* **2015,** *209,* 337–342.

24. Yan, Y.; Huang, L. B.; Zhou, Y.; Han, S. T.; Zhou, L.; Sun, Q. J.; Zhuang, J. Q.; Peng, H. Y.; Yan, H.; Roy, V. A. L. Surface Decoration on Polymeric Gate Dielectrics for Flexible Organic Field-Effect Transistors via Hydroxylation and Subsequent Monolayer Self-Assembly. *ACS Appl. Mater. Interfaces.* **2015,** *7,* 23464–23471.

25. Ha, Y. G. Low Temperature Solution-Processed Gate Dielectrics for Low-Voltage Organic Field-Effect Transistors. *J. Nanosci. Nanotechnol.* **2015,** *15,* 6617–6620.

26. She, X. J.; Liu, J.; Zhang, J. Y.; Gao, X.; Wang, S. D. Operational Stability Enhancement of Low-Voltage Organic Field-Effect Transistors Based on Bilayer Polymer Dielectrics. *Appl. Phys. Lett.* **2013,** *103,* 133303.

27. Veres, J.; Ogier, S. D.; Leeming, S. W.; Cupertino, D. C.; Khaffaf, S. M. Low-k Insulators as the Choice of Dielectrics in Organic Field-Effect Transistors. *Adv. Funct. Mater.* **2003,** *13,* 199–204.

28. Ruckenstein, E.; Hong, L. Polymers with Hydrophilic and Hydrophobic Polymeric Side-Chains. *Macromolecules.* **1993,** *26,* 1363–1368.

29. Virkar, A. A.; Mannsfeld, S.; Bao, Z. A.; Stingelin, N. Organic Semiconductor Growth and Morphology Considerations for Organic Thin-Film Transistors. *Adv. Mater.* **2010,** *22,* 3857–3875.

30. Subbarao, N. V. V.; Gedda, M.; Iyer, P. K.; Goswami, D. K. Enhanced Environmental Stability Induced by Effective Polarization of a Polar Dielectric Layer in a Trilayer Dielectric System of Organic Field-Effect Transistors: A Quantitative Study. *ACS Appl. Mater. Interfaces.* **2015,** *7,* 1915–1924.

31. Zhang, M. H.; Tiwari, S. P.; Kippelen, B. Pentacene Organic Field-Effect Transistors with Polymeric Dielectric Cnterfaces: Performance and Stability. *Org. Electron.* **2009,** *10,* 1133–1140.

32. Hwang, D. K.; Fuentes-Hernandez, C.; Kim, J.; Potscavage, W. J.; Kim, S. J.; Kippelen, B. Top-Gate Organic Field-Effect Transistors with High Environmental and Operational Stability. *Adv. Mater. (Weinheim, Ger.)* **2011,** *23,* 1293–1298.

33. Ng, T. N.; Daniel, J. H.; Sambandan, S.; Arias, A. C.; Chabinyc, M. L.; Street, R. A. Gate Bias Stress Effects Due to Polymer Gate Dielectrics in Organic Thin-Film Transistors. *J. Appl. Phys.* **2008,** *103,* 044506.

34. Choi, H. H.; Lee, W. H.; Cho, K. Bias-Stress-Induced Charge Trapping at Polymer Chain Ends of Polymer Gate-Dielectrics in Organic Transistors. *Adv. Funct. Mater.* **2012,** *22,* 4833–4839.

35. Choi, C. G.; Baez, B. S. Effects of Hydroxyl Groups in Gate Dielectrics on the Hysteresis of Organic Thin Film Transistors. *Electrochem. Solid-State Lett.* **2007,** *10,* 347–350.

36. Gu, G.; Kane, M. G.; Mau, S. C. Reversible Memory Effects and Acceptor States in Pentacene-Based Organic Thin-Film Transistors. *J. Appl. Phys.* **2007,** *101,* 014504.

37. Park, J.; Do, L. M.; Bae, J. H.; Jeong, Y. S.; Pearson, C.; Petty, M. C. Environmental Effects on the Electrical Behavior of Pentacene Thin-Film Transistors with a Poly(methyl methacrylate) Gate Insulator. *Org Electron.* **2013,** *14,* 2101–2107.

38. Hong, S.; Kim, D.; Kim, G. T.; Ha, J. S. Effect of Humidity and Thermal Curing of Polymer Gate Dielectrics on the Electrical Hysteresis of SnO2 Nanowire Field-Effect Transistors. *Appl. Phys. Lett.* **2011,** *98,* 102906.

39. Tsai, T. D.; Chang, J. W.; Wen, T. C.; Guo, T. F. Manipulating the Hysteresis in Poly(vinyl alcohol)-Dielectric Organic Field-Effect Transistors Toward Memory Elements. *Adv. Funct. Mater.* **2013,** *23,* 4206–4214.

40. Kim, S. H.; Jang, J.; Jeon, H.; Yun, W. M.; Nam, S.; Park, C. E. Hysteresis-Free Pentacene Field-Effect Transistors and Inverters Containing Poly(4-vinyl Phenol-co-methyl methacrylate) Gate Dielectrics. *Appl. Phys. Lett.* **2008,** *92,* 183306.

41. Hwang, D. K.; Lee, K.; Kim, J. H.; Im, S.; Park, J. H.; Kim, E. Comparative Studies on the Stability of Polymer Versus SiO2 Gate Dielectrics for Pentacene Thin-Film Transistors. *Appl. Phys. Lett.* **2006,** *89, 093507.*

42. Xu, W. T.; Rhee, S. W. Hysteresis-Free Organic Field-Effect Transistors with High Dielectric Strength Cross-Linked Polyacrylate Copolymer as a Gate Insulator. *Org. Electron.* **2010,** *11,* 836–845.

43. Guo, T. F.; Tsai, Z. J.; Chen, S. Y.; Wen, T. C.; Chung, C. T. Influence of Polymer Gate Dielectrics on N-channel Conduction of Pentacene-Based Organic Field-Effect Transistors. *J. Appl. Phys.* **2007,** *101,* 124505.

44. Nimmakayala, V. V. S.; Murali, G.; Suresh, V.; Parameswar, K. I.; Goswami, D. K. Effect of Thickness of Bilayer Dielectric on 1,7-Dibromo-N,N'-Dioctadecyl -3,4,9,10-Perylene-tetracarboxylic Diimide Based Organic Field-Effect Transistors. *Phys. Status Solidi. A.* **2014,** *211,* 2403–2411.

45. Klauk, H. Organic Thin-Film Transistors. *Chem. Soc. Rev.* **2010,** *39,* 2643–2666.

46. Singho, N. D.; Lah, N. A. C.; Johan, M. R.; Ahmad, R. FTIR Studies on Silver-Poly(Methylmethacrylate) Nanocomposites via In-Situ Polymerization Technique. *Int. J. Electrochem. Sci.* **2012,** *7,* 5596–5603.

47. Arumugam, S.; Cortizo-Lacalle, D.; Rossbauer, S.; Hunter, S.; Kanibolotsky, A. L.; Inigo, A. R.; Lane, P. A.; Anthopoulos, T. D.; Skabarat, P. J. An Air-Stable DPP-Thieno-TTF Copolymer for Single-Material Solar Cell Devices and Field Effect Transistors. *ACS Appl. Mater. Interfaces.* **2015,** *7,* 27999–28005.

48. Noh, Y. H.; Park, S. Y.; Seo, S. M.; Lee, H. H. Root Cause of Hysteresis in Organic Thin Film Transistor with Polymer Dielectric. *Org. Electron.* **2006,** *7,* 271–275.

49. Padma, N.; Sen, S.; Sawant, S. N.; Tokas, R. A Study on Threshold Voltage Stability of Low Operating Voltage Organic Thin-Film Transistors. *J. Phys. D. Appl. Phys.* **2013,** *46.*

50. Di Girolamo, F. V.; Aruta, C.; Barra, M.; D'Angelo, P.; Cassinese, A. Organic Film Thickness Influence on the Bias Stress Instability in Sexithiophene Field Effect Transistors. *Appl. Phys. A. Mater.* **2009,** *96,* 481–487.

51. Hashim, A. A.; *Polymer Thin Films;* Intech Publishers: Ausrtia, 2010.

52. Williams, G.; Watts, D. C.; Dev, S. B.; North, A. M. Further Considerations of Non Symmetrical Dielectric Relaxation Behaviour Arising from a Simple Empirical Decay Function. *Trans. Faraday Soc.* **1971,** *67,* 1323–1335.

53. Gezo, J.; Lui, T. K.; Wolin, B.; Slichter, C. P.; Giannetta, R. Stretched Exponential Spin Relaxation in Organic Superconductors. *Phys. Rev. B.* **2013,** *88,* 140504.

CHAPTER 3

NONLINEAR THERMAL INSTABILITY IN A FLUID LAYER UNDER THERMAL MODULATION

PALLE KIRAN*

Department of Mathematics, Rayalaseema University, Kurnool 518007, Andhra Pradesh, India.

Corresponding author. E-mail: kiran40p@gmail.com

CONTENTS

ABSTRACT

Using the complex non-autonomous Ginzburg–Landau model, oscillatory mode of thermal instability is investigated for oscillatory convective flow. The non-Newtonian fluid is considered for the existence of oscillatory mode of convection. The perturbation technique is used to simplify nonlinear problem. The Nusselt number is obtained numerically in terms of the system parameters, to study heat transfer in the medium. The effect of viscoelastic fluid parameters, stress relaxation λ_1 and strain retardation λ_2 times on heat transfer have been discussed. Thermal modulation is considered to control the heat transfer in the medium. This modulation destabilizes at low frequencies and stabilizes at high frequencies, but this trend is reverse for amplitude of modulation. Finally, it is found that heat transfer enhances for oscillatory mode than in stationary case.

3.1 INTRODUCTION

For an oscillatory mode of convection, it is required to consider an external phenomenon like rotation, magnetic field, and viscoelastic fluids. As such a candidate here, we consider a viscoelastic fluid layer to study Rayleigh–Bénard problem. In general, industrial fluids are basically non-Newtonian; particularly viscoelastic fluids have been potentially important with a lot of industrial applications. The characteristics of heat transfer in viscoelastic fluids layer are important in chemical processing industries. Proper understanding of convective motion and its behavior is necessary for controlling many processes, e.g., geothermal reservoirs, filtration, enhanced oil recovery, polymer filament package, manufacturing processes, such as injection molding, crystal growth, transport of chemical substances, petroleum industry, and composite impregnations.

Malkus and Veronis[1] investigated a nonlinear stability analysis of the Rayleigh–Bénard convection by employing the power series method. The nonlinear stability analysis is an important concept to understand the structure of turbulence; it becomes one of the most active research fields. Herbert[2] considered the stability of viscoelastic liquids in heated plane Couette flow and found that the presence of elasticity has a destabilizing effect on the flow. Green III[3] found that an oscillatory mode of convective motion is possible at the onset of instability. Vest and Arpaci[4] reported the occurrence of overstability for the typical Bénard–Rayleigh convection of a horizontal layer of homogeneous Maxwellian fluid heated from below.

For the viscoelastic fluid, Rosenbalt[5] performed a nonlinear stability analysis of Rayleigh–Bénard convection. He treated the bifurcation problem corresponding to the exchange of stabilities and overstability. Hamabata[6] further considered the effect of internal heat generation on the overstability of a viscoelastic liquid layer. For an Oldroyd B fluid the critical condition to mark the oscillatory motion was predicted by Kolkka and Ierley[7] with a two-dimensional fluid layer heated from below. Martinez-Mardones et al.[8] expanded the investigation by including the binary aspect to the viscoelastic fluids. Yoon et al.[9] analyzed the onset of thermal convection in a horizontal porous layer saturated with viscoelastic liquid using a linear theory. Laroze et al.[10] have analyzed the effect of viscoelastic fluid on bifurcations of convective instability; he analyzed that the nature of the convective solution depends largely on the viscoelastic parameters. Malashetty et al.[11] studied the onset of convection in a binary viscoelastic fluid layer using linear and weak nonlinear stability analyses. The Oldroyd B model was utilized by Sheu[12] to describe the rheological behavior of a viscoelastic nanofluid layer, incorporating the effects of thermophoresis and Brownian motion, and onset criterion for stationary and oscillatory convection was derived analytically using linear stability analysis. Comissiong et al.[13] investigated the Bénard–Marangoni thermal instability problem in a viscoelastic Jeffreys fluid layer with internal heat generation, the onset of oscillatory convection was studied using linear stability analysis and the dependence of critical Rayleigh number for oscillatory convection on the internal heat generation, relaxation and retardation times was derived. Kumar and Bhadauria[14] have studied the effect of rotation on the onset of double diffusive convection in a horizontal saturated porous layer of a viscoelastic fluid using linear and nonlinear analyses. Kumar and Bhadauria[15] investigated the effect of rotation on thermal instability in an anisotropic porous layer saturated with a viscoelastic fluid and found the onset criteria for steady and non-steady cases considering linear and nonlinear stability analyses. Rajib and Layek[16] analysis revealed that the onset of oscillatory convective motion strongly depends on relaxation times of viscoelastic fluid and the relaxation time must be greater than the retardation time of the fluid. Also they discussed the dependence of relaxation parameters and Prandtl number on the onset of convection of different viscoelastic fluid models.

The effect of modulation on the dynamical system has interest because stabilization or destabilization may occur in the presence of modulation, which thus enhances the mass, momentum, and heat transport, Davis.[17] The effects of temperature modulation on the thermal instability were studied by: Venezian,[18] Rosenblat and Herbert,[19] Yih and Li,[20] Finucane and Kelly,[21]

Bhatia and Bhadauria,[22,23] Bhadauria and Bhatia,[24] and of gravity modulation by Gresho and Sani,[25] Clever et al.,[26] Bhadauria et al.,[27] Bhadauria,[28] and Bhadauria et al.[29] The gravity modulation on the thermal instability of Newtonian and non-Newtonian fluids is investigated by Yang,[30] the linear stability theory has been used and found that for both fluids modulation has a destabilization effect at low frequencies and a slight stabilization effect at high frequencies, which increases with increasing the amplitude of modulation. Kim et al. [31] used modified Darcys law and employed linear stability theory to find the critical condition of the onset of convective motion. Using linear stability analysis, they showed that overstability is a preferred mode for a certain parameter range. Based on linear stability, a nonlinear stability analysis was also conducted, and Landau equations and Nusselt number variations are derived for steady and oscillatory modes of convection.

Recently, Bhadauria et al.[32,33] investigated nonlinear thermal instability for an oscillatory mode convection considering porous media. They are the first who investigated this nonlinear convection for oscillatory mode of convection using Ginzburg–Landau model. Bhadauria et al.[34] presented the same results for non-Newtonian fluid layer under gravity modulation. But, the same model[34] has not been extended in the direction of temperature modulation. In which one can show that phase angle of the modulation play a role along with modulation parameters to control heat transfer. With this, this chapter presents the missing part that is a weak nonlinear stability analysis for the oscillatory mode of convection under temperature modulation. Deriving a non autonomous complex Ginnzburg–Landau model heat transfer results are presented.

3.2   GOVERNING EQUATION

An infinitely extended horizontal viscoelastic fluid layer, confined between two stress-free boundaries at $z = 0$ and $z = d$, is considered. The stress-free boundaries are maintained at constant temperature and the fluid layer is heated from below. A physical configuration of the model is given in Figure 3.1.

The hydrodynamic equations are simplified by assuming Oberbeck–Boussinesq approximation. The constitutive equations for non-Newtonian viscoelastic fluid model with the relaxation time $\overline{\lambda_1}$ and retardation time $\overline{\lambda_2}$ may be represented as[16,34]:

$$\nabla . \vec{q} = 0 \qquad (3.1)$$

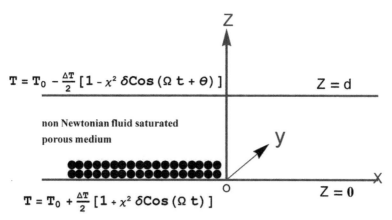

FIGURE 3.1 Physical configuration of the problem.

$$\left(\lambda_1 \frac{\partial}{\partial t} + 1\right)\left(\frac{\partial \vec{q}}{\partial t} + (\vec{q}.\nabla)\vec{q} - \frac{1}{\rho_o}\nabla P + \frac{\rho}{\rho_o}\vec{g}\right) - \nu\left(\lambda_2 \frac{\partial}{\partial t} + 1\right)\nabla^2 \vec{q} = 0,$$

$$(3.2)$$

$$\frac{\partial T}{\partial t} + (\vec{q}.\nabla)T = \kappa_T \nabla^2 T,$$

$$(3.3)$$

$$\rho = \rho_0\left[1 - \alpha_T (T - T_0)\right],$$

$$(3.4)$$

where the physical variables have their usual meanings and are given in nomenclature. The externally imposed thermal boundary conditions are considered as (Venezian[18])

$$T = T_0 + \frac{\Delta T}{2}\left[1 + \chi^2 \delta\cos(\Omega t)\right], \qquad at\ z = 0$$

$$= T_0 - \frac{\Delta T}{2}\left[1 - \chi^2 \delta\cos(\Omega t + \theta)\right], \qquad at\ z = d \qquad (3.5)$$

where ΔT is the temperature difference across the fluid layer, δ and Ω are amplitude and frequency of temperature modulation, and θ is the phase angle.

3.3   BASIC STATE

The conduction state is assumed to be quiescent, and the quantities in this state are given by

$$\vec{q}_b = 0,\ p = p_b(z,t),\ T = T_b(z,t),\ p = p_b(z,t). \tag{3.6}$$

Substituting the Eq 3.6 in Eqs 3.1–3.4, we get the following the relations which helps us to define basic state pressure and temperature:

$$\frac{\partial p^b}{\partial_z} = -p_b\vec{g}, \tag{3.7}$$

$$\frac{\partial p^b}{\partial_z} = -p_b\vec{g}, \tag{3.8}$$

$$\rho_b = \rho_0\left[1 - \alpha_T\left(T_b - T_b\right)\right]. \tag{3.9}$$

The solution of the Eq 3.8, subjected to the boundary conditions (3.5), is given by

$$T_b(z,t) = T_s(z) + \chi^2\,\delta\mathrm{Re}\left[T_1(z,t)\right], \tag{3.10}$$

where
$$T_S(Z) = T_O + \frac{\Delta T}{2}\left(1 - \frac{2z}{d}\right), \tag{3.11}$$

$$T_1(z,t) = \left(\left\{a(\varsigma)e^{\frac{\varsigma z}{d}} + a(-\varsigma)e^{-\frac{\varsigma z}{d}}\right\}e^{i\Omega t}\right), \tag{3.12}$$

and $a(\varsigma) = \dfrac{\Delta T}{2}\cdot\dfrac{\left(e^{-i\theta} - e^{-\varsigma}\right)}{\left(e^{\varsigma} - e^{-\varsigma}\right)}$ and $\varsigma^2 = \dfrac{-i\Omega d^2}{\kappa_T}$. Here $T_s(z)$ is the study part, while $T_1(z,t)$ is the oscillatory part of the basic state temperature field $T_b(z,t)$. The finite amplitude perturbations on the basic state are superposed in the form:

$$\vec{q} = \vec{q}_b + q',\ p = p_b + p',\ p = p_b + p',\ T = T_b + T', \tag{3.13}$$

We introduce the Eq 3.13, the basic state temperature field given by Eq 3.10, and then use the steam function ψ as $u' = \dfrac{\partial \psi}{\partial z}$, $\omega' = \dfrac{\partial \psi}{\partial x}$, for two-dimensional flow, in the Eqs. 3.1—3.4.

The equations are then non-dimensionalized using the physical variables: The resulting non-dimensionalized system of

$$(x,y,z) = d(x^*, y^*, z^*), t = \frac{d^2}{\kappa_T} t^*, \psi = \kappa_T \psi^*, T' = \Delta T T^*, \lambda_1 = \frac{\kappa_T \overline{\lambda_1}}{d^2}, \lambda_2 = \frac{\kappa_T \overline{\lambda_2}}{d^2}, \text{ and } \Omega = \frac{\kappa_T}{d^2} \Omega^*$$

equations can be expressed as (dropping the asterisk)

$$\left(\lambda_1 \frac{\partial}{\partial t} + 1 \right) \left(\frac{1}{Pr} \frac{\partial}{\partial t} \nabla^2 \psi - \frac{1}{Pr} \frac{\partial (\psi, \nabla^2 \psi)}{\partial (x,z)} + Ra \frac{\partial T}{\partial x} \right) - \left(\lambda_2 \frac{\partial}{\partial t} + 1 \right) \nabla^4 \psi = 0, \quad (3.14)$$

$$-\frac{dT_b}{dz} \frac{\partial \psi}{\partial x} + \left(\frac{\partial}{\partial t} - \nabla^2 \right) T = \frac{\partial (\psi, T)}{\partial (x,z)}, \quad (3.15)$$

where $Ra = \dfrac{\alpha_T g \Delta T d^3}{\nu \kappa_T}$ is thermal Rayleigh number, $\nu = \dfrac{\mu}{\rho_0}$ is kinematic viscosity. The above system will be solved by considering the stress free and isothermal boundary conditions as given below

$$\psi = \frac{\partial^2 \psi}{\partial z^2} = T = 0 \quad \text{on} \quad z = 0 \quad z = 1. \quad (3.16)$$

We introduce a small perturbation parameter χ that shows deviation from the critical state of onset convection, then the variables of weak nonlinear state may be expanded as power series of χ as[34–42]:

$$Ra = R_0 + \chi^2 R_2 + \chi^4 R_4 + \ldots,$$
$$\psi = \chi \psi_1 + \chi^2 \psi_2 + \chi^3 \psi_3 + \ldots,$$
$$T = \chi T_1 + \chi^2 T_2 + \chi^3 T_3 + \ldots, \quad (3.17)$$

where R_0 is the critical value of the critical Rayleigh number at which the onset of convection takes place in the absence of temperature modulation.

3.4 BIFURCATION OF PERIODIC SOLUTION

In order to allow for anticipated frequency shift along the bifurcation solution, we introduce the fast time scale τ and the slow time scale s. Therefore, we scale the time variable such that, the form of time derivative will be taken by $\dfrac{\partial}{\partial t} = \dfrac{\partial}{\partial \tau} + \chi^2 \dfrac{\partial}{\partial s}$. In the first order problem, the nonlinear term in energy equation will vanish. Therefore, the first order problem reduces to the linear stability problem for overstability. At the lowest order, we have

$$
\begin{bmatrix}
\dfrac{1}{\Pr}\left(\lambda_1 \dfrac{\partial}{\partial \tau}+1\right)\dfrac{\partial}{\partial \tau}\nabla^2 - \left(\lambda_2 \dfrac{\partial}{\partial \tau}+1\right)\nabla^4 & R_0\left(\lambda_1 \dfrac{\partial}{\partial \tau}+1\right)\dfrac{\partial}{\partial x} \\[4mm]
\dfrac{\partial}{\partial x} & \left(\dfrac{\partial}{\partial \tau}-\nabla^2\right)
\end{bmatrix}
\begin{bmatrix} \psi_1 \\[2mm] T_1 \end{bmatrix}
=
\begin{bmatrix} 0 \\[2mm] 0 \end{bmatrix}
\tag{3.18}
$$

The solution of the lowest order system subject to the boundary conditions Eq 3.16, is assumed to be

$$
\psi_1 = \left(\mathbb{B}(s)e^{i\omega\tau} + \overline{\mathbb{B}}(s)e^{-i\omega\tau}\right)\sin ax \sin \pi z, \tag{3.19}
$$

$$
T_1 = \left(\mathbb{A}(s)e^{i\omega\tau} + \overline{\mathbb{A}}(s)e^{-i\omega\tau}\right)\cos ax \sin \pi z. \tag{3.20}
$$

The undetermined amplitudes are functions of slow time scale and are related by the following relation:

$$
\mathbb{B}(s) = -\frac{c+i\omega}{a}\mathbb{A}(s), \tag{3.21}
$$

where $c = a^2 + \pi^2$. The values of the critical Rayleigh number and the corresponding wave number for stationary mode of convection are

$$
R_0 = \frac{c^3}{a^2}, \tag{3.22}
$$

$$
a = \frac{\pi}{\sqrt{2}}, \tag{3.23}
$$

which are the classical results of Chandrasekhar.[35] We find critical Rayleigh number for oscillatory convection as:

$$
R_0 = \frac{c^3}{a^2} - \frac{\left(\left(\lambda_1 + \lambda_2 \Pr\right)c+1\right)c\omega^2}{a^2 \Pr}, \tag{3.24}
$$

which is same as obtained by Rajib et al.[16] Here we calculate the corresponding critical wave number while minimizing critical Rayleigh number with respect to the square of wave number. The critical Rayleigh number and corresponding wave number does not depend on (λ_1, λ_2) in stationary mode but in oscillatory mode. Also we see that, the overstability can occur for a particular wave number a only, if the following inequality holds

$$\lambda_1 > \lambda_2 + \frac{1 + \Pr}{c \Pr}. \tag{3.25}$$

In the second order, we get the following relations

$$\psi_2 = 0, \tag{3.26}$$

$$\left(\frac{\partial}{\partial \tau} - \nabla^2\right) T_2 = \frac{\partial \psi_1}{\partial x}\frac{\partial T_1}{\partial z} - \frac{\partial \psi_1}{\partial z}\frac{\partial T_1}{\partial x}. \tag{3.27}$$

From the above relation, according to Kim et al.,[31] we can deduce that the velocity and temperature fields have the terms having frequency 2ω and independent of fast time scale. Thus, we write the second order temperature term as follows:

$$T_2 = \{T_{20} + T_{22}e^{2i\omega\tau} + \overline{T}_{22}\,e^{-2i\omega\tau}\}\sin 2\pi z, \tag{3.28}$$

where T_{22} and T_{20} are temperature fields having the terms having the frequency 2ω and independent of fast time scale, respectively. The solutions of the second order problems are:

$$T_{20} = \frac{a}{8\pi}\{\mathbb{A}(s)\overline{\mathbb{B}}(s) + \overline{\mathbb{A}}(S)\mathbb{B}(S)\}, \tag{3.29}$$

$$T_{22} = \frac{\pi a}{8\pi^2 + 4i\omega}\mathbb{A}(s)\mathbb{B}(s). \tag{3.30}$$

The horizontally averaged Nusselt number, $N u(\tau)$, for the oscillatory mode of convection is given by:

$$Nu(s) = 1 - \chi^2 \left(\frac{\partial T_2}{\partial z}\right)_{z=0} \tag{3.31}$$

Using the expression of T_2, given in Eq 3.28, one can simplify Eq 3.31 as

$$Nu(s) = 1 + \left(\frac{c}{2} + 2\pi^2 \frac{\sqrt{c^2 + \omega^2}}{\sqrt{64\pi^4 + 16\omega^2}} \right) |A(s)|^2.$$ (3.32)

It is clear that the thermal modulation is effective at third order and affects N $u(s)$ through $A(s)$ which is evaluated at third order. We calculate the mean value of Nusselt number (MNu) for better understanding of the results on heat transport. However, a representative time interval that allows a clear comprehension of the effect needs to be chosen. The interval $(0, 2\pi)$ seems to be an appropriate interval to calculate MNu. The time-averaged Nusselt is defined as:

$$MNu = \frac{1}{2\pi} \int_0^{2\pi} Nu(\tau) d\tau.$$ (3.33)

At the third order, we get the following relation with undetermined coefficients on right hand side

$$\begin{bmatrix} \frac{1}{Pr} \left(\lambda_1 \frac{\partial}{\partial \tau} + 1 \right) \frac{\partial}{\partial \tau} \nabla^2 - \left(\lambda_2 \frac{\partial}{\partial \tau} + 1 \right) \nabla^4 & R_0 \left(\lambda_1 \frac{\partial}{\partial \tau} + 1 \right) \frac{\partial}{\partial x} \\ \frac{\partial}{\partial x} & \left(\frac{\partial}{\partial \tau} - \nabla^2 \right) \end{bmatrix} \begin{bmatrix} \psi_6 \\ T_3 \end{bmatrix} = \begin{bmatrix} R_{31} \\ R_{32} \end{bmatrix},$$ (3.34)

where the expressions for R_{31} and R_{32} are given in the appendix. Now under the stability condition for the existence of third order solution, we obtain the following Landau equation that describes the temporal variation of the amplitude $A(s)$ of the convection cell

$$\frac{dA(s)}{ds} - \gamma^{-1} F(s) A(s) + \gamma^{-1} k |A(s)|^2 A(s) = 0,$$ (3.35)

where the coefficients γ, $F(s)$, and k are given in the Appendix. Writing $A(s)$ in the phase-amplitude form, we get

$$A(s) = |A(s)| e^{i\phi}.$$ (3.36)

Now substituting the expression Eq 3.36 in Eq 3.35, we get the following equations for the amplitude $|A(s)|$:

$$\frac{d|A(s)|^2}{ds} - 2p_r |A(s)|^2 + 2l_r |A(s)|^4 = 0,$$ (3.37)

$$\frac{d\left(ph\left(\mathbb{A}(s)\right)\right)}{ds} = pi - li\left|\mathbb{A}(s)\right|^2, \tag{3.38}$$

where $\gamma^{-1}F(s) = p_r + ip_i$, $\gamma^{-1}k = l_r + il_i$ and $ph(.)$ represents the phase shift.

3.5   RESULTS AND DISCUSSION

We have investigated a local nonlinear thermal instability in a layer of visco-elastic fluid, considering oscillatory mode of convection, under temperature modulation. The amplitude equation has been derived by using complex non-autonomous Ginzburg–Landau equation. The study is limited to a certain range of values of system parameters. We take δ values around 0.3, since we consider small amplitude of convection. The effect of time on Nusselt number is investigated with respect to the other physical quantities of unsteady problem. The temperature modulation of the boundaries has been considered in following three cases:

1. In-phase modulation (IPM) ($\theta = 0$),
2. Out-of-phase modulation (OPM) ($\theta = \pi$), and
3. Modulation of only the lower boundary (LBMO) ($\theta = -i\infty$), the upper boundary is kept at constant temperature.

In order to illustrate the effects of relaxational parameters λ_1, λ_2, the frequency Ω and the amplitude δ of modulation on heat transport, we plot the curves of Nusselt number vs. time s. It is observed that the relation Eq 3.25 leads to an interesting result; that for a viscoelastic fluid layer heated underneath; the oscillatory type of instability is possible only when the relaxation parameter λ_1 is greater than the retardation parameter λ_2.

Also, it is clear from the relation Eq 3.24 that the oscillatory convection depends on both relaxation and retardation times.

The results corresponding to the temperature modulation have been depicted in Figures 3.2–3.6, where we have plotted Nu with respect to the slow time s. It is found that the value of Nu starts with 1, thus showing the conduction state. Initially, heat transfer across the fluid layer takes place through conduction when s is small. The values of Nu increases for intermediate values of s, thus showing that convection is in progress. The effect of the Prandtl number is important, because many practical available visco-elastic fluids have large Prandtl numbers. It is quite interesting to note that when the Prandtl number is small, the critical value of the Rayleigh number decreases significantly for increasing Prandtl number so that the Prandtl number has a tendency to destabilize the system, compatible with results obtained by Wenchang Tan et al.[36] and Kim et al.[31] In Figure 3.2a IPM case, as $P r$ increases, there is an increment in heat transfer, compatible with the results obtained by Bhadauria et al.[29] for considering low viscous fluids. The similar effect of $P r$ can be seen in the case of OPM given in Figure 3.3a but, the behavior of the plots varying sinusoidal, where the effect of amplitude and frequency of modulation enters. In Figures 3.2b and 3.3b, the effect of an increment in the value of relaxation parameter λ_1 is destabilizing as the value of $N u$ increases on increasing λ_1. Further, the effect of retardation parameter λ_2 is found to stabilize the system as the heat transfer decreases on increasing λ_2 given in Figures 3.2c and 3.3c.

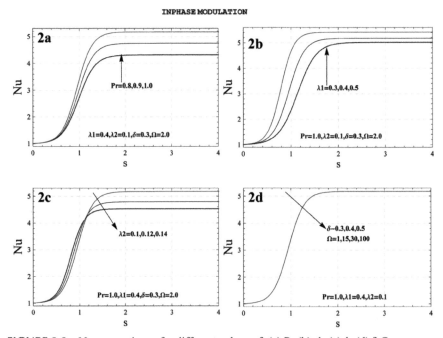

FIGURE 3.2 Nu versus time s for different values of (a) Pr (b) λ_1 (c) λ_2 (d) δ Ω.

The effects of frequency Ω and the amplitude δ of modulation on heat transport are clearly visible in the cases OPM, given in Figure 3.3d,e; however, no effect is found in IPM case in Figure 3.2d.

This is also clear due to the fact that in phase modulation of the boundary temperature does not substantially modify the temperature gradient across the fluid layer, therefore not much effect on heat transfer. Thus, the results in the case of IPM (Fig. 3.3a–d) are same as that are in unmodulated case. However in cases of OPM, the effect of temperature modulation on heat transfer is quite visible in Figures 3.3a–f, and is oscillatory in nature. Here we can find certain frequencies where the value of Nu is high thus destabilizing the system and low thus stabilizing the system From Figure 3.3d, one can see that the amplitude of modulation enhances the heat transfer as δ increases but opposite in the case of Ω given in Figure 3.3e. From Figure 3.3e, we find that the effect of temperature modulation decreases as the frequency of modulation increases, and finally when Ω is very large, the effect of modulation disappears altogether, thus confirming the results of Venezian[18] and Yang.[30]

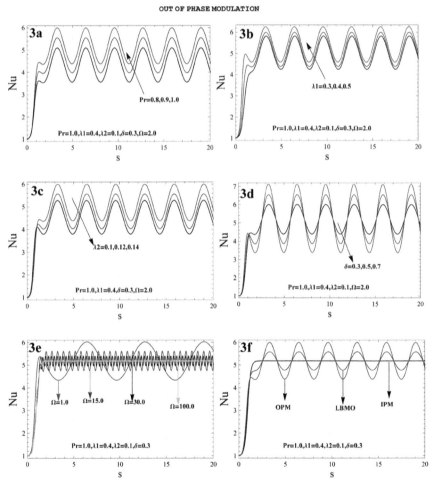

FIGURE 3.3 Nu versus time s for different values of (a) Pr (b) λ_1 (c) λ_2 (d) δ (e) Ω (f) Comparision.

Here we are not presenting the results of lower boundary modulation case as they are similar to the results obtained in case of OPM. On comparing the results in Figure 3.3f, it is found that Nu^{LBMO} is lower than Nu^{OPM} but higher than Nu^{IPM} as given below:

$$Nu^{IPM} < Nu^{LBMO} < Nu^{OPM}$$

In Figure 3.4 we compare the results of oscillatory and stationary mode of convection for OPM. It is found that heat transfer is more in oscillatory mode of convection than in stationary mode. It can be observed that

$$Nu^{st} < Nu^{osc}$$

for the same wave number. This implies that oscillatory instabilities can set in before stationary mode. Similar results have also been obtained by Rajib and Layek[16] and Kim et al.[31]

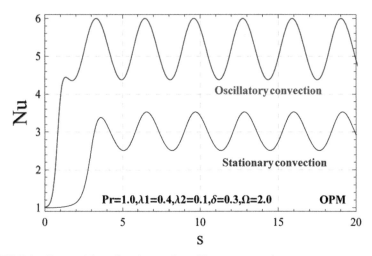

FIGURE 3.4 Comparision of stationary & oscillatory convection.

In Figures 3.5 and 3.6, as special cases, we plot the amplitude of convection $A(s)$ vs. time s, it is found that amplitude enhances the heat transfer as δ increases but opposite in frequency Ω, thus conforms the results obtained by Siddheshwar et al.[37] These parameters effects have been investigated in the porous medium case by Siddheshwar et al.[38] and Bhadauria et al.[39-41]

A better way of presenting our results according to Siddheshwar et al.[42] the effect of modulation on mean Nusselt number depends on both the phase difference θ and frequency Ω of modulation than only on the choice of the small amplitude modulation. From Figures 3.7 and 3.8, it is evident that for a given frequency of modulation there is a range of θ in which MNu increases with increasing θ and another range in which MNu decreases.

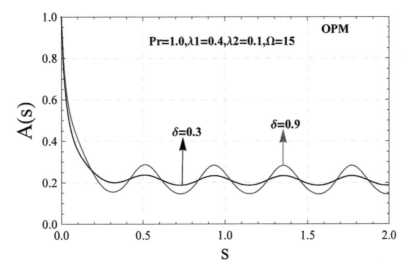

FIGURE 3.5   Effect of amplitude of modulationon amplitude of convection.

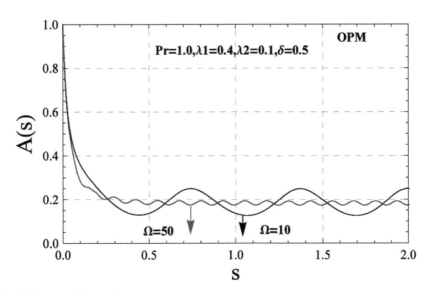

FIGURE 3.6   Effect of frequency of modulation on amplitude of convection.

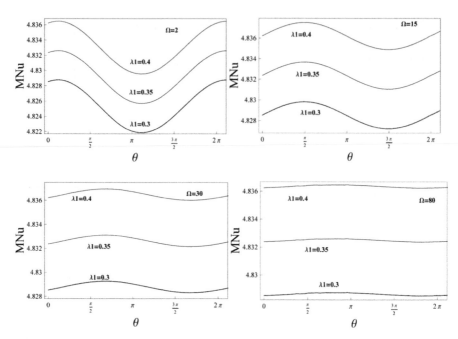

FIGURE 3.7   Effect of θ on MNu for different values of λ_1 and Ω.

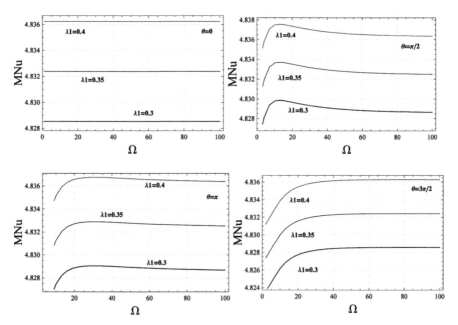

FIGURE 3.8   Effect of Ω on MNu for different values of λ_1 and θ.

Thus, one can conclude that, the combination of choices of Ω and θ can be made depending on the demands on heat transport in an application situation. Heat transfer can be regulated (enhanced or reduced) with the external mechanism of temperature modulation effectively. We also can observe our results in Figures 3.7 and 3.8 are the results which are similar to Siddheshwar et al.[42] for the Newtonian fluid case. It is clear that for temperature modulation the boundary temperatures should not be in in-phase modulation (synchronized), where the effect of modulation is negligible on heat transport.

3.6 CONCLUSIONS

The effect of temperature modulation on overstability of Rayleigh–Bénard convection investigated by performing a weakly nonlinear stability analysis resulting in the complex Ginzburg–Landau model. The following conclusions are made by the previous analysis:

1. Effect of IPM is negligible on heat transport in the system.
2. In the case of IPM, the effect of δ and Ω are also found to be negligible on heat transport.
3. Effect of λ_1 is to enhance the heat transport for all three types of modulations.
4. Effect of λ_2 is to decrease the heat transport for all three types of modulations.
5. In the case of IPM, Nu increase steadily for intermediate value of time s and ultimately becomes constant when s is large.
6. In the case of OPM and LBMO, Nu shows an oscillatory nature.
7. It is found that Nu^{LBMO} is lower than Nu^{OPM} but higher than Nu^{IPM}.
8. From the results of MNu, we conclude that, depending on the choice of Ω, θ heat transfer regulate effectively.

ACKNOWLEDGMENTS

The author acknowledges the support and encouragement from his parents (father P. Thikkanna and mother P. Sugunamma). He dedicates this work to Mr. A. Kumar (a Ph.D. student of Prof. B. S. Bhadauria during the period 2008–2011 at BHU). Finally I acknowledge the support of Department of Atomic Energy, New Delhi, Govt of India, for providing financial support in the form of NBHM-PDF.

APPENDIX

The dimensionless frequency of the neutral oscillatory mode is

$$\omega^2 = \frac{cPr\,(\lambda_1 - \lambda_2) - (1+Pr)}{\lambda_1 - \lambda_2) - (1+Pr)}.$$

The basic state solution which appears in Eq 3.15, influences the stability problem through the factor $\dfrac{\partial T_b}{\partial z}$ which is given by

$$\frac{dT_b}{dz} = -1 + \chi^2 \delta\,(f_2(z,t)),$$

where

$$f_2\,(z,t) = Re\,f(z)e^{(-i\Omega t)},$$

$$f(z) = \left(A(\varsigma)e^{\varsigma z} + A(-\varsigma)e^{-\varsigma z}\right), \quad A(\varsigma) = \frac{\varsigma}{2}\left(\frac{e^{-i\theta} - e^{-\varsigma}}{e^{\varsigma} - e^{-\varsigma}}\right) \& \varsigma = (1-i)\sqrt{\frac{\Omega}{2}}.$$

The expressions given in Eq 3.34 are

$$R_{31} = \lambda_2 \frac{\partial}{\partial s}(\nabla^4 \psi_1) - R_0 \lambda_1 \frac{\partial}{\partial s}\left(\frac{\partial T_1}{\partial x}\right) - R_2\left(\lambda_1 \frac{\partial}{\partial \tau} + 1\right)\left(\frac{\partial T_1}{\partial x}\right)$$
$$-\frac{1}{Pr}\left(\lambda_1 \frac{\partial}{\partial \tau} + 1\right)\frac{\partial}{\partial s}(\nabla^2 \psi_1) - \frac{1}{Pr}\lambda_1 \frac{\partial}{\partial s}\left(\frac{\partial}{\partial \tau}\nabla^2 \psi_1\right),$$

$$R_{32} = \frac{\partial \psi_1}{\partial x}\frac{\partial T_2}{\partial s} - \frac{\partial T_1}{\partial s} + \delta f_2\,(z,s)\frac{\partial \psi_1}{\partial x}.$$

The coefficients given in Eq 3.35 are

$$\gamma = \left[1 - a\Delta_1 R_0 \lambda_1 + \frac{c^2\Delta_1\lambda_2\,(c+i\omega)}{a} + \frac{c\Delta_1\,(c+i\omega)(1+2i\omega\lambda_1)}{a\,Pr}\right],$$

$$F(s) = \left[2\delta(c+i\omega)I_1\,(s)\right],$$

$$I_1\,(s) = \int_0^1 f_2\,(s)\sin^2\,\pi z dz,$$

$$k = \left(\frac{c^2 + ic\omega}{4} + \frac{\pi^2 c^2 + \pi^2 \omega^2}{8\pi^2 + 4i\omega}\right),$$

and $\Delta_1 = \dfrac{a\,Pr}{i\alpha c\,Pr(1+i\omega\lambda_1) + (1+i\omega\lambda_2)c^2}.$

NOMENCLATURE

Latin Symbols

A	Amplitude of convection
θ	Phase angle
a	Wave number
δ	Amplitude of temperature modulation
d	Depth of the fluid layer
g	Acceleration due to gravity
Nu	Nusselt number
p	Reduced pressure
Ra	Thermal Rayleigh number, $Ra = \dfrac{\beta_T g \Delta T d^3}{\nu \kappa_T}$
R_0	Critical Rayleigh number
T	Temperature
ΔT	Temperature difference across the fluid layer
t	Time
(x, z)	Horizontal and vertical co-ordinates

Greek Symbols

α	T Coefficient of thermal expansion
χ	Perturbation parameter
κ_T	Effective thermal diffusivity
Ω	Frequency of modulation
ω	Dimensionless oscillatory frequency
λ_1	Stress relaxation time
λ_2	Strain retardation time
μ	Viscosity of the fluid
ν	Kinematic viscosity, $\left(\dfrac{\mu}{\rho_0} \right)$
ρ	Fluid density
ψ	Stream function
s	Slow time $s = \chi^2 t$
T'	Perturbed temperature
$\hat{k}$	Vertical unit vector

Other symbols

$$\nabla^2_1 \qquad \frac{\partial^2}{\partial x^2} + \frac{\partial^2}{\partial y^2}$$

$$\Delta^2 \qquad \nabla^2_1 + \frac{\partial^2}{\partial z^2}$$

KEYWORDS

- nonlinear stability
- thermal modulation
- overstability
- viscoelastic fluids

REFERENCES

1. Malkus, W. V. R.; Veronis, G. Finite Amplitude Cellular Convection. *J. Fluid. Mech.* **1958,** *4,* 225–260.
2. Herbert, D. M. On the Stability of Viscoelastic Liquids in Heated Plane Couette Flow. *J. Fluid. Mech.* **1963,** *17,* 353–359.
3. Green, T. III. Oscillating Convection in an Elasticoviscous Liquid. *Phys. Fluids.* **1968,** *11,* 1410.
4. Vest, C. M.; Arpaci, V. S. Overstability of a Viscoelastic Fluid Layer Heated from Below. *J. Fluid. Mech.* **1969,** *36,* 613–623.
5. Rosenblat, S. Thermal Convection of a Viscoelastic Fluid. *J. Nonnewton. Fluid. Mech.* **1986,** *21,* 201–223.
6. Hamabata, H. Overstability of a Viscoelastic Liquid Layer with Internal Heat Generation. *Int. J. Heat Mass Transf.* **1986,** *29,* 645–647.
7. Kolkka, R. W.; Ierley, G. R. On the Convected Linear Stability of a Viscoelastic Oldroyd B Fluid Heated from Below. *J. Nonnewton. Fluid Mech.* **1987,** *25,* 209–237.
8. Martinez-Mardones, J.; Tiemann, R.; Walgraef, D. Thermal Convection Thresholds in Viscoelastic Solutions. *J. Nonnewton. Fluid Mech.* **2000,** *93,* 1–15.
9. Yoon, D. Y.; Kim, M. C.; Choi, C. K. The Onset of Oscillatory Convection in a Horizontal Porous Layer Saturated with Viscoelastic Liquid. *Transp. Porous. Media.* **2004,** *55,* 275–284.
10. Laroze, D.; Martinez-Mardones, J.; Bragard, J. Thermal Convection in a Rotating Binary Viscoelastic Liquid Mixture. *Eur. Phys. J. Spec. Top.* **2007,** *146,* 291–300.
11. Malashetty, M. S.; Swamy, M. The Onset of Double Diffusive Convection in a Viscoelastic Fluid Layer. *J. Nonnewton. Fluid Mech.* **2010,** *165,* 1129–1138.

12. Sheu, L. J. Thermal Instability in a Porous Medium Layer Saturated with a Viscoelastic Nanofluid. *Transp. Porous Media.* **2011,** *88,* 461–477.

13. Comissiong, D. M. G.; Dass, T. D.; Ramkissoon, H.; Sankar, A. R. On Thermal Instabilities in a Viscoelastic Fluid Subject to Internal Heat Generation. *World Acad. Sci. Eng. Tech.* **2011,** *56,* 8–24.

14. Kumar, A.; Bhadauria, B. S. Double Diffusive Convection in a Porous Layer Saturated with Viscoelastic Fluid Using a Thermal Non-equilibrium model. *Phys. Fluids.* **2011,** *23,* 054101.

15. Kumar, A.; Bhadauria, B. S. Nonlinear Two Dimensional Double Diffusive Convection in a Rotating Porous Layer Saturated by a Viscoelastic Fluid. *Transp. Porous Med.* **2011,** *87,* 229–250.

16. Rajib, B.; Layek, G. C. The Onset of Thermo Convection in a Horizontal Viscoelastic Fluid Layer Heated Underneath. *Thermal Energy Power Eng.* **2012,** *1,* 1–9.

17. Davis, S. H. The Stability of Time-Periodic flows. *Ann. Rev. Fluid Mech.* **1976,** *8,* 57–74.

18. Venezian, G. Effect of Modulation on the Onset of Thermal Convection. *J. Fluid Mech.* **1969,** *35,* 243–254.

19. Rosenblat, S.; Herbert, D. M. Low-Frequency Modulation of Thermal Instability. *J. Fluid Mech.* **1970,** *43,* 385–398.

20. Yih, C. S.; Li, C. H. Instability of Unsteady Flows or Configurations 2. Convective Instability. *J. Fluid Mech.* **1972,** *54,* 143–152.

21. Finucane, R. G. Kelly, R. E. Onset of Instability in a Fluid Layer Heated Sinusoidally from Below. *Int. J. Heat Mass Transf.* **1976,** *19,* 71–85.

22. Bhatia, P. K.; Bhadauria, B. S. Effect of Modulation on Thermal Convection Instability. *Z. Natur. Forsch.* **2000,** *55,* 957–966.

23. Bhatia, P. K.; Bhadauria, B. S. Effect of Low Frequency Modulation on Thermal convection Instability. *Z. Natur. Forsch.* **2001,** *56,* 507–522.

24. Bhadauria, B. S.; Bhatia, P. K. Time Periodic Heating of Rayleighn-Bénard Convection. *Phys. Scripta.* **2002,** *66,* 59–65.

25. Gresho, P. M.; Sani, R. L. The Effects of Gravity Modulation on the Stability of a Heated Fluid Laer. *J. Fluid Mech.* **1970,** *40,* 783–806.

26. Clever, R.; Schubert, G.; Busse, F. H. Two Dimensional Oscillatory Convection in a Gravitationally Modulated Fluid Layer. *J. Fluid Mech.* **1993,** *253,* 663–680.

27. Bhadauria, B. S.; Bhatia, P. K.; Debnath, L Convection in Hele-Shaw Cell with Para-Metric Excitation. *Int. J. Non Linear Mech.* **2005,** *40,* 475–484.

28. Bhadauria, B. S. Gravitational Modulation of Rayleigh-Bénard Convection. *Proc. Natl. Acad. Sci. India.* **2006,** *76,* 61–67.

29. Bhadauria, B. S.; Hashim, I.; Siddheshwar, P. G. Effect of Internal Heating on Weakly Nonlinear Stability Analysis of Rayleigh-Bénard Convection under G-jitter. *Int. J. Nonlinear Mech.* **2013,** *54,* 35–42.

30. Wen-Mei, Yang. Stability of Viscoelastic Fluids in a Modulated Gravitational Field. *Int. J. Heat Mass Transf.* **1997,** *40,* 1401–1410.

31. Kim, M. C.; Lee, S. B.; Kim, S.; Chung, B. J. Thermal Instability of Viscoelastic Fluids in Porous Media. *Int. J. Heat Mass Transf.* **2003,** *46,* 5065–5072.

32. Bhadauria, B. S.; Kiran, P. Weakly Nonlinear Oscillatory Convection in a Viscoelastic Fluid Saturating Porous Medium under Temperature Modulation. *Int. J. Heat Mass Transf.* **2014,** *77,* 843–851.

33. Bhadauria, B. S.; Kiran, P. Weak Nonlinear Oscillatory Convection in a Viscoelastic Fluid Saturated Porous Medium under Gravity Modulation. *Transp. Porous Media.* **2014,** *104,* 451–467.

34. Bhadauria, B. S.; Kiran, P. Heat and Mass Transfer For Oscillatory Convection in a Binary Viscoelastic Fluid Layer Subjected to Temperature Modulation at the Boundaries. *Int. Commun. Heat Mass Transf.* **2014,** *58,* 166–175.

35. Chandrasekhar, S. *Hydrodynamic and Hydromagnetic Stability*; Oxford University Press: Oxford, 1961.

36. Wenchang, T.; Takashi, M. Stability Analysis of a Maxwell Fluid in a Porous Medium Heated from Below. *Phys. Lett. A.* **2007,** *360,* 454–460.

37. Siddheshwar, P. G.; Bhadauria, B. S.; Pankaj, M.; Srivastava, A. K. Study of Heat Transport by Stationary Magneto-Convection in a Newtonian Liquid Under Temperature or Gravity Modulation Using Ginzburg-Landau Model. *Int. J. Non Linear Mech.* **2012,** *47,* 418–425.

38. Siddheshwar, P. G.; Bhadauria, B. S.; Srivastava. A. An Analytical Study of Nonlinear Double Diffusive Convection in a Porous Medium with Temperature Modulation/ Gravity Modulation. *Transp. Porous Med.* **2012,***91,* 585–604.

39. Bhadauria, B. S.; Siddheshwar, P. G.; Jogendra, K.; Suthar, Om. P. Non-Linear Stability Analysis of Temperature/gravity Modulated Rayleigh-Bénard Convection in a Porous Medium. *Transp. Porous Med.* **2012,** *92,* 633–647.

40. Bhadauria, B. S.; Hashim, I.; Siddheshwar, P. G. Study of Heat Transport in a Porous Medium under G-jitter and Internal Heating Effects. *Transp. Porous Med.* **2013,** *96,* 21–37.

41. Bhadauria, B. S.; Palle, K. Heat Transport in an Anisotropic Porous Medium Saturated with Variable Viscosity Liquid under Temperature Modulation. *Transp. Porous Med.* **2013,** *100,* 279–295.

42. Siddheshwar, P. G.; Bhadauria, B. S.; Suthar, Om. P. Synchronous and Asynchronous Boundary Temperature Modulations of Bnard-Darcy Convection. *Int. J. Non Linear Mech.* **2013,** *49,* 84–89.

CHAPTER 4

APPLICATION OF A RANDOM SEQUENTIAL ALGORITHM FOR THE MODELING OF MACROMOLECULAR CHAINS' CROSS-LINKING EVOLUTION DURING A POLYMER SURFACE MODIFICATION

STANISLAV MINÁRIK[1*], VLADINÍR LABAŠ[2], ONDREJ BOŠÁK[2], and MARIÁN KUBLIHA[2]

[1]*Research Centre of Progressive Technologies, Slovak University of Technology, Bratislava, Slovak Republic*

[2]*Institute of Materials Science, Faculty of Materials Science and Technology in Trnava, Slovak University of Technology, Bratislava, Slovak Republic*

Corresponding author. E-mail: stanislav.minarik@stuba.sk

CONTENTS

ABSTRACT

The aim of this work is to introduce a model for the numerical study of cross-linking process during some chemical reactions in composites with macromolecular structure. We define a parameter that quantifies the degree of cross-linking and indicates the rate of bridge formation between macromolecular chains. The mentioned parameter is determined by the number of positions free for cross-link connections in the model of macromolecular structure. Then the evolution of this parameter is calculated, provided that random and sequential mechanism of bridge formation is preferred. The result of the calculation is compared with experimental data characterizing evolution of AC electric conductivity of styrene-butadiene rubber (SBR) rubber mixtures during silane cross-linking phase of treatment. Good agreement of calculated results with experimental data was observed. This work offers a new unconventional view on kinetics of cross-linking chemical reactions in polymer structures based on non-chemical concept. We believe that our results should create a basis for novel theoretical models and simulations of such reactions to understand their nature. The presented contribution contains only brief mathematical statements, not exact proofs. We try to make accessible mathematically rigorous results for a wide range of researchers mainly in the field of the modeling and simulation of chemical reactions.

4.1 INTRODUCTION

Cross-linking is a process of the joining of polymer chains with bonds that occurs during polymer synthesis with addition of atoms or molecules. It is a type of polymerization reaction that branches out from the main molecular chain to form a network of chemical links, and cross-linking agents are added to the material to enable this process. The cross-linking results in a random three-dimensional network in the structure of interconnected chains. The degree of cross-linking process affects many properties of polymeric materials. The covalent or ionic bond that links one polymer chain to another is called "cross-link." Cross-links can be formed by chemical reactions that are usually initiated by heat.

Cross-linking of macromolecular chains phenomena takes part in various natural or technological processes in divers areas. Many modern technological procedures cover cross-linking polymerization reactions to achieve suitable monolayer surface coverage of products. There may be mentioned uses such as silanization of metal, glass or rubber[1-4] and vulcanization of rubber[5-7] which are used to increase strength and decrease degradation rate of elastomeric polymers. Cross-linking process has been studied in biology where cross-linking theory of aging was proposed.[8-10] According to this theory, accumulation of cross-linked proteins damages cells and tissues, slowing down bodily processes, which results in aging. Recent studies show that cross-linking reactions are involved in the age-related changes in the studied proteins. The cross-linking process in protein systems is investigated also in the medicine[11] where protein-coupled receptors excite drug targets.

Our work is oriented on the problem of the macromolecular cross-linking evolution modeling and simulation. At the beginning the sequential mechanism of cross-link formation was confirmed by means of a simple probabilistic analysis of the time scaling of cross-linking formation phenomena. Next we continue with mathematics based on a simple calculation of the number of cross-links formed during polymer surface modification. We search for a formula enabling determination of the number of possible cross-links which may be currently formed in macromolecular structure. This number of cross-links is used for the quantification of degree of cross-linking evolution. In the final section we compare results of the calculation with experimental data of AC electrical conductivity of styrene-butadiene rubber (SBR) mixtures measured during silanization. Satisfying agreement between calculated and experimental data were observed just in the temperature range where cross-link formation is expected.

4.2   RANDOM SEQUENTIAL FORMATION OF CROSS-LINKS

For our purpose the cross-linking evolution is quantified by the total number of possible cross-links P which may be formed between all pairs of macromolecular chains in the model of macromolecular structure as shown in the Figure 4.1. We suppose that the number P characterizes current level of the cross-linking process. This number decreases in time when a cross-linking process is in the progress because positions for possible cross-link formation are occupied gradually. We expect random and sequential character of the filling of these possible positions. In this section we intent to calculate evolution of the quantity P under this assumption.

Let N be the total number of macromolecular chains in the system. We expect that in the beginning of cross-linking process there are just n_i positions for the cross-link connections in any i-th macromolecular chain

(i = 1, 2, ..., N). We consider that the cross-linking is a sequential process. This means that cross-links are formed one after the other in time. Only one bond can be formed at any moment namely at a position which is free for the connection (unoccupied by bond formed in one of the previous step). More bonds cannot be formed at the same time. Let P_k be total number of possible cross-links which may be formed in the k-th step of sequential cross-linking process. The probability of cross-link formation in the k-th step depends precisely on P_k.

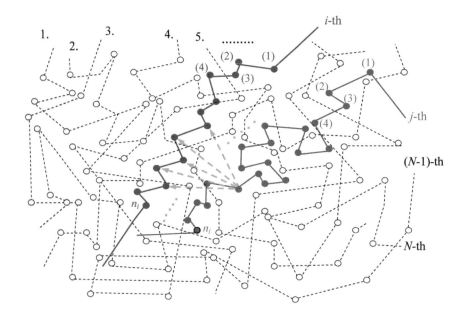

FIGURE 4.1 Scheme illustrating the model of macromolecular structure. Possible cross-linking bonds formation between i-th and j-th macromolecular chains are shown. Green dashed arrows demonstrate the variability of connections (cross-links). Number of posible connections in the whole system of N macromolecular chains is discussed in the text.

In the first step (k = 1) the total number of possible cross-links is:

$$2P_1 = 0 + n_1 n_2 + n_1 n_3 + n_1 n_4 + n_1 n_5 + n_1 n_6 + n_1 n_7 + n_1 n_8 + n_1 n_9 + \quad \ldots + n_1 n_{N-1} + n_1 n_N +$$
$$+ n_2 n_1 + 0 + n_2 n_3 + n_2 n_4 + n_2 n_5 + n_2 n_6 + n_2 n_7 + n_2 n_8 + n_2 n_9 + \quad \ldots + n_2 n_{N-1} + n_2 n_N +$$
$$\vdots$$
$$+ n_N n_1 + n_N n_2 + n_N n_3 + n_N n_4 + n_N n_5 + n_N n_6 + n_N n_7 + n_N n_8 + \quad \ldots + n_N n_{N-1} + 0$$

Assume that just one cross-link between i-th and j-th macromolecular chains was formed in the first step. This means that in the second step must be considered $n_i \to n_i-1, n_j \to n_j-1$ and the total number of possible cross-links formation for $k = 2$ can be found by the same a way as:

$$2P_2 = 0+n_1n_2+.... + n_1(n_i-1)+ ....+ n_1(n_j-1) ... + n_1n_{N-1}+ n_1n_N+ .... + (n_i-1)$$
$$n_1+ (n_i-1)n_2+ .... + (n_i-1)(n_j-1)+ ... + (n_i-1)n_N+ .... +(n_j-1)n_1+ (n_j-1)n_2+ ....$$
$$(n_j-1)(n_i-1) + ... + (n_5-1)n_N+ ....+n_Nn_1+ n_Nn_2+.... n_N(n_i-1)+ ....+ n_N(n_j-1)+...$$
$$+ n_Nn_{N-1}+ 0$$

It can be easily shown that there is another relationship between P_1 and P_2:

$$2P_2 = 2P_1 - 4\sum_{m=1}^{N} n_m +2\left(n_i +n_j\right)+2, \quad \text{i.e.,} \quad P_2 = P_1 - 2\sum_{m=1}^{N} n_m +\left(n_i +n_j\right)+1. \quad (4.1)$$

On the basis of a generalization of the eq 4.1 next formula for any k can be found:

$$P_{k+1} = P_k - 2\sum_{m=1}^{N} n_m^{(k)} +\left(n_p^{(k)} +n_q^{(k)}\right)+1, \quad P_{k+2} = P_{k+1} - 2\sum_{m=1}^{N} n_m^{(k+1)} +\left(n_w^{(k+1)} +n_s^{(k+1)}\right)+1, \quad (4.2)$$

where $n_i^{(j)}$ is number of positions for the cross-link connection in i-th macro-molecular chain at j-th step of calculation. We note that:

$$s^{(k)} = \sum_{m=1}^{N} n_m^{(k)}, \quad s^{(k+1)} = \sum_{m=1}^{N} n_m^{(k+1)} \quad \text{i.e.,} \quad s^{(k+1)} = s^{(k)} - 2. \quad (4.3)$$

Therefore it holds:

$$P_{k+1} = P_k - 2s^{(k)} +\left(n_l^{(k)} +n_m^{(k)}\right)+1, \quad (4.4)$$

$$P_{k+2} = P_{k+1} - 2\left(s^{(k)} -2\right)+\left(n_{l'}^{(k+1)} +n_m^{(k+1)}\right)+1 = P_k - 4s^{(k)} +\left(n_{l'}^{(k)} +n_{m'}^{(k)}\right)$$
$$+\left(n_l^{(k+1)} +n_m^{(k+1)}\right)+6, \quad (4.5)$$

$$P_{k+3} = P_k - 4s^{(k)} +\left(n_{l'}^{(k)} +n_{m'}^{(k)}\right)+\left(n_l^{(k+1)} +n_m^{(k+1)}\right)+6 - 2s^{(k+2)} +\left(n_l^{(k+2)} +n_m^{(k+2)}\right)+1. \quad (4.6)$$

Because $s^{(k+2)} = s^{(k+1)} -2 = s^{(k)} -4$, eq 4.6 can be rewritten to the form:

$$P_{k+3} = P_k - 6s^{(k)} +\left(n_{l'}^{(k)} +n_{m'}^{(k)}\right)+\left(n_l^{(k+1)} +n_m^{(k+1)}\right)+\left(n_{l''}^{(k+2)} +n_{m''}^{(k+2)}\right)+15. \quad (4.7)$$

If we continue the calculation in the same a way, we get:

$$P_{k+4} = P_{k+3} - 2s^{(k+3)} + \left(n_{l'''}^{(k+3)} + n_{m'''}^{(k+3)}\right) + 1 \tag{4.8}$$

where $s^{(k+3)} = s^{(k+2)} - 2 = s^{(k-1)} - 4 = s^{(k)} - 6$ and

$$P_{k+4} = P_k - 8s^{(k)} + \left(n_{l'}^{(k)} + n_{m'}^{(k)}\right) + \left(n_l^{(k+1)} + n_m^{(k+1)}\right) + \left(n_{l''}^{(k+2)} + n_{m''}^{(k+2)}\right)$$
$$+ \left(n_{l'''}^{(k+3)} + n_{m'''}^{(k+3)}\right) + 28. \tag{4.9}$$

So in the next step:

$$P_{k+5} = P_k - 8s^{(k)} + \left(n_{l'}^{(k)} + n_{m'}^{(k)}\right) + \left(n_l^{(k+1)} + n_m^{(k+1)}\right) + \left(n_{l''}^{(k+2)} + n_{m''}^{(k+2)}\right) + \left(n_{l'''}^{(k+3)} + n_{m'''}^{(k+3)}\right) +$$
$$+ 28 - 2s^{(k+4)} + \left(n_l^{(k+4)} + n_m^{(k+4)}\right) + 1$$

where $s^{(k+4)} = s^{(k+3)} - 2 = s^{(k+2)} - 4 = s^{(k+1)} - 6 = s^{(k)} - 8$ and

$$P_{k+5} = P_k - 10s^{(k)} + \left(n_{l'}^{(k)} + n_{m'}^{(k)}\right) + \left(n_l^{(k+1)} + n_m^{(k+1)}\right) + \left(n_{l''}^{(k+2)} + n_{m''}^{(k+2)}\right) + \left(n_{l'''}^{(k+3)} + n_{m'''}^{(k+3)}\right) +$$
$$+ \left(n_l^{(k+4)} + n_m^{(k+4)}\right) + 45 \ldots$$

Therefore after v steps we can expect:

$$P_{k+v} = P_k - 2 v s^{(k)} + \sum_{m=1}^{v} (4m-3) + \left(n_{l'}^{(k)} + n_{m'}^{(k)}\right) + \left(n_l^{(k+1)} + n_m^{(k+1)}\right) + \left(n_{l''}^{(k+2)} + n_{m''}^{(k+2)}\right) + \ldots . \tag{4.10}$$

If we consider that:, eq (4.10) can be rewritten to the form:

$$P_{k+v} = P_k + 2v^2 - \left(2s^{(k)} + 1\right)v + \left(n_{l'}^{(k)} + n_{m'}^{(k)}\right) + \left(n_l^{(k+1)} + n_m^{(k+1)}\right) + \ldots + \left(n_\sigma^{(k+v-1)} + n_\mu^{(k+v-1)}\right). \tag{4.11}$$

Then the total number of possible cross-links which may be formed in the v -th step of cross-linking process calculated from this process beginning (this beginning is characterized by $k = 0$) is:

$$P_v = P_0 + 2v^2 - \left(2s^{(0)} + 1\right)v + \left(n_{l'}^{(0)} + n_{m'}^{(0)}\right) + \left(n_l^{(1)} + n_m^{(1)}\right) + \ldots + \left(n_\sigma^{(v-1)} + n_\mu^{(v-1)}\right). \tag{4.12}$$

where $s^{(0)} = \displaystyle\sum_{i=1}^{N} n_i^{(0)} = const$ and $P_0 = \dfrac{1}{2} \displaystyle\sum_{\substack{i=1 \\ (i \neq j)}}^{N} \sum_{j=1}^{N} n_i^{(0)} n_j^{(0)} = const$.

4.3   CONTINUUM LIMIT IN THE FIRST APPROXIMATION

We assume random and sequential character of cross-link formation during macromolecular chains cross-linking evolution. This character is given by gradual filling of free positions for the cross-link connections in the macromolecular system. Let cross-linking bonds are formed gradually and separately with the mean frequency of their formation ϕ. In this case the continuum limit of the process can be considered by substitution $v = v(t)$ in eq 4.12, where t is time measured from the beginning of the process. Then the cross-linking evolution as a function of time can be estimated as follows:

$$P(t) = 2v^2(t) - \left\{2\sum_{i=1}^{N} n_i^{(0)} + 1\right\} v(t) + \frac{1}{2}\sum_{\substack{i=1 \\ (i \neq j)}}^{N}\sum_{j=1}^{N} n_i^{(0)} n_j^{(0)} + \left(n_{i'}^{(0)} + n_{m'}^{(0)}\right)$$

$$+ \left(n_{i}^{(1)} + n_{m}^{(1)}\right) + \ldots + \left(n_{\sigma}^{(v-1)} + n_{\mu}^{(v-1)}\right). \tag{4.13}$$

In terms of the cross-linking undergoing for sufficiently large values v we can apply:

$$v \gg 0 \quad \Leftrightarrow \quad n_i^{(0)} \gg n_i^{(v-1)}. \tag{4.14}$$

This means that double summation of $n_i^{(0)} n_j^{(0)}$ is dominant in eq 4.13 under conditions in eq 4.14. Therefore in the first approximation we can assume:

$$\frac{1}{2}\sum_{\substack{i=1 \\ (i \neq j)}}^{N}\sum_{j=1}^{N} n_i^{(0)} n_j^{(0)} + \left(n_{i'}^{(0)} + n_{m'}^{(0)}\right) + \left(n_{i}^{(1)} + n_{m}^{(1)}\right) + \ldots + \left(n_{\sigma}^{(v-1)} + n_{\mu}^{(v-1)}\right) \approx const = B \tag{4.15}$$

when the cross-linking process is running sufficiently long time t. Then the time dependence of parameter P results from the continuum limit of eq 4.12 as:

$$P(t) = 2v^2(t) - Av(t) + B(t), \quad \text{where } B(t) \approx B = const, \text{ and } A = 2\sum_{i=1}^{N} n_i^{(0)} + 1 \tag{4.16}$$

The proposed model of macromolecular chains' cross-linking process is based on the assumption that certain type of macromolecular structure relaxation to a cross-linked form occurs. Such process is then characterized by relaxation time τ that characterizes cross-linking rate, i.e., determines a mean frequency ϕ of macromolecular cross-links formation. In order to determine the time dependence of parameter P we have to calculate $v(t)$ and substitute it to the eq 4.16.

Let the probability of formation of the single cross-link connection at any given time be proportional to the total number of possible cross-links which may be currently formed in macromolecular system. This number is determined just by the parameter P and therefore next formula can be written as:

$$\frac{dP_v}{dt} = -\left(\frac{1}{\tau}\right)P_v. \tag{4.17}$$

If eq 4.16 is considered:

$$\frac{dP_v}{dv}\frac{dv}{dt} = -\left(\frac{1}{\tau}\right)P_v \tag{4.18}$$

and the time dependence $v(t)$ can be calculated as:

$$(4v-A)\frac{dv}{dt} = -\left(\frac{1}{\tau}\right)(2v^2 - Av + B) \quad \dots \quad \text{i.e.} \quad v^2 - \frac{A}{2}v + \frac{B}{2} = \frac{\Omega}{2}e^{-\frac{t}{\tau}} \tag{4.19}$$

where Ω is constant resulting from initial conditions. In the beginning of the cross-linking process, i.e., in the time $t = 0$ the value $v = 0$ and $\Omega = B$ results from eq 4.19 for that reason and next formula can be written:

$$v(t) = \frac{A}{4} - \frac{1}{\sqrt{2}}\sqrt{Be^{-\frac{t}{\tau}} + \left(\frac{A^2}{8} - B\right)} \tag{4.20}$$

Let N_c is the finite number of cross-linking bonds in the macromolecular system, then:

$$\lim_{t\to\infty} v(t) = \lim_{t\to\infty}\left\{\frac{A}{4} - \frac{1}{\sqrt{2}}\sqrt{Be^{-\frac{t}{\tau}} + \left(\frac{A^2}{8} - B\right)}\right\} = \frac{A}{4} - \frac{1}{\sqrt{2}}\sqrt{\left(\frac{A^2}{8} - B\right)} = N_c. \tag{4.21}$$

Constant B can be determined from eq 4.21 as $B = N_c(A - 2N_c)$ and substituted into eq 4.20. Subsequently the $v(t)$ takes the form:

$$v(t) = \alpha\left\{1 - \sqrt{\beta\left(e^{-\frac{t}{\tau}} - 1\right) + 1}\right\} \quad \dots \text{where}: \alpha = \frac{A}{4}, \beta = 8\left(\frac{N_c}{A}\right)\left(1 - 2\frac{N_c}{A}\right),$$

$$\text{and } A = \frac{1}{4}\left\{2\sum_{i=1}^{N} n_i^{(0)} + 1\right\}. \tag{4.22}$$

Substituting eq. 4.22 into eq 4.16, the time dependence of parameter P can be determined in the next form:

$$P(t) = 2\alpha^2 \beta e^{-\frac{t}{\tau}} \tag{4.23}$$

As can be seen from eq 4.23 the quantity P characterizing cross-linking evolution shows a exponential decrease in time in the first approximation. Frequency of macromolecular cross-links formation can be easy calculated as:

$$\phi(t) = \frac{d\,v(t)}{dt} = \frac{\alpha\beta}{2\tau}e^{-\frac{t}{\tau}}\left\{\beta\left(e^{-\frac{t}{\tau}}-1\right)+1\right\}^{-\frac{1}{2}} \tag{4.24}$$

The quantity in eq 4.24 determines the number of cross-links formed per time unit (1 sec) in the macromolecular system, i.e., the rate of cross-linking process. Results in eqs 4.22 and 4.24 for different values of τ are shown in Figures 4.3 and 4.4 in Section 4.5.

4.4 COMPARISON OF THE MODEL RESULTS WITH EXPERIMENTAL DATA

Silanization is the covering of a surface through self-assembly with oregano-functional alkoxysilane molecules. Mineral components like rubber mixtures, metal oxide or glass surfaces can all be silanized, because they contain silanol or hydroxyl groups. These groups attack and displace the alkoxy groups on the silane thus forming a covalent –Si–O–Si– bond. This means that the main goal of silanization is to form bonds across the interface between mineral (inorganic) components and organic components present in the system of macromolecular structure. We intuitively encourage the idea that formation of these bonds is realized by the same mechanism as it is described in Section 4.2 (see Fig. 4.4).

Silanization of resin is the process of covering the resin surface through silane coupling agents. Silane or silica treatment of resin composite is technology step during rubber mixture preparation which is oriented on cross linking and reinforcement of silica or silane-filled rubber compounds. To make a silane bridge between inorganic fillers and the polymer matrices, a bi-functional molecule which reacts with both the fillers and the polymer is required. Generally the silanization takes place in four steps:

I.   hydrolysis of silanes into silanols;
II.   condensation of silanols into oligomers;
III. formation of hydrogen bonds between oligomers and hydroxyl groups;
IV. the reaction of Si-OH groups with OH groups to build silane cross-linking bonds.

Temperature dependences of the AC electric conductivity $\sigma(t)$ of prepared SBR mixtures (used in the manufacture of tyres) during silanization were measured at the linear increasing of temperature ($\eta = 1°C$ min^{-1}) in the temperature range from 60°C up to 170°C. Measurements were realized using equipment Good Will LCR 819 at the frequency 1 kHz. Measured rubber mixtures differed from each other by the content of their constituents. Results of these measurements are shown in Figure 4.2 (blue points). Details of the experimental measurement can be found in our work.[12] The existence of a local maximum in the temperature dependence $\sigma(t)$ has been identified experimentally as a response of silanization reaction. However this electrical response covers all silanization steps (I–IV.) mentioned above. Results of the random sequential algorithm calculation described in previous section are shown in Figure 4.2 (red line) for the purpose of the presented model verification. If the electrical conductivity of rubber mixtures σ reflects the silane cross-linking evolution we can expect that:

$$\sigma \approx \gamma P, \tag{4.25}$$

where γ is a constant and P is the parameter introduced in one of the previous sections and determined by eq 4.23. We point out that we have not investigated yet the relationship between σ and parameter P in detail. Equation 4.25 is still our preliminary hypothesis. However if the standard mechanism of electrical conductivity is considered, we believe that the linear relationship between σ and parameter P is acceptable option. Afterwards when the temperature t increases linearly we can substitute $t = (T - T_0)/\eta$ to eq 4.23 and find expected temperature dependence of AC electric conductivity σ in the silane cross-linking bonds formation temperature range in next form:

$$\sigma(t) = \sigma_0 e^{-\frac{T}{\vartheta}}, \tag{4.26}$$

where σ_0 and ϑ are constants. So exponential decrease of AC electric conductivity $\sigma(t)$ is expected in the silane cross-linking bonds formation temperature range, i.e., in the last step of silanization reaction corresponding to the point IV, mentioned above.

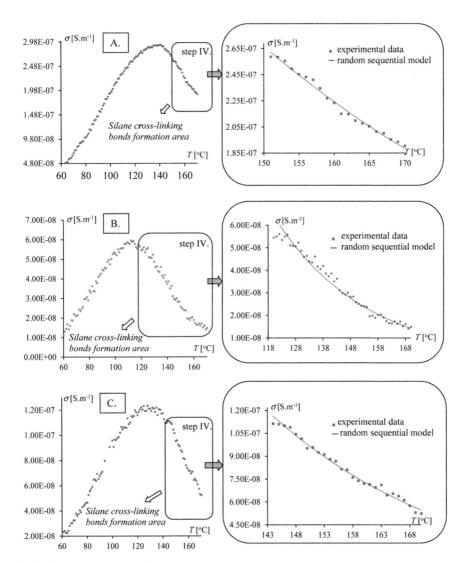

FIGURE 4.2 AC electrical conductivity of rubber mixtures σ as a function of temperature T measured during silanization. Peaks A. B. and C. on the left represent electrical response of whole silanization proces (they cover all steps I–IV.). On the right side there are segments of these graphs corresponding to silane cross-linking formation (step IV).

TABLE 4.1 Coefficients σ_0 and ϑ in the Formula (4.26) Calculated by Least Squares Method. Graphs of the Calculated Dependence Obtained using these Coefficients in (4.26) are Illustrated in Fig. 4.2 (Red Lines on the Right Side). Values in the Last Column in this Table Result from Correlation between Experimental and Calculated Data and Demonstrate the Reliability of the Random Sequential Algorithm in the Modeling of Macromolecular Chains Cross-linking Evolution.

Rubber Mixture	Coefficient σ_0 [S.m$^{1/2}$.s^{-1}]	Coefficient ϑ [$^\circ$C^{-1}]	Correlation Coefficient
A	4×10^{-6}	58,8235	0.9893
B	2×10^{-6}	3,3333	0.9784
C	8×10^{-6}	34,4828	0.9867

4.5   DISCUSSION AND CONCLUSION

Major motivation of our work was to verify the applicability of random sequential algorithm for description of kinetics of macromolecular chains cross-linking process in some specific polymer structures. As we can see from results presented above experimental data correspond quite well to the model calculation just in the last area of the measured temperature range, i.e., in the last step of silanization. This fact can be interpreted in accordance with assumption that random sequential character of silane cross-linking bonds formation is dominant in the corresponding time interval (i.e., just in step IV). The agreement achieved between theoretical results and experiment was quite satisfying for measured samples prepared from SBR rubber mixtures. A question to ask is: what is dependence of parameters σ_0 and ϑ in eq 4.26 on rubber mixture composition and conditions of silanization reaction. Further systematic analysis of a great number of experimental data is necessary to find answer to this question.

Random sequential character of cross-linking bond formation is the most important feature of the presented model that leads to its major assertion that number of possible cross-links which may be formed between all pairs of macromolecular chains in system decreases exponentially in time (see eqs. 4.23 and 4.26) while parameters of exponential function satisfy:

$$\alpha = \frac{A}{4}, \quad \beta = 8\left(\frac{N_c}{A}\right)\left(1 - 2\frac{N_c}{A}\right), \quad \vartheta \approx \eta\tau \quad \text{where } A = \frac{1}{4}\left\{2\sum_{i=1}^{N} n_i^{(0)} + 1\right\}.$$

Of course one could rightly say that a two parameter exponential equation fits to the data shown is not surprising. Maybe a two-parameter linear fit would not be much worse. This means it will be necessary to search for a stronger support for the model in the experimental data in the future research. Concerning the interpretation of our results it should be noted that the possible discrepancy of model results with experimental data may be due to the fact that relaxation time slightly varies with temperature, i.e., $\tau = \tau(T)$. Change of the τ value greatly affects functions $v(t)$ and $\phi(t)$ as it is shown in Figures 4.3 and 4.4.

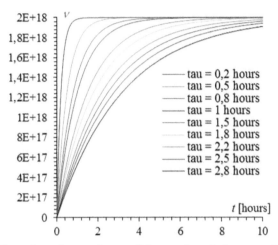

FIGURE 4.3 Time dependence of cross-links number during cross-linking evolution calculated from (4.22) for $N_c = 2.10^{18}$ and $A = 2.10^{20}$ and various values of τ.

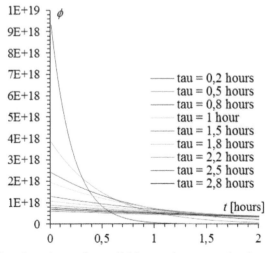

FIGURE 4.4 Time dependence of cross-linking rate in macromolecular structure calculated from (4.24) for $N_c = 2.10^{18}$ and $A = 2.10^{20}$ and various values of τ.

Transformations from the time to temperature dependences $v(t) \rightarrow v(T)$ and $\phi(t) \rightarrow \phi(T)$ are more complicated in that case. For example the frequency of macromolecular cross-links formation can be determined as:

$$\phi(t,T) = \frac{d\,v(t,T)}{dt} = \frac{\partial v(t,T)}{\partial t} + \frac{\partial v(t,T)}{\partial T}\frac{\partial T}{\partial t} = \frac{\alpha\beta}{2\tau(T)}\frac{e^{-\frac{t}{\tau(T)}}}{\sqrt{\beta\left(e^{-\frac{t}{\tau(T)}}-1\right)+1}}\left\{1-\left(\frac{t}{\tau}\right)\frac{\partial\tau}{\partial T}\eta\right\}. \quad (4.27)$$

However, considering values of correlation coefficients shown in Table 4.1 it can be concluded that the temperature dependence of τ is not significant in case of measured samples.

Moreover it is clear from eq 4.24 for values of $\tau \gg t$ the frequency ϕ can be considered constant:

$$\lim_{\substack{t \to 0 \\ \tau}} \phi(t) = \frac{\alpha\beta}{2\tau} = const. \quad (4.28)$$

This fact leads to a quadratic decrease of the quantity P in time in the case of macromolecular structures with large values of τ.

ACKNOWLEDGMENT

This research was supported by Slovak Grant Agency under projects VEGA 1/0356/13 and 1/0184/14. This work was also funded by the ERDF - Research and Development Operational Programme under the project "University Scientific Park Campus MTF STU - CAMBO" ITMS: 26220220179.

KEYWORDS

- random sequential algorithm
- cross-linking
- macromolecular chain
- silanization
- electric conductivity

REFERENCES

1. Varghese, S.; Karger-Kocsis, J. Natural Rubber-based Nanocomposites by Latex Compounding with Layered Silicates. *Polymer.* **2003,** *44,* 4921–4927.
2. Xie,Y.; Hill, C, A, S.; Xiao, Z.; Militz, H.; Mai, C. Silane Coupling Agents Used for Natural Fiber/polymer Composites: A Review.*Compos. Part A. Appl. Sci. Manuf.* **2010,** *41,* 806–819.
3. Allara, D, L.; Parikh, A, N.; Rondelez, F. Evidence for a Unique Chain Organization in Long Chain Silane Monolayers Deposited on Two Widely Different Solid Substrates. *Langmuir.* **1995,** 2357–2360.
4. Sombatsompop, N.; Wimolmala, E.; Markpin, T. Fly-ash Particles and Precipitated Silica as Fillers in Rubbers. II. Effects of Silica Content and Si69-Treatment in Natural Rubber/ Styrene-Butadiene Rubber Vulcanizates. *J. Appl. Polym. Sci.* **2007,** *104,* 3396–3405.
5. Salgueiro, W.; Somozaa, A.; Marzocca, A. J.; Consolati, G.; Quasso, F. Evoltulion of the Crosslink Structure in the Elastomers NR and SBR. *Radiat. Phys. Chim.* **2007,** *76,*142–145.
6. Saville, B.; Watson, A. A. Structural Characterization of Sulfur-Vulcanized Rubber Networks. *Rubber Chem. Technol.* **1967,** *40,* 100–148.
7. Ding, R. Leonov, A. I. A Kinetic Model for Sulfur Accelerated Vulcanization of a Natural Rubber Compound. *J. Appl. Polym. Sci.***1996,** *61,* 455–463.
8. Bjorksten, J.; Tenhu, H. The Crosslinking Theory of Aging—Added Evidence. *Exp. Gerontol.* **1990,** *25,* 91.
9. Kuntal Pal; Karsten Melcher, H.; Eric Xu. Structure Modeling Using Genetically Engineered Crosslinking. *Cell.* **2013,** *155,* 1207–1208.
10. Michael, F. Butler; Yiu-Fai Ng; Paul, D. A. Pudney. Mechanism and Kinetics of the Crosslinking Reaction Between Biopolymers Containing Primary Amine Groups and Genipin. *J. Polym. Sci.,* **2003,** *41,* 3941–3953.
11. Cheung, D. T.; Nimni, M. E. Mechanism of Crosslinking of Proteins by Glutaraldehyde I: Reaction with Model Compounds. *Connect Tissue. Res.* **1982,** *20,* 187–99.
12. Emil Seliga; Ondrej Bošák; Stanislav Minárik; Marián Kubliha; Vladimír Labaš; Juraj Slabeycius. Electrical Response of Silanization of Rubber Mixtures. *Adv. Mater. Phys. Chem.* **2013,** *3,* 105–111.

CHAPTER 5

POLYANILINE-COATED INORGANIC OXIDE COMPOSITES FOR BROADBAND ELECTROMAGNETIC INTERFERENCE SHIELDING

MUHAMMAD FAISAL*

Department of Physics, PES Institute of Technology, Bangalore South Campus, Bangalore 560100, Karnataka, India

Corresponding author. E-mail: faismuhammad@gmail.com; muhammadfaisal@pes.edu

CONTENTS

ABSTRACT

In recent years, there is a growing interest in conducting polymer-based composites because of their unique and fascinating properties and economic viability. This chapter aims to provide the details of studies on electromagnetic attributes of synthesized polyaniline (PAni)–inorganic oxide composites. These composites are studied from the application point of view to optimize as broadband electromagnetic interference (EMI) shielding materials. There is an ever growing need for adjustable EMI shielding materials operating in a wide range of frequencies.

5.1 INTRODUCTION

In the domain of composite materials research, intrinsically conducting polymers (ICPs) and their composites have been intensively studied for uses in a wide range of technological applications. In particular, polyaniline, one of the most promising conducting polymers (CPs), is considered to be a low cost material which is environmentally, thermally, and chemically stable. This chapter reports the studies on polyaniline/inorganic oxide composite materials that offer modified electromagnetic properties. These organic/inorganic hybrids form a promising class of new materials, owing to the advantageous properties of the conducting polymer matrix and the embedded inorganic particles. Three different inorganic oxides—stannic oxide (SnO_2), yttrium oxide (Y_2O_3), and manganese dioxide (MnO_2)—are selected due to their attractive properties. The major objective of this work is to develop cost-effective conducting composites having high electromagnetic interference (EMI) shielding properties and microwave absorbing capacity in the broadband frequency of 8–18 GHz, covering X- and Ku-bands of practical relevance.

5.1.1 CONDUCTING POLYMERS

Polymer composite systems with special properties is a field of increasing scientific and technical interest, offering the opportunity to synthesize a broad variety of promising materials, with a wide range of technological applications. With the discovery of ICPs in 1960, an attractive subject of research was initiated because of the interesting properties and numerous application possibilities of ICPs.[1] Polymers are generally insulators, and the low conductivity of polymers has been extensively used for insulating copper wires and as outer structure of electrical appliances, which prevent people from direct contact with electricity. However, some polymers have been synthesized with remarkable ability to conduct electricity. An organic polymer that possesses the electronic, magnetic, electrical, and optical properties of a metal is called an ICP.[2] During the past four decades, research in CPs and conducting polymer composites (CPCs) have witnessed an increase in interest from academic and industrial research laboratories world-wide, aiming at design and synthesis of new materials in combination with the characterization and application of their novel functional properties.[2,3]

ICPs find their potential applications in multidisciplinary areas such as electrical, electronics, thermoelectric, electrochemical, electromagnetic, electromechanical, electroluminescence, electro-rheological, chemical, membrane, and sensors.[4-15] Their use as new materials in value-added industrial and consumer products continues in opening up entirely new domains of polymeric application. Most of the scientists working on CPs are attracted by the fact that their conductivity can shift, in a reversible way, over several orders of magnitude by oxidation/reduction (also called doping/dedoping) processes. The availability of these new organic semiconductors has opened up possibilities to rebuild electronics and microelectronics producing flexible devices.[16-18] Among the large variety of conducing polymers, polyaniline (PAni) has emerged as the most promising one because of its diverse properties (Fig. 5.1).

Conducting polymer–inorganic oxide composites are expected to improve or complement the electrical, electromagnetic, chemical, and structural properties over their single components, to achieve maximum efficiency required on the different processes taking place in specific applications.[19-24] In view of this, this work aims to investigate the preparation and use of polyaniline composites with inorganic oxide dispersants to synthesize CPCs and to evaluate its broad band microwave properties.

The ability of conjugated polymers to carry delocalized electronic charges is used in many applications in which a metallic, semi-conducting, or electrically tunable medium is involved.[25] The conductivity of ICPs can be tuned by chemical manipulation of the polymer backbone, by the process of doping and blending with other dispersants. These ICPs can be blended into traditional polymers to form electrically conductive blends with various modified properties.

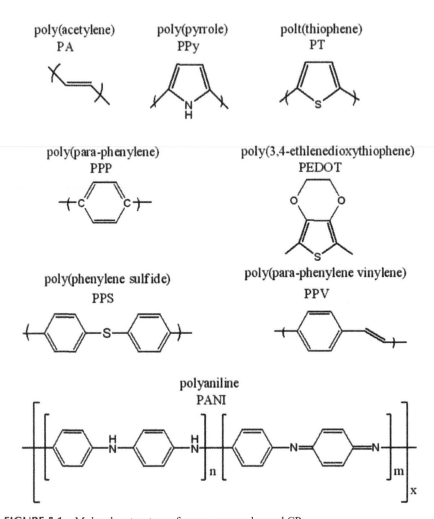

FIGURE 5.1   Molecular structure of some commonly used CPs.

The importance of the field of research of CPs has been recognized by the attribution of the chemistry Nobel Prize in 2000 to A. J. Heeger, A. G. MacDiarmid, and H. Shirakawa.[26] The field of conducting polymers has opened a new frontier in interdisciplinary research and development. Excitement about these CPs is evidenced by the fact that these synthetic metals have attracted scientists from such diverse areas of interest as synthetic chemistry, electrochemistry, solid state physics, materials science, polymer science, biology, optics, electronics, and electrical engineering.[27–30] The common electronic feature of pristine (undoped) CPs is the π-conjugated system which is formed by overlapping of carbon pz orbitals and alternating carbon–carbon bond length. In polyaniline, nitrogen pz orbitals and C6 rings also are part of the conjugation path. The extensive main chain π-conjugation strongly determines its physical properties.[31] Thus, a key requirement for a polymer to become intrinsically electrically conducting is that there should be an overlap of molecular orbitals to allow the formation of delocalized molecular wave function. Besides this, molecular orbitals must be partially filled so that there is a free movement of electrons throughout the lattice.[32] In general, CPs form less ordered structures with many defects and distortion. Major works dealing with CPs are devoted to the problems of their synthesis and possible applications.[33–37]

5.1.2   POLYANILINE

Among the CPs, polyaniline (PAni) has emerged as the material of choice for many applications. The interest in this conducting polymer stems from the fact that many different ring and nitrogen-substituted derivatives can be readily synthesized and that each of the derivatives can exist in several different oxidation states which can in principle be "doped" by a variety of different dopants either by non-redox processes or by partial chemical or electrochemical oxidation. These properties, combined with the fairly high conductivity after doping, their ease of synthesis and processing, satisfactory environmental stability, and the relatively low cost of the starting materials, strongly suggest their significant potential technological applicability.

Polyaniline and its related derivatives form a diverse family of CPs in terms of both their electronic and structural characteristics. These properties are intimately coupled with the suitable molecular level intra-chain and inter-chain interactions. Amorphous polyaniline exhibits relatively poor inter-chain electron transport behavior, thus rending these materials quasi-one-dimensional. Crystalline PAni salts, depending on the primary dopant, the presence of water and/or secondary dopants, and the sample processing history, can display vastly improved inter-chain ordering which results in an increase in the overall conductivity. This leads to a three-dimensional band picture and the development of a metallic-like state. Understanding the central structure/property issues underlying this broad diversity is a formidable goal and requires a detailed knowledge of both the nascent polymer structure and its subsequent evolution. Also, when they are in the composite form, their electrical as well as dielectric properties are altered from those of basic materials.[38–43]

5.1.3 CONDUCTING POLYMER COMPOSITES

Generally composite materials can be defined as materials consisting of two or more components with different properties and distinct boundaries between the components. The idea of combining several components to produce a new material with new properties that are not attainable with individual components has been the inspiring reason toward the development of composite science technology. Composites differ from traditional materials in a sense that interesting properties can result from the complex interaction of the nanostructured heterogeneous phases.

Polymer-based composites are important materials in the development of systems exhibiting various functional properties resulting from the synergistic interaction between the matrix phase and the dispersed phase. In the case of CPCs, the many properties of CPs, such as non-corrosiveness, light weight, mechanical strength, and the possibility to tune electrical conductivity and dielectric behavior, can be utilized along with the properties of the selected dispersed phase to make multifunctional materials of technological relevance.

Polyaniline has been extensively studied in the past due to its unique electronic, redox and acid–base properties as well as numerous potential applications in many fields. In order to improve and extend its functions, the preparation of polyaniline composites has attracted a great deal of attention. Composites consisting of PAni and one or more components, which can be metals, metal oxides, metalloids, nonmetals, inorganic and organic/bioorganic compounds, as well as biological materials and natural products, were prepared and characterized by numerous research groups. Preparation, properties and applications of PAni composites have been an area of immense interest as such materials can be tuned for various technologically important applications.[44–49] Numerous applications of PAni composites in various domains of technology, including electronic nanodevices, chemical, and biological sensors, catalysts, and electrocatalysts, energy storage, microwave absorption and electromagnetic interference (EMI) shielding,[50–52] static electricity dissipation,[53,54] electrorheological (ER) fluids, and various applications in biomedicine have been identified.[48] It is reported that the PAni–metal nanoparticle composites not only retained the original respective intrinsic performances, but also exhibited coadjutant effect that is beneficial for improving the properties of PAni matrix in composites.[55] PAni–metal nanoparticle composites exhibited a wide range of morphologies, ranging from nanosphere, nanorod, and complex nanosheet assemblies, by varying the initial concentration and molar ratio aniline to the dispersant in a low-temperature oxidative polymerization process.[56] Dispersity and morphology of these composites decide many of the physical properties of the resulting system. The PAni–metal nanoparticle composites have shown excellent activity, stability, reproducibility and durability for many applications. Synergism of action of PAni and the dispersed phase is one of the most characteristic features of these composites.[57] The composites of polyaniline have been well documented for a number of inorganic dispersants including TaS_2,[58] $FeOCl$,[59] V_2O_5,[60] MoO_3[61] $CdPS_3$,[62] and $VOPO(4).2H_2O$[63] This research work is inspired by the fact that the detailed analysis of microwave properties of conducting polyaniline composites with inorganic dispersants in the broad microwave frequency range of 8–18 GHz, encompassing both X- and Ku-bands, is very limited.

The choice of the best method to produce composites with specified characteristics remains an unresolved problem, because the processing method may significantly determine the properties of the resulting composite materials. Many composite materials are composed of just two phases; one is termed the matrix, which is continuous and surrounds the other phase, often called the dispersed phase. Popular methods to produce PAni-containing composites[36] include synthetic methods based on aniline polymerization in the presence of or inside a matrix polymer, and blending methods to mix a previously prepared PAni with a matrix polymer. The synthetic direction is probably preferable if it is necessary to produce inexpensive conducting composites, due to the use of inexpensive aniline instead of more expensive PAni. The various synthetic methods include colloidal dispersions,[64] electrochemical encapsulation,[65] coating of inorganic polymers, and in-situ polymerization with nanoparticles,[66] which have opened new avenues for composite material synthesis. The experimental composition followed in this study is the in-situ chemical oxidative polymerization of aniline monomer over the fine graded dispersants, which is simple, cost effective, and economical. It is one of the most widely accepted methods for the preparation of composites with polyaniline as the matrix as reported in the literature.[67–75]

5.1.4 IMPORTANCE OF COMPOSITES

Polymeric nanocomposites consisting of inorganic nanoparticles and organic polymers represent a new class of materials that exhibit improved performance compared to their microparticle counterparts.[76] It is therefore expected that they will advance the field of engineering applications. The properties of composites are a function of the properties of the constituent phases, their relative amounts, and the geometry of the dispersed phase. Incorporation of inorganic nanoparticles into a polymer matrix can significantly affect the properties of the matrix. Polymer matrices reinforced with modified inorganic particles combine the functionalities of polymer matrices, which include low weight and easy formability, with the unique features of the inorganic nanoparticles. The composites obtained by incorporation of these types of materials can lead to improvements in several areas, such as optical, mechanical, electrical, magnetic, rheological, and fire retardancy properties.[77,78] Hence, the development of organic–inorganic nanocomposites, often achieved by adding modified inorganic particles into polymer matrices, is intended to produce composite materials with various improved properties. The properties of polymer composites depend on the type of nanoparticles that are incorporated, their size and shape, their concentration and their interactions with the polymer matrix. The main problem with polymer composites is the prevention of particle aggregation. It is difficult to produce monodispersed nanoparticles in a polymer matrix because the dispersants agglomerate due to their specific surface area and volume effects. An additional problem found in these composites is a lower impact strength than that found in the organic precursor alone due to the stiffness of the inorganic material, leading to the use of elastomeric additives to increase the toughness of the composites.[79]

Polymer–inorganic dispersant composites have been in a close relationship with the design of advanced electronic and optoelectronic devices. Su and Kuramoto[80] reported the synthesis of PAni–TiO_2 nanocomposites by in-situ polymerization of PAni in the presence of TiO_2 nanoparticles. The as-synthesized nanocomposite films showed appreciable conductivity (1–10 S/cm), which was further increased by thermal treatment at 80 °C for 1 h. Mo et al.[81] also carried out the synthesis of PAni–TiO_2 nanocomposites using TiO_2 nanoparticles and colloids. The dielectric constant and loss also increased with increased TiO_2 loading. The conductivity of the nanocomposites also gradually increased as the amount of TiO_2 increased from 1 to 5 wt%. Ma et al.[82] reported that the surface resistivity of polystyrene resin–ZnO nanocomposites synthesized by melt-blending decreases as the concentration of ZnO increases. The addition of 30 wt% of either ZnO spherical particles or whiskers also caused a reduction in the surface resistivity of the composite materials from 1.0×10^{16} to 8.98×10^{12} Ω/cm^2 and 9.57×10^{10} Ω/cm^2, respectively. The amount of ZnO in the polystyrene resin can be gradually increased to form a conductive network. Ma et al.[83] observed that functionalization of TiO_2 nanoparticles improves the electrical properties of polyethylene–TiO_2 nanocomposites. Singha and Thomas[84] investigated the dielectric properties of epoxy nanocomposites containing TiO_2, ZnO, and Al_2O_3 nano-fillers at low filler concentrations by weight and observed some unique electrical properties that could be advantageous in several existing and potential electrical systems.

Organic–inorganic composite materials possess unique properties as new materials and compounds for academic research as well as for the development of innovative industrial applications. These nanocomposites combine the unique properties of organic and inorganic components in one material. The basic multifunctional feature of these nanocomposite materials makes them potentially applicable in various areas in high added-value applications such as smart coatings for corrosion protection and abrasion resistance; artificial membranes for ultra and nanofiltration, pervaporation and gas separation; catalysts and nanoscopic reactors; adsorbents of toxic metal ions; biomaterials for osteo-reconstructive surgery; or ophthalmic devices with optoelectronic and magnetic properties for telecommunications or information displays.

The combination of the inorganic particles with conducting polymer leads to formation of conducting polymer composite possessing unique combination of electrical and dielectric properties. This property of the composite can be used as an electromagnetic shielding material with microwave attenuation properties. The electromagnetic wave consists of an electric (E) and the magnetic field (H) right angle to each other. This electromagnetic energy has been suppressed in the shielding purpose. The conducting polyaniline–inorganic oxides in this study are expected to be effective in electromagnetic shielding and microwave attenuation. The dielectric permittivity ε ($\varepsilon = \varepsilon' + i\varepsilon''$), depending upon the pulsation ($\omega = 2\pi f$), is one of the most important physical parameters that characterize the material as a microwave absorber. Hence, a promising combination of polyaniline with inorganic oxide particles can be a better choice of composite materials with modified dielectric properties to be used for microwave attenuation. However, high dispersant contents are required from 10 to 30 wt% to reach a percolation level leading to enhanced microwave losses. It is reported that material performances are dependent on the filler content as well as particle aggregation phenomena in the composites.[85–89] In the formation of the composites, as monomers usually have a low molecular weight by their nature, the inorganic dispersants can penetrate and the interstitial volume inside the polymer becomes partially filled with inorganic particles. It is important that the formation of interpenetrating networks between inorganic and organic dispersants in the composites improves the compatibility between constituents and builds strong interfacial interaction between the two phases. In-situ polymerization is a method in which inorganic particles are first dispersed in a monomer, and the resulting mixture is polymerized using a technique similar to bulk polymerization.

The inorganic oxide particles identified for this study are stannic oxide (SnO_2), yttrium trioxide (Y_2O_3), and manganese dioxide (MnO_2). Stannic oxide is a crystalline powder which is produced thermally from high grade tin metal and widely employed for technological applications including electronic ceramics, capacitors, conductive coatings, chemicals, glaze opacifiers, pigments, special refractories, lapidary, and lens polishing. Incorporation of SnO_2 particles with high dielectric constant and various other properties, the electromagnetic shielding characteristics of PAni can be altered. Y_2O_3 facilitates excellent microwave and dielectric properties.[90,91] Y_2O_3 has good thermal stability up to 2200 °C with high dielectric constant and only small deviations from stoichiometry under normal conditions of temperature and pressure.[92] Manganese dioxide (MnO_2) is a low-band gap, high optical constant semiconductor that exhibits ferroelectric properties and holds a high dielectric constant, which is a function of frequency.[93,94] MnO_2 has already been recognized as one of the most promising electrode materials for various energy-storage technologies with its natural abundance and environmental compatibility and literatures show that MnO_2 is a good electromagnetic wave absorber.[95-98]

5.2 ELECTROMAGNETIC SHIELDING

Today's world heavily relies on the technologies based on electromagnetic radiation, one of the most fundamental phenomena in nature. The extensive study of electromagnetism pioneered by James Clerk Maxwell in the 19th century brought us radio, television, wireless communication, global navigation, radars, and other devices that sustain the needs of our rapidly evolving society. However, as our technological system grows in complexity, the unwanted interactions start to happen between its constituents. The phenomenon known as EMI is a disturbance that affects an electrical circuit of a device due to an electromagnetic radiation from another electrical circuit. EMI shielding and microwave attenuation materials are receiving enormous interest to protect electronics, instrumentation, and environment in commercial, industrial, health care, and defense applications in the full range of the EMI frequency spectrum.[99-103] The interference across communication channels, automation, and processing controls result in the loss of valuable time, energy, and resources. Hence, EMI has been addressed under electromagnetic compatibility (EMC) regulations by governments and organizations around the world (IEC-TC77, IEC-CISPR, FCC-20780, MIL-STD-461/462, etc.)[104–106] to focus on the key issues related to EMI shielding. One of the important parameters in characterizing EMI shielding materials is the electromagnetic interference shielding effectiveness (EMI SE) values in the frequency range of interest.[100–107] A total of −15 to −30 dB of shielding effectiveness (SE) is considered acceptable for most industrial and consumer applications. A good shielding material should prevent both incoming and outgoing EMI. More precisely, an EMI shield in electronic equipments controls the excessive self-emission of electromagnetic waves and also ensures the undisturbed functioning of the device in presence of external electromagnetic fields.[108] Furthermore, the ongoing downscaling trend in electronics requires the shielding material to be thin, lightweight, easy to manufacture, and cheap.

A shield is conceptually a barrier to the transmission of electromagnetic energy. For efficient shielding action, shield should possess either mobile charge carriers (electrons or holes) or electric and/or magnetic dipoles which interact with the electric (E) and magnetic (H) vectors of the incident electromagnetic energy. In the recent past, a wide variety of materials[109–118] have been used for EMI shielding with a broad range of electrical conductivity (σ), good electromagnetic attributes such as permittivity (ε), or permeability (μ) and engineered morphological features. The optimization of EMI shielding performance with certain level of required attenuation, meeting a set of physical criteria, maintaining economics and regulating the involved shielding mechanism is not a straight forward task and involves complex interplay of intrinsic properties (σ, ε, and μ) of the shield and the logical selection of extrinsic parameters.

Among the numerous applications of CPs, their use to reduce EMI is receiving great attention from the scientific and technological communities. ICPs are promising materials for shielding electromagnetic radiation and reducing or eliminating EMI because of their relatively high "σ" and "ε" values and their ease of control through chemical processing. Also, they are relatively lightweight compared to standard metals, are flexible, and do not corrode. More specifically, polyaniline (PAni) characteristics of frequency agility, light weight, simple tuning of functional properties, and non-corrosive nature, make this polymer a suitable candidate for the development of electromagnetic shielding and microwave attenuation materials where protection against EMI without degradation of device performance is the main aim. Tuning of conductive and dielectric properties in PAni-based composites can be achieved by suitable selection of polymerization conditions and controlled addition of suitable dispersants.

Polyaniline composites with modified electrical and dielectric properties could be used as an effective EMI shielding material. For such an application typical materials include copper or aluminum, which have high enough conductivity (σ) and dielectric constant (ε), which contribute to high EMI SE. Despite EMI shielding systems based on metals having good mechanical and shielding properties; they are heavy, are susceptible to corrosion, and possess limited forming capability with major shielding component as reflection. The use of polyaniline–inorganic oxide composite for EMI reduction and microwave attenuation would represent a promising alternative to overcome the above described limitation of metal-based systems. Accordingly, polyaniline (PAni)–inorganic oxide composite could be considered potential candidates to replace or to use together with metallic materials for electromagnetic shielding applications. An effective shielding material should have a low reflectivity in a wide frequency range and it should be lightweight, mechanically stable, cheap and easy to deposit. Polyaniline composites with modified morphology and electrical properties would have a unique set of electromagnetic properties and can cause higher microwave attenuation over a broad range of frequencies. These heterogeneous composite structures with high surface-to-volume ratio and adjustable electromagnetic properties can be potentially used for EMI-shielding applications.

5.2.1 SHIELDING PHENOMENON

In EMI shielding, the overall attenuation is achieved by the phenomena of reflection loss or return loss (caused by the reflection at the surface of the shield, R_{dB}), absorption loss (caused by the absorption effect as the waves proceed through the shield, A_{dB}) and multiple reflections (caused by the additional effects of multiple reflections and transmissions in the interior of the shield, M_{dB}),[107–119,120] the remaining energy will be transmitted for interference (Fig. 5.2). Reflection depends on the charge carriers; absorption depends on the presence of dipoles; whereas multiple reflections are the total internal reflections within the material. These three mechanisms contribute toward overall attenuation with SE_R, SE_A, and SE_M as corresponding SE components due to reflection, absorption, and multiple reflections, respectively. Among these three mechanisms, reflection and absorption suppress most of the electromagnetic fields. Shielding by absorption is important for various practical applications as compared to reflection.

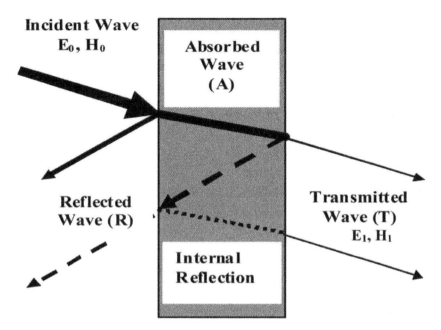

FIGURE 5.2   Schematic representation of mechanism of EMI shielding.

The EMI SE of a material is defined as the attenuation of the prop-agating electromagnetic waves produced by the shielding material. It is generally measured in terms of reduction in magnitude of incident power/field upon transition across the shield. SE can be expressed in logarithmic scale as:[111, 120–124]

$$SE \text{ (dB)} = SE_R + SE_A + SE_M$$

$$= 10 \log(P_0/P_1) = 20 \log(E_0/E_1) = 20 \log(H_0/H_1) = 20 \log(F_0/F_1),$$

where P_0 (E_0, H_0, or F_0) and P_1 (E_1, H_1, or F_1) are the power (electric, magnetic, or plane-wave field intensity) of incident and transmitted electro-magnetic waves, respectively.

5.2.2   EXPERIMENTAL TECHNIQUE

Experimentally, shielding is measured using network analyzer. Scalar network analyzer (SNA) measures only the amplitude of signals whereas vector network analyzer (VNA) measures magnitude as well as phases of various signals. Consequently, SNA cannot be used to measure complex signals (e.g., complex permittivity or permeability) and therefore, VNA is the most widely used measurement technique.[125] In this study, microwave characteristics of the polyaniline–inorganic oxide composite samples were measured using the wave guide transmission line method.[126–128] Waveguide measurements are considered as a very effective way to characterize small quantities of new materials over broad frequency ranges. Other microwave characterization techniques such as contactless cavity perturbation method simplify the sample size and shape but yield data at a single frequency that is dependent upon the resonant frequency of the cavity. This technique is not practical for a broad frequency response study. Also, another popular method, namely, free-space technique, has been used on processable materials using relatively large sample area.

The measurement scheme is shown in Figure 5.3. In the X and Ku-band rectangular wave guide sample holder, since the outer conductor is grounded, the alternating current in the inner conductor created by the VNA produces a transverse electromagnetic wave of a certain frequency. As the wave travels through the sample holder part of the wave gets reflected, because of the impedance mismatch, and the other gets transmitted. The relationship between the incident, reflected and the transmitted powers can be expressed in terms of scattering (S) parameters.

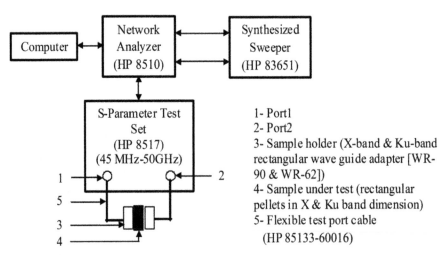

1- Port1
2- Port2
3- Sample holder (X-band & Ku-band rectangular wave guide adapter [WR-90 & WR-62])
4- Sample under test (rectangular pellets in X & Ku band dimension)
5- Flexible test port cable (HP 85133-60016)

FIGURE 5.3 Schematic of the VNA setup used for the microwave measurements in the X- and Ku-bands.

5.2.3 SCATTERING (S) PARAMETERS

The scattering matrix (S-matrix) is the mathematical concept that fully describes the propagation of an electromagnetic wave through a multi-port network. For a signal incident on one port, some fraction of the signal bounce back out of that port, some of it scatter and exit from other ports and some of it disappear as heat. The S-matrix for a two-port device has four coefficients known as S-parameters that represent all possible input–output signal paths. The first number in the subscript of the S-parameter refers to the responding port, while the second represents the incident port (Fig. 5.4). Each S-parameter (S_{11}/S_{22} and S_{21}/S_{12}) is a complex number that represents magnitude and angle, without any unit.

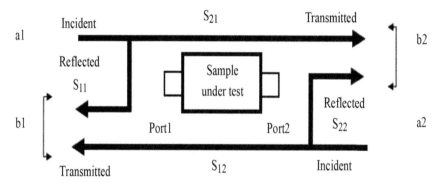

FIGURE 5.4 Sketch of complex S-parameters of a two-port device.

The scattering parameters (S-parameters) correspond to the transmission (S_{12}/S_{21}) and reflection (S_{11}/S_{22}) of the incident electromagnetic waves were measured. These S-parameters were used to compute the complete EMI shielding and dielectric properties of the composites in the entire X-and Ku-band microwave spectrum. Scattering parameters connect the input and output circuit quantities using the reflection and transmission parameters normally adopted in microwave analysis.

5.3 SYNTHESIS OF POLYANILINE AND ITS COMPOSITES

Polyaniline has been known for many years since the synthesis of the so-called aniline blacks that enjoyed early use as cotton dyes in the 1860s. Only sporadic studies of polyaniline's were undertaken until the investigations by MacDiarmid and co-workers in the mid-1980s, and the discovery of electrical conductivity for its emeraldine salt (ES) form led to an explosion of interest in this fascinating conducting polymer.[129–131] Polyaniline is one of the most technologically important materials among all the CPs, with its remarkable stability, good conductivity, redox properties, and processibility.[132,133] Polyaniline has been prepared by a number of methods, the most common being chemical and electrochemical oxidation. Other techniques include solid state polymerization, plasma polymerization, precursor polymer route, and template polymerization.

Chemical oxidative methods are preferred to electrochemical polymerization because of its cost-effectiveness and bulk quantity of the polymer that can be prepared during the onset of the reaction. To date it has been the major commercial method of producing CPs. CPs are made conducting, or "doped," by the reaction of conjugated semiconducting polymer with an oxidizing agent, a reducing agent or a protonic acid, resulting in highly delocalized polycations or polyanions.[134,135] The mobile charge carriers are introduced by oxidation or reduction and counterions by "doping." The conductivity of PAni can be tuned by chemical manipulation of the polymer backbone, by the nature of the dopant, by the degree of doping, and by blending with other polymers/dispersants.[136–139] Among the reagents for the synthesis of polyaniline, dopant is the most important reagent, because it is mainly attributed to electrical properties of PAni. In general, inorganic acids (e.g., HCl, H_2SO_4, $H3PO_4$, and HF) and organic acids are widely used as dopants for doping PAni which strongly affect the properties of the PAni.[140]

In this work, polyaniline has been synthesized by chemical polymerization technique, and the polyaniline–inorganic oxide composites are prepared by in-situ polymerization technique. We have selected hydrochloric acid (HCl) in equimolar proportion to aniline and ammonium persulfate $((NH)_4)_2S_2O_8)$ APS) as the oxidizing agent. The most widely employed chemical oxidant has been aqueous ammonium persulfate $(NH_4)_2S_2O_8$, leading to the incorporation of HSO_4^-/SO_4^{2-} as the dopant anions (A–) in the obtained PAni/HCl. Acidic conditions (pH ≤ 3) are usually required to assist the solubilization of the aniline in water and to avoid excessive formation of undesired sub products.[141] To minimize the presence of residual aniline and to obtain the best yield of polyaniline, the stoichiometric persulpahate/aniline ratio of 1.25 is recommended.[142] The scheme of chemical oxidative polymerization of aniline can be represented as shown in Figure 5.5.

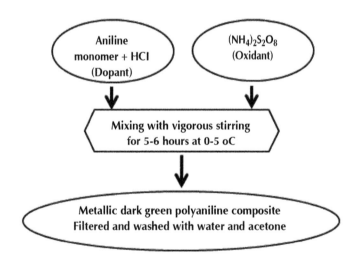

$$+2\,n\,HCl\ +5\,n\,H_2SO_4\ +5\,n\,(NH_4)_2SO_4$$

FIGURE 5.5 Chemical synthesis of PAni via oxidative polymerization by peroxydisulfate.

All the conducting PAni composites were synthesized by in-situ polymerization using HCl as dopant and APS in water as oxidant[35–38] and is schematically represented below (Fig. 5.6). First, a certain weight percentage (10, 20, and 30 wt%) dispersant particles (SnO_2, Y_2O_3, and MnO_2) were suspended in 1 M HCl solution and stirred for 1 h to get well dispersed. The 2 ml aniline monomer was added to the suspension and stirred for 30 min. The oxidant APS (the molar ratio of APS to aniline was 1:1) was then slowly added dropwise to the suspension mixture with a constant vigorous stirring. The polymerization was allowed to proceed for 10 h at a controlled temperature 0–5 °C. The composite was obtained by filtering and washing the suspension with deionized water and acetone and dried under vacuum at 60 °C for 24 h.

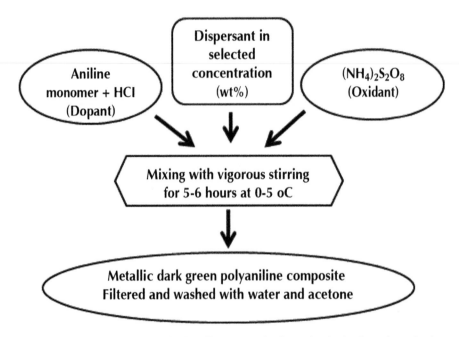

FIGURE 5.6 Representation of polyaniline composite formation by in-situ polymerization.

The composites, i.e., PAni–SnO$_2$ composites (PSN1: PAni with 10 wt% SnO$_2$; PSN$_2$: PAni with 20 wt% SnO$_2$; and PSN3: PAni with 30 wt% SnO$_2$), PAni–Y$_2$O$_3$ composites (PYO1: PAni with 10 wt% Y$_2$O$_3$; PYO2: PAni with 20 wt% Y$_2$O$_3$; and PYO3: PAni with 30 wt% Y$_2$O$_3$), and PAni–MnO$_2$ composites (PMN1: PAni with 10 wt% MnO$_2$; PMN2: PAni with 20 wt% MnO$_2$; and PMN3: PAni with 30 wt% MnO$_2$), were compressed in the form of rectangular pellets for microwave property studies in both the X- and Ku-bands. Rectangular samples of thickness 2.0 mm of standard X- and Ku-band rectangular waveguide dimensions (X-band waveguide standard: WR-90 with inside dimensions 2.286 cm × 1.016 cm and Ku-band waveguide standard: WR-62 with inside dimensions 1.579 cm × 0.789 cm) were compacted using homemade die (Fig. 5.7) under a pressure of 9 tons in a table-top hydraulic press at room temperature. The quantity of the composite sample and pressing time were optimized and maintained for X-band as well as Ku-band samples.

FIGURE 5.7 Homemade die and plunger for processing circular and rectangular pellets.

5.4 CHARACTERIZATION

5.4.1 MORPHOLOGICAL PROPERTIES: SCANNING ELECTRON MICROSCOPY (SEM) AND TRANSMISSION ELECTRON MICROSCOPY (TEM)

The surface morphology of the composites investigated using SEM and TEM are shown in Figure 5.8(a–e) for PAni–SnO$_2$ composites, Figure 5.9(a–f) for PAni–Y$_2$O$_3$ composites, and Figure 5.10(a–d) for PAni–MnO$_2$ composites. Polyaniline (Figs. 5.8(a), 5.9(a), and 5.10(a)) prepared under acidic medium exhibits a characteristic aggregated globular morphology, which is in agreement with those of literature.[143–145] The SEM micrograph of pure SnO$_2$ (Fig. 5.8(b)) shows heterogeneous cubic flaky grains, and pure Y$_2$O$_3$ particles showed polyhedral flaky aggregates (Fig. 5.9(b)) as reported by elsewhere.[146,147] Such kind of morphology is very important to facilitate electromagnetic absorbing properties of various shielding materials, as they act as electric field/magnetic field trap centers leading to electric/magnetic loss.[148] Also, the dielectric losses at microwave frequencies can be tailored with the inclusion of flake-like dispersants in the composites.[149] The micrograph of MnO$_2$ (Fig. 5.10(b)) exhibited flaky strips with agglomeration. The SEM and TEM of all the composites indicate the formation of PAni layer over SnO$_2$, Y$_2$O$_3$, and MnO$_2$ particles, which will have unique morphological modifications leading to novel functionalities. The SEM micrographs (Figs. 5.8(c), 5.9(c), and 5.10(c)) reveal the formation of composites with the dispersant particles being entrapped into PAni chains. The morphological modifications of these composites show great anisotropy compared to pure PAni and the inorganic dispersants. As a consequence, these electrically conducting composites with high anisotropy are expected to be strongly polarizable. Electrical conduction and various polarization mechanisms contribute to modified electromagnetic properties of the materials.[149–151] The encapsulation of the dispersants by the PAni matrix was confirmed by the SEM micrographs.

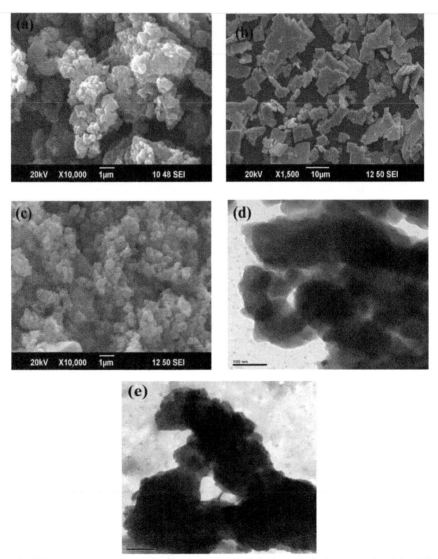

FIGURE 5.8   SEM images of (a) PAni, (b) SnO_2, and (c) PAni–SnO_2 composite (30 wt%); TEM images of (d) PAni, and (e) PAni–SnO_2 composite (30 wt%).

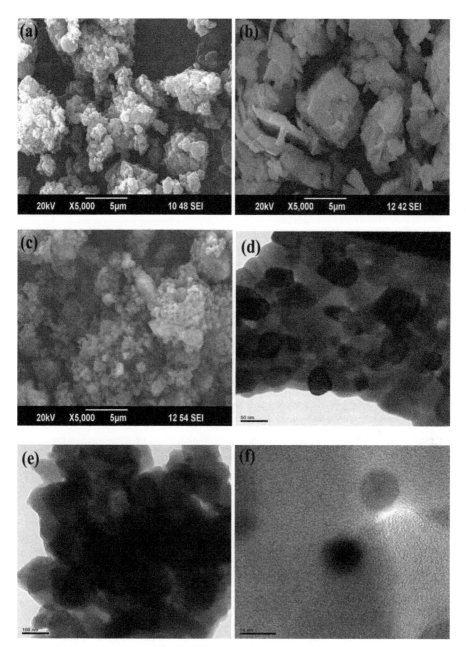

FIGURE 5.9 SEM photographs of (a) PAni, (b) Y_2O_3, and (c) PAni–Y_2O_3 composite (20 wt%); low magnification HRTEM images of (d) Y_2O_3, (e) PAni–Y_2O_3 composite, and (f) high magnification HRTEM image of PAni–Y_2O_3 composite.

FIGURE 5.10   SEM images of (a) PAni, (b) MnO_2, and (c) PAni–MnO_2 composite (30 wt%); TEM image of (d) PAni–MnO_2 composite (20 wt%).

5.4.2 X-RAY DIFFRACTION STUDIES

In this study, X-ray diffraction was carried out by X-ray powder diffraction method at an ambient temperature for the structural characterization of the samples. X-ray diffractograms of pristine PAni, SnO_2, and PAni–SnO_2 (20 wt%, PSN2) composites are shown in Figure 5.11(a, b, and c). In Figure 5.11(a), the broad peak in the range $2\theta = 20°$ to $30°$ is a characteristic semi crystalline peak of polyaniline. This indicates the presence of short-range π-conjugation in PAni. The XRD pattern of SnO_2 in Figure 5.11(b) indicates the presence of tetragonal structure (JCPDS file number 01-088-0287) with the crystalline peaks appearing at $2\theta = 18°$, 29°, 33°, 36°, 47°, 50°, 57°, 62°, and 69°. The crystalline peaks of SnO_2 are retained in the XRD pattern of the PAni–SnO$_2$ (20 wt%) composite (Fig. 5.11(c)) with a slight decrease in intensity of peaks. The strong interaction between PAni backbone chain and SnO$_2$ particles during the polymerization reaction did influence in the reduction of peak intensity corresponds to PAni–SnO$_2$ composite. Since the characteristic peaks of SnO$_2$ also appears in PAni–SnO$_2$ composite, suggests that SnO$_2$ has retained its structure after the formation of composite with PAni.

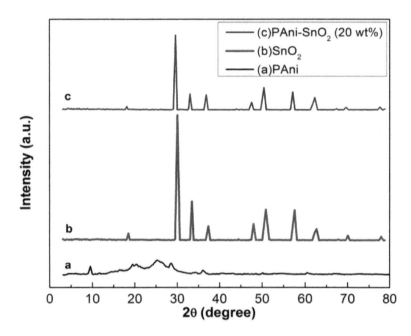

FIGURE 5.11 XRD spectra of (a) PAni, (b) SnO$_2$, and (c) PAni–SnO$_2$ composite (20 wt%).

The XRD patterns of polyaniline, Y$_2$O$_3$, and PAni–Y$_2$O$_3$ composite are shown in Figure 5.12(a–c). The XRD pattern of polyaniline (Fig. 5.12(a)) shows amorphous features with broad semi-crystalline diffraction peaks centered at around $2\theta = 21°$ and $28°$. These peaks correspond to the scattering from bare polymer chains of the protonated polyaniline. It has been reported that the structure of polyaniline doped with common counter ions are predominately amorphous. Figure 5.12(b) shows the diffraction peaks corresponding to single phase of cubic Y$_2$O$_3$ (JCPDS 41-1105). For PAni–Y$_2$O$_3$ composite (Fig. 5.12(c)), the diffraction peaks of Y$_2$O$_3$ can be clearly observed in the composite, with a slight variation of intensity. This confirms that Y$_2$O$_3$ retained its structure even after the formation of composite.

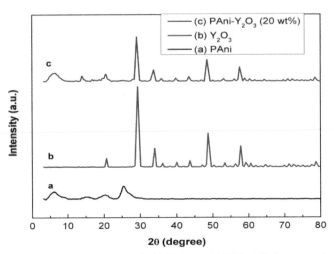

FIGURE 5.12   XRD spectra of (a) PAni, (b) Y_2O_3, and (c) PAni–Y_2O_3 composite (20 wt%).

Figure 5.13(a, b, and c) shows the X-ray diffraction patterns of the pure PAni, MnO_2, and composite having 20 wt% of MnO_2 in polyaniline. The XRD pattern of polyaniline suggests that it has amorphous features with broad semi-crystalline diffraction peaks centered at around $2\theta = 25.1°$, $26.2°$, $28.8°$, and $31.3°$, respectively. Figure 5.13 shows the diffraction peaks corresponding to orthorhombic structure of MnO_2 (JCPDS 14-0644) with prominent peaks located at $2\theta = 18.9°$, $27.5°$, $35°$, and $42.2°$. By comparing

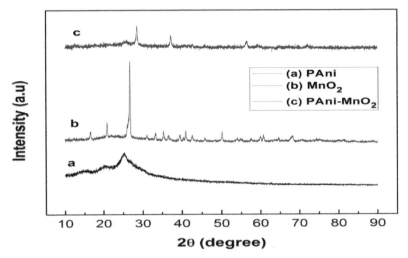

FIGURE 5.13   XRD spectra of (a) PAni, (b) MnO_2, and (c) PAni–MnO_2 composite (20 wt%).

the X-ray diffraction pattern of the composite (Fig. 5.13(c)) with that of standard MnO_2 XRD pattern, it is confirmed that there is no change in the structure of MnO_2 due to its dispersion in polyaniline during the polymerization reaction. Hence through this X-ray diffractogram, it can be concluded that the MnO_2 has retained the structure even though it is dispersed in polyaniline during polymerization reaction.

5.5 EMI SHIELDING STUDIES

5.5.1 EMI SHIELDING EFFECTIVENESS

EMI SE is the ability of a shielding material to attenuate the incident electromagnetic field. EMI SE values of PAni–SnO_2, PAni–Y_2O_3, and PAni–MnO_2 composites in the frequency range of 8–12 GHz (X-band) are presented in Figure 5.14(a–c). The composites exhibit significantly improved SE compared to that of pure polyaniline. The combination of SnO_2, Y_2O_3, and MnO_2 with PAni leads to synergistic effect toward the improvement in EMI SE. Figure 5.14(a–c) also represents the contribution toward improvement in EMI SE of each of the composite materials with different concentration (wt%) of the chosen dispersants. For PAni–SnO_2 composites, the SE values increased up to 30 wt% of SnO_2 in PAni. With increasing Y_2O_3 content in PAni from 10 to 20 wt%, there is an improvement in EMI SE from 22.7 to 24.1 dB and reduced from 23.9 to 20.3 dB for higher concentrations of Y_2O_3 in PAni. In the case of PAni–MnO_2 composites, 20 wt% of MnO_2 concentration in the PAni matrix shows the highest EMI SE in the range of 22.3–20.4 dB.

The EMI SE values for PAni–SnO_2, PAni–Y_2O_3, and PAni–MnO_2 composites in the Ku-band (12–18 GHz) are presented in Figure 5.15(a–c). The EMI SE of all the composites shows significant improvement as compared to that of pure polyaniline. In the case of PAni–SnO_2 composites, it is observed that the EMI SE is maximum for the composite with 30 wt% of SnO_2 in PAni as compared to other samples. It can be observed from Figure 5.15 that the SE increases with increase of Y_2O_3 concentration up to 20 wt% and decreases further with 30 wt% Y_2O_3 loading. The EMI SE is maximum for the composite with 20 wt% of MnO_2 in PAni as compared to other PAni–MnO_2 composites. The difference in the respective EMI SE values of these PAni–inorganic oxide composites with varying concentration of the dispersants is quite significant. This behavioral shift is due to the percolation in the composites, and the maximum EMI SE indicates the percolation

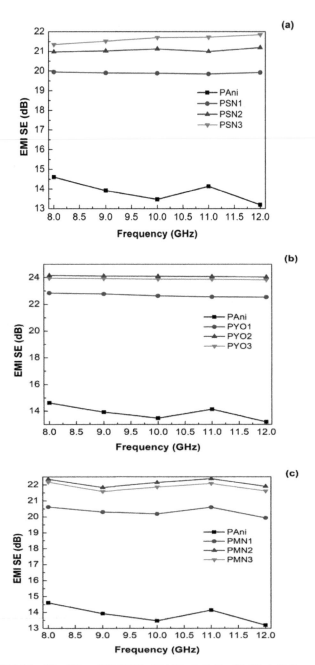

FIGURE 5.14   The X-band EMI SE of (a) PAni–SnO$_2$, (b) PAni–Y$_2$O$_3$, and (c) PAni–MnO$_2$ composites.

threshold value of the dispersant concentration for SE. The EMI SE values of of PAni–Y_2O_3 and PAni–MnO_2 composites show optimum results for the percolation threshold at 20 wt%. However, for PAni–SnO_2 composites the percolation threshold was observed at 30 wt% loading of SnO_2 in PAni. In the Ku-band, the variations of SE values for all three composite series with varying concentration (wt%) of the dispersants were similar to that observed in the X band. In general, shielding effectiveness values observed to be maximum around the optimal loading of dispersants corresponding to the percolation threshold for all the composites in this study. The results show that varying the concentration of the dispersants in PAni matrix can provide better tuning of the conducting path resulting in the enhancement of electromagnetic functionality. The observed EMI shielding behavior is very similar to that reported in the literature for different polymers and their composites[152–157] with improved EMI SE values for practical applications in both X- and Ku-bands.

The observed higher values of SE of all these composites in the broad frequency range 8–18 GHz are most significant in view of the possible technological applications of these composites. Improved SE of all these composites can be attributed to the strong interaction between the PAni matrix and the inorganic oxides together with resulting morphological modifications. The higher EMI SE of these composites can also be attributed to the modified dielectric properties of these composites. The electromagnetic waves can be absorbed and attenuated in the composite materials which have a high permittivity.[158–161] The attenuation is also contributed due to the mobile charge carriers (polarons/bipolarons) in the materials leading to strong polarization and relaxation effects. The presence of more mobile charges can effectively interact with the incoming electromagnetic fields in the radiation. The inorganic dispersant can act as interconnecting bridge between the various conducting chains of polyaniline resulting in long range charge transport. The charge transport as well as a number of possible relaxation modes can lead to increased shielding efficiency. It is known that addition of dispersant to PAni matrix creates large surface area or interface area as a result of their dispersion in the polymer matrix.[162–164] Hence the exhibited effective EMI SE of PAni–SnO_2, PAni–Y_2O_3, and PAni–MnO_2 composites is also resulted from the absorption of the incident electromagnetic energy entering the composite and its dissipation through the interface boundaries.[165] The experimental results of EMI SE show the synergistic effect of the two phases (deposition of PAni matrix over $SnO_2/Y_2O_3/MnO_2$) toward improved electromagnetic properties. The analysis revealed that PAni–Y_2O_3 composites show the highest EMI SE followed by samples PSN and PMN successively.

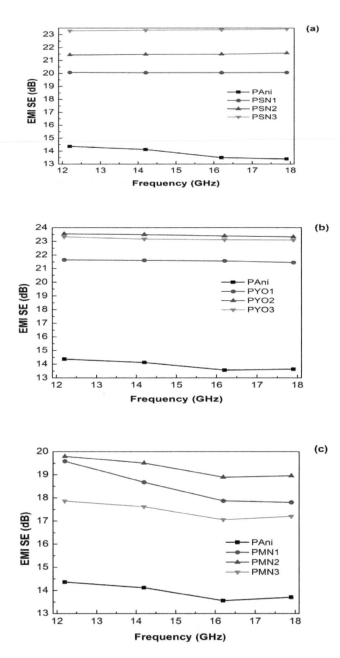

FIGURE 5.15   The Ku-band EMI SE of (a) PAni–SnO$_2$, (b) PAni–Y$_2$O$_3$, and (c) PAni–MnO$_2$ composites.

The shielding efficiency target for a shielding material in commercial applications is expected to be around –20 dB.[166–168] All the polyaniline–inorganic oxide ($SnO_2/Y_2O_3/MnO_2$) composites in this study showed proper EMI shielding effect for commercial applications and it is expected that these EMI shielding materials may be optimized for various applications in a broad frequency range of 8–18 GHz.

5.5.2 COMPLEX PERMITTIVITY

The shielding efficiency of a material depends on the electromagnetic attributes of the material, such as complex permittivity.[169,170] The complex permittivity ($\varepsilon^* = \varepsilon' - i\varepsilon''$) values of the composites are measured based on Nicolson–Ross–Weir (NRW) procedure.[171,172] The real (ε') and imaginary (ε'') relative permittivity values represent the stored energy and dissipated energy, from an external electric field by a material, respectively.[173,174] When the electromagnetic waves strike the material, it induces displacement currents (indicating polarization capability of the material, represented by real permittivity ε') and conduction electrical current (indicating effectiveness of mobile charge transport, represented by dielectric loss ε'') within the material.[175]

Figures 5.16 and 5.17 show the dielectric constant (ε') and dielectric loss (ε'') in PAni–SnO_2, PAni–Y_2O_3, and PAni–MnO_2 composites in the X-band. The dielectric permittivity, ε' and ε'', of PAni–SnO_2 composites were found to have average values of 84.9–74.8 and 97.7–87.5, respectively. The complex permittivity of PAni–Y_2O_3 composites shows average values of 97.4–84.7 and 118.6–106.8, respectively. The dielectric constant and dielectric loss of PAni–MnO_2 composites show average values of 49.6–40 and 55.2–46.3, respectively. The ε' spectra of PAni–SnO_2 composites show marginal variation in the whole X-band range. The ε' values have increased for PSN composites with increasing SnO_2 content from 10 to 30 wt% in PAni matrix. Increasing SnO_2 content in polyaniline matrix leads to a rise in the permittivity values up to 30 wt% indicating the modified electromagnetic attributes of these composites. A similar variation of the permittivity was observed in the case of PAni–Y_2O_3 and PAni–MnO_2 composites, with the percolation behavior. The complex permittivity values for PAni–Y_2O_3 composite samples show that both the ε' and ε'' values reached maximum as the Y_2O_3 concentration is increased to 20 wt%. The ε^* values for the PAni–MnO_2 composites show decreasing trend with increasing concentration (30 wt%) of the dispersant MnO_2. The PMN samples exhibited higher values of ε' and

ε'' for 20 wt% loading of MnO_2 in PAni, indicating the optimum values at the percolation threshold of 20 wt%.

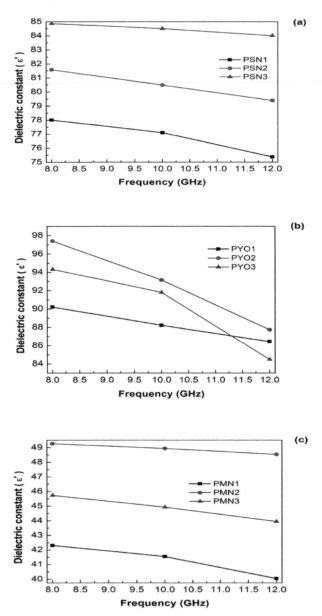

FIGURE 5.16 Dielectric constant (ε') of (a) PAni–SnO_2, (b) PAni–Y_2O_3, and (c) PAni–MnO_2 composites at X-band frequencies.

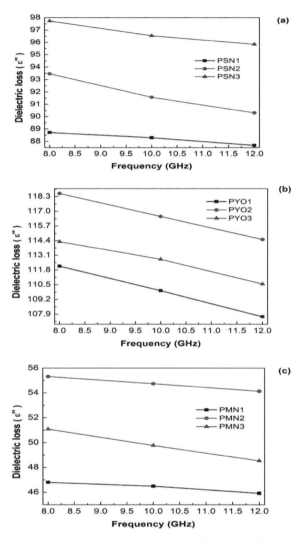

FIGURE 5.17   Dielectric loss (ε'') of (a) PAni–SnO$_2$, (b) PAni–Y$_2$O$_3$, and (c) PAni–MnO$_2$ composites at X-band frequencies.

Figures 5.18(a–c) and 5.19(a–c) show the complex permittivity spectra, real (ε') and imaginary (ε'') parts, respectively, for the composite samples PSN, PYO, and PMN in the Ku-band. Figures 5.18(a) and 5.19(a) show that ε' and ε'' of PAni–SnO$_2$ composites are stable throughout the Ku-band and show a marginal decreasing trend with increasing frequency. Both ε' and ε'' values increase as the SnO$_2$ percentage is increased in the composite up to

30 wt% and decreased marginally for higher wt% of PAni–SnO$_2$ composites. On the other hand, ε' and ε" increased with Y$_2$O$_3$ loading in PAni (Figures 5.18b and 5.19b) up to 20 wt% loading and thereafter exhibited minimal decrement. As shown in Figures 5.18c and 5.19c, ε' and ε" increased with increasing concentration of MnO$_2$ content for PAni–MnO$_2$ composites; however, the complex permittivity decreased for higher wt% (PMN3)

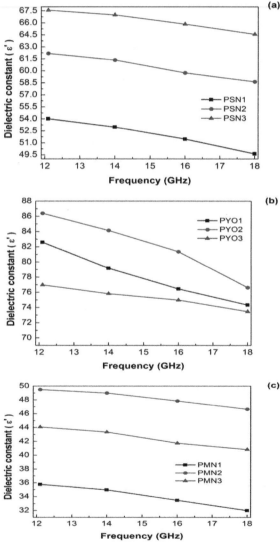

FIGURE 5.18 Dielectric constant (ε') of (a) PAni–SnO$_2$, (b) PAni–Y$_2$O$_3$, and (c) PAni–MnO$_2$ composites at Ku-band frequencies.

composite sample. Similar percolative trend was observed for PAni–SnO$_2$, PAni–Y$_2$O$_3$, and PAni–MnO$_2$ composites for the whole frequency range of 8–18 GHz, with different percolation threshold values. These complex permittivity results pointed out that all the PAni–inorganic oxide composites in this study show better dielectric response and resultant effective shielding properties throughout the 8–18 GHz broadband frequency range: 20 wt% is the optimum concentration of Y$_2$O$_3$ for PAni–Y$_2$O$_3$ composites; 20 wt% is the effective loading of MnO$_2$ for PAni–MnO$_2$; and 30 wt% is the optimum

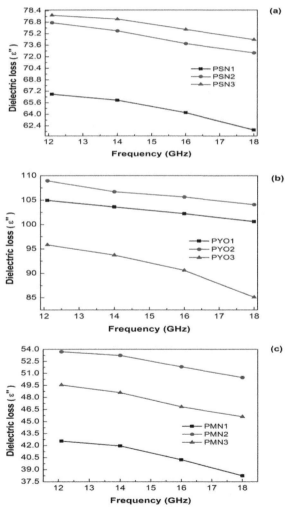

FIGURE 5.19   Dielectric loss (ε''') of (a) PAni–SnO$_2$, (b) PAni–Y$_2$O$_3$, and (c) PAni–MnO$_2$ composites at Ku-band frequencies.

loading of SnO_2 for PAni–SnO_2 composites in achieving higher EMI SE and better dielectric response.

The exhibited larger permittivity values of all these composites can be attributed to the modified space charge polarization effect, dielectric relaxation, and electrical conductivity. Also, these composites are expected to generate dielectric losses linked to their modified morphology. In the case of polyaniline–inorganic oxide composites, morphological modifications contribute to dielectric constant and dielectric loss due to interfacial polarization and its relaxation, as the semiconducting inorganic particles dispersed in the PAni matrix give rise to heterogeneity. Different relaxation frequencies of various dipoles formed in the PAni–inorganic oxide composite structure, hopping of mobile charge carriers, and relaxation due to interfacial polarization together with the effects of space charge polarization all are responsible for the observed dielectric attributes and the efficient shielding properties. The observed decrement in the complex permittivity values of all three sets of PAni composites followed the similar variations in the EMI SE. In conclusion, permittivity data were in agreement with EMI SE results for PAni–SnO_2, PAni–Y_2O_3, and PAni– MnO_2 composites. The complex permittivity of the PAni–inorganic oxide composites depends on the texture of the percolation, and higher concentration of the dispersants leads to a high level of particle aggregation leading to deterioration of the electromagnetic properties.[176]

5.6 CONCLUSIONS

EMI shielding properties of three different PAni composites were tested in the frequency range of 8–18 GHz, and results indicate better EMI SE and dielectric properties. From the analysis of SEM images, it was found that SnO_2, Y_2O_3, and MnO_2 are well incorporated into PAni matrix in their respective composites. Using SnO_2 as the dispersant, increased SE was observed at 30 wt% concentration, especially at the percolation threshold. The combination of PAni and Y_2O_3 shows higher SE values for all composites, with observed maximum corresponding to 20 wt%. The observed shielding and dielectric properties of these composites can be attributed to different mechanisms acting synergistic for the modified structures composed of PAni and inorganic dispersant. These light weight composites are favorable for practical broadband EMI shielding and microwave absorbing applications.

ACKNOWLEDGMENT

The author gratefully acknowledges the management of PES Institute of Technology, Bangalore South Campus, for their support and encouragement toward carrying out this research work.

KEYWORDS

- **intrinsically conducting polymers**
- **polyaniline**
- **composites**
- **inorganic oxides**
- **electromagnetic interference**
- **EMI shielding materials**
- **dielectric permittivity**

REFERENCES

1. Bhadra, S.; Khastgir, D.; Singha, N. K.; Lee, J. H. *Prog. Polym. Sci.* **2009**, *34,* 783.
2. Skotheim, T. A.; Elsenbaumer, R. L.; Reynolds, J. R. *Hand Book of Conducting Polymers*, 2nd ed.; Marcel Dekker: New York, 1998; Vols. 1–2.
3. Nalwa, H. S. *Organic Conductive Molecules and Polymers*; Wiley: England, 1977; Vol. 2.
4. Unsworth, J.; Lunn, B. A.; Innis, P. C.; Jin, Z.; Kaynak, A.; Booth, N. G. *J. Intel. Mat. Syst. Str.* **1992**, *3,* 380.
5. Schoch Jr, K. F. *IEEE Electr. Insul. Mag.* **1994**, *10,* 29.
6. Angelopoulos, M. *IBM J. Res. Develop.* **2001**, *45,* 57.
7. Gospodinova, N.; Terlemezyan, L. *Prog. Polym. Sci.* **1998**, *23,* 1443.
8. Saxena, V.; Malhotra, B. O. *Curr. Appl. Phys.* **2003**, *3,* 293.
9. MacDiarmid, A. G. *Angew. Chem. Int. Ed.* **2001**, *40,* 2581.
10. Saxena, V.; Shirodkar, V. *Appl. J. Polym. Sci.* **2001**, *77,* 1050.
11. Ahuja, T.; Mir, I. A.; Kumar, D.; Rajesh, *Biomater.* **2007**, *28,* 791.
12. Sharma, A. L.; Malhotra, B. D. *J. Appl. Polym. Sci.* **2001**, *81,* 1460.
13. Friend, R. H. *Pure Appl. Chem.* **2001**, *73,* 425.
14. Touwslager, F. J.; Willard, N. P.; de Leeuw, D. M. *Appl. Phys. Lett.* **2002**, *81,* 4556.
15. Bhakshi, A. K.; Bhalla, G. *J. Sci. Ind. Res.* **2004**, *63,* 715.
16. Guo, Y.; Yu, G.; Liu, Y. *Adv. Mater.* **2010**, *22,* 4427.
17. Klauk, H. Ed. *Organic Electronics*: *Materials, Manufacturing, and Applications;* Wiley-VCH Weinheim, 2006.

18. So, F. Ed. Organic Electronics : *Materials, Processing, Devices and Applications* CRC Press: Boca Raton, USA, 2010.
19. Parvatikar, N.; Jain, S.; Kanamadi, C. M.; Chougule, B. K.; Bhoraskar, S. V.; Ambika Prasad, M. V. N. *J. Appl. Polym. Sci.* **2007**, *103,* 653.
20. Makeiff, D. A.; Huber, T. *Synth. Met.* **2006**, *156,* 497.
21. Hatchett, D. W.; Josowicz, M. *Chem. Rev.***2008**, *108,* 746.
22. Prakash, S.; Kale, B. B.; Amalnerkar, D. P. *Synth. Met.* **1999**, *106,* 53.
23. Sathiyanarayanan, S.; Syed Azim, S.; Venkatachari, G. *Synth. Met.* **2007**, *157,* 205.
24. Gangopadhyay, R.; De, A. *Chem. Mater.* **2000**, *12,* 608.
25. McGhee, M. D.; Miller, E. K.; Moses, D.; Heeger, A. J. *Advances in Synthetic Metals: Twenty Years of Progress in Science and Technology;* ed. by P. Bernier, S. Lefrant and G. Bidan, Elsevier, Amsterdam, 1999.
26. Heeger, A. J. *Rev. Mod. Phys.* **2001**, *73,* 681.
27. Teles, F. R. R.; Fonseca, L. R. Mater. *Sci. Eng.:C* **2008**, *28,* 1530.
28. Heeger, P. S.; Heeger, A. J. *Proc. Natl. Acad. Sci.* USA. **1999**, *96,* 12219.
29. McQuade, D. T.; Pullen, A. E.; Swager, T. M. *Chem. Rev.* **2000**, *100,* 2537.
30. Lu, J.;and Toy, P. *Chem. Rev.* **2009**, *109,* 815.
31. Salancek, W. R.; Clark, D. T.; Samuelsen (eds.), E. J. *Science and Applications of Conducting Polymers;* Adam Hilger: Bristol, 1991.
32. Bakhsgum, A.; Kaur, A.; Arora, V. *Indian J. Chem.* **2012**, *5,* 57.
33. Scrosati, B. *Prog. Solid State Chem.* **1988**, *18,* 1.
34. Bakhshi, A. K. Bull. *Mater. Sci.***1995**, *18,* 469.
35. Novak, P.; Muller, K.; Santhanam, K. S. V.; Haas, O. *Chem. Rev.* **1997**, *97,* 207.
36. Anand, J.; Palaniappan, S.; Sathyanarayana, D. N. *Prog. Polym. Sci.* **1998**, *23,* 993.
37. Pud, A.; Ogurtsov, N.; Korzhenko, A.; Shapoval, S. *Prog. Polym. Sci.* **2003**, *28,* 1701.
38. Ginder, J. M.; Epstein, A. J. *Phys. Rev. B.* **1990**,*41,* 10674.
39. Wu, H. L.; Philips, P. *Phys. Rev. Lett.* **1991**, *66,* 1366.
40. Lavarda, F. C.; dos Santos, D. A.; Galvao, D. S.; Laks, Phys, B. *Rev. Lett.* **1994**, *73,* 1267.
41. Alvarez, A. V.; Sordo, J. A. J. *Chem. Phys.* **2008**, *126,* 1747061.
42. Zhang, D. *Polym. Testing* **2007**, *26,* 9.
43. Freund, M. S.; Deore, B. A. *Self-Doped Conducting Polymers;* John Wiley and Sons Ltd: UK, 2007.
44. Sih, B. C.; Wolf, M. O. *Chem. Commun.* **2005**, 3375.
45. Tsakova, V. in: A. Eftekhari (Ed.), *Nanostructured Conductive Polymers;* Wiley: London, Chapter7, 2010.
46. Bissessur, R.; in: Eftekhari (Ed.), A. *Nanostructured Conductive Polymers*; Wiley: London, Chapter 6, 2010.
47. Baibarac, M.; Baltog, I.; Lefrant, S.; in: Eftekhari (Ed.), A. *Nanostructured Conductive Polymers*; Wiley: London, Chapter 5, 2010.
48. Lu, X.; Zhang, W.; Wang, C.; Wen, T. C.; Wei, Y. *Prog. Polym. Sci.* **2011**, *36,* 671.
49. Snook, G. A.; Kao, P.; Best, A. S. J. *Power Sources.* 2011, *196,* 1.
50. Bhattacharya, A.; De, A. *Prog. Solid State. Chem.* **1996**, *24,* 141.
51. Cottevieille, D.; Le Mehaute, A.; Challioui, C.; Mirebeau, P.; Demay, J. N. *Synth. Met.* **1999**, *101,* 703.
52. Makela, T.; Sten, J.; Hujanen, A.; Isotalo, H. *Synth. Met.* **1999**, *101,* 707.
53. Jonas, F.; Heywang, G. *Electrochim. Acta.* **1994**, *39,* 1345.

54. Stenger-Smith, J. D. *Prog. Polym. Sci.* **1998,** *23,* 57.

55. Li, X.; Sun, J.; Huang, M. *Prog. Chem.* **2007,** *19,* 787.

56. Huang, Y. F.; Park, Y. I.; Kuo, C. Y.; Xu, P.; Williams, D. J.; Wang, J.; Lin, C. W.; Wang, H. L. J. *Phys. Chem. C.* **2012,** *116,* 11272.

57. Marjanovic, G. C. *Synth. Met.* **2013,** *170,* 31.

58. Dutta, K.; De, S. K. J. *Nanoparticle Res.* **2007,** *9,* 631.

59. Whittingham, M. S. *Prog. Solid State Chem.* **1978,** *12,* 41.

60. Wu, C. G.; DeGroot, D. C.; Marcy, H. O. *J. Am. Chem. Soc.***1995,** *117,* 9229.

61. Liu, Y. J.; DeGroot, D. C.; Schindler, J. L.; Kannewurf, C. R.; Kanatzidis, M. G. *Adv. Mater.* **1993,** *5,* 369.

62. Pan, T. M Yeh, W. W. *IEEE Trans. Electron Devices* **2008,** *55,* 2354.

63. Hill, P. G.; Foot, P. J. S.; Budd, D.; Davis, R. *Mater. Sci. Forum* **1993,** *122,* 185.

64. Jarjeyes, O.; Fries, P. H.; Bidan, G. *Synth. Met.* **1995,** *69,* 43.

65. Huang, C. L.; Matijevic, E. *J. Mater. Res.* **1995,** *10,* 1327.

66. Wu, K. H.; Shin, M.; Yang, C. C. J. *Polym. Sci. Part A-Polym. Chem.* **2006,** *44,* 2657.

67. Roy, A. S.; Anilkumar, K. R.; Prasad, M. V. N. A. *J. Appl. Polym. Sci.* **2012,** *123,*1928.

68. Syed Khasim, Raghavendra, S. C.; Revanasiddappa, M.; Sajjan, K. C.; Lakshmi, M.; Faisal, M. Bull. *Mater. Sci.* **2011,** *34,* 1557.

69. Pandey, M.; Srivastava, A.; Anchal Srivastava, Shukla, R. K. *Sensor. Trans. J.* **2010,** *113,* 33.

70. Saini, P.; Choudhary, V.; Singh, B. P.; Mathur, R. B.; Dhawan, S. K. *Mater. Chem. Phys.* **2009,** *113,* 919.

71. Parvatikar, N.; Jain, S.; Khasim, S.;Revanasiddappa, M.; Bhoraskar, S. V.; Prasad, M. V. N. A. *Sensor. Actuat.:B.* **2006,** *114,* 599.

72. Ravikiran, Y. T.; Lagare, M. T.; Sairam, M.; Mallikarjuna, N. N.; Sreedhar, B.; Manohar, S.; MacDiarmid, A. G.; Aminabhavi, T. M. *Synth. Met.* **2006,** *156,* 1139.

73. Parvatikar, N.; Jain, S.; Bhoraskar, S. V.; Ambika Prasad, M. V. N. *J. Appl. Polym. Sci.* **2006,** *102,* 5533.

74. Syed Khasim, Raghavendra, S. C.; Revanasiddappa, M.; Ambika Prasad, M. V. N. *Ferroelectrics.* **2005,** *325,* 111.

75. Raghavendra, S. C.; Syed Khasim.; Revanasiddappa, M.; Ambika Prasad, M. V. N.; Kulkarni, A. B; Bull. *Mater. Sci.* **2003,** *26,* 733.

76. Jeon, I. Y.; Baek, J. B. *Materials.* **2010,** *3,* 3654.

77. Breiner, J. M.; Mark, J. E. *Polym.* **1998,** *39,* 5486.

78. Rong, M. Z.; Zhang, M. Q.; Zheng, Y. X.; Zeng, H. M.; Friedrich, K. *Polym.* **2001,** *42,* 3301.

79. Bao, S. P.; Tjong, S. C. Composites Part A: Appl. *Sci. Manuf.* **2007,** *38,* 378.

80. Su, S. J.; Kuramoto, N. *Synth. Met.* **2000,** *114,* 147.

81. Mo, T. C.; Wang, H. W.; Chen, S. Y.; Yeh, Y. C. *Ceram. Int.* **2008,** *34,* 1767.

82. Ma, C. C. M.;Chen, Y. J.; Kuan, H. C. *J. Appl. Polym. Sci.* **2006,** *100,* 508.

83. Ma, D.; Hugener, T. A.; Siegel, R. W.; Christerson, A.; Martensson, E.; Onneby, C.; Schadler, L. S. *Nanotechnology.* **2005,** *16,* 724.

84. Singha, S.; Thomas, M. J. *IEEE Trans. Dielectr. Electr. Insul.* **2008,** *15,* 12.

85. Park, J. G.; Louis, J.; Cheng, Q.; Bao, J.; Smithyman, J.; Liang, R.; Wang, B.; Zhang, C.; Brooks, J. S.; Kramer, L.;. Fanchasis, P.; Dorough, P. *Nanotechnology.* **2009,** *20,* 415702.

86. Bi, H.; Kou, K. C.; Ostrikov, K. K.; Yan, L. K.; Wang, W. C. *J. Alloy. Compd.* **2009,** *478,* 796.

87. Zhao, D. L.; Li, X.; Shen, Z. M. *J. Alloy. Compd.* **2009,** *471,* 457.

88. Shi, S. L.; Liang, J. *Nanotechnology.* **2008,** *19,* 255707.

89. Zhang, X. F.; Dong, X. L.; Huang, H; Lv, B.; Lei, L. P.; Choi, C. J. *J. Phys. D: Appl. Phys.* **2007,** *40,* 5383.

90. Cranton, W. M.; Spink, D. M.; Stevens, V.; Thomas, C. B. *Thin Solid Films.* **1993,** *226,* 156.

91. Ding, D.; Xiao, Y.; Han, P.; Zhang, Q. *J. Rare Earth.* **2010,** *28,* 765.

92. David, S.; Richard, D.; Joan, R.; Siu, W. C. *Biochem. Biophys. Res. Commun.* **2006,** *342,* 86.

93. Folgueras, L. C.; Alves, M. A.; Martin, I. M.; Rezende,M. C. *PIERS Proc.* **2009,** *23–27,* 301.

94. Li, X.; Chen, W.; Bian, C.; He, J.; Xu, N.; Xue, G. *Appl. Surf. Sci.* **2003,** *217,* 16.

95. Guan, H.; Liu, S.; Zhao, Y.; Duan, Y. *J. Electron. Mater.* **2006,** *35,* 892.

96. Pang, S. C.; Anderson, M. A; Chapman, T. W. *J. Electrochem. Soc.* **2000,** *147,* 444.

97. Guan, H.; Zhao, Y.; Liu, S.; Lv, S. *Eur. Phys. J.-Appl. Phys.* **2006,** *36,* 235.

98. Fryz, C. A.; Shui, X.; Chung, D. D. L. *J. Power Sources.* **1996,** *18,* 41.

99. Das, A.; Hayvac, H. T.; Tiwari, M. K.; Bayer, I. S.; Erricolo, D.; Constantine, M. *J. Colloid Interface Sci.* **2011,** *353,* 311.

100. Celozzi, S.; Araneo, R.; Lova, G. *Electromagnetic Shielding*; John Wiley & Sons, Inc, Hoboken, New Jersey, 2008.

101. Yang, Y. L.; Gupta, M. C.; Dudley, K. L.; Lawrence, R. W *Nano Lett.* **2005,** *5,* 2131.

102. Bagwell, R. M.; McManaman, J. M.; Robert, C. *Compos. Sci. Technol.* **2006,** *66,* 522.

103. Li, N.; Huang, Y.; Du, F.; He, X.; Lin, X.; Gao, H.; Ma, Y.; Li, F.; Chen, Y.; Eklund, P. C. *Nano Lett.* **2006,** *6,* 1141.

104. Kim, M. S.; Kim, H. K.; Byun, S. W.; Jeong, S. H.; Hong, Y. K.; Joo, J. S.; Song, K. T.; Kim, J. K.; Lee, C. J.; Lee, J. Y. *Synth. Met.* **2002,** *126,* 233.

105. Al-Saleh, M. H.; Sundararaj, U. *Carbon.* **2009,** *47,* 2.

106. Fugetsu, B.; Sano, E.; Sunada, M.; Sambongi, T.; Shibuya, X.;Wang, T.; Hiraki, *Carbon.* ***2008,*** *46,* 1256.

107. Kaiser, K. L. *Electromagnetic Shielding*; CRC Press, Boca Raton: USA, 2006.

108. Abbas, S. M.; Dixit, A. K.; Chatterjee, R.; Goel, T. C. *J. Magn. Magn. Mater.* **2007,** *309,* 20.

109. Abbas, S. M.; Chatterjee, R.; Dixit, A. K.; Kumar, A. V. R.; Goel T. C. *J. Appl. Phys..***2007,** *101,* 074105.

110. Colaneri, N. F.; Shacklette, L. W. *IEEE Trans. Instrum. Meas.* **1992,** *41,* 291.

111. Joo, J.; Epstein, A. J. *Appl. Phys. Lett.* **1994,** *65,* 2278.

112. Ott, H. W. *Electromagnetic Compatibility Engineering*; John Wiley & Sons: New Jersey, 2009.

113. Saini, P.; Choudhary, V.; Singh, B. P.; Mathur, R. B.; Dhawan, S. K. *Mater. Chem. Phys.* **2009,** *113,* 919.

114. Saini, P.; Choudhary, V.; Dhawan, S. K. *Polym. Adv. Technol.* **2010,** *21,*1.

115. Saini, P.; Choudhary, V.; Singh, B. P.; Mathur, R. B.; Dhawan, S. K. *Synth. Met.* **2011,** *161,* 1522.

116. Schulz, R. B.; Plantz, V. C.; Brush, D, R. *IEEE Trans. Electromagn. Compat.* ***1988,*** *30,* 187.

117. Singh, P.; Babbar, V. K.; Razdan, A.; Srivastava, S. L.; Puri, R. K. *Mater. Sci. Eng. B.* **1999,** *67,* 132.
118. Singh, P.; Babbar, V. K.; Razdan, A.; Srivastava, S. L.; Goel, T. C. *Mater. Sci. Eng. B.* **2000,** *78,* 70.
119. Chung, D. D. L. *Composite Materials: Science and Applications* ;Springer: London, 2003.
120. Jana, P. B.; Mallick, A. K.; De, S. K. *IEEE Trans. Electromagn. Compat.* **1992,** *34,* 478.
121. Hong, Y. K.; Lee, C. Y.; Jeong, C. K.; Lee, D. E.; Kim, K.; Joo, J. *Rev. Sci. Instrum.* **2003,** 74, 1098.
122. Zhang, X.; Zhang, J.; Song, W.; Liu, Z. *J. Phys. Chem. B.* **2006,** *110,* 1158.
123. Kaynak, A. *Mater. Res. Bull.* **1996,** *31,* 845.
124. Singh, K.; Ohlan, A.; Saini, P.; Dhawan, S. K. *Polym. Adv. Technol.* **2008,** *19,*229.
125. Ailton De Souza Gomes (ed.), *New Polymers for Special applications*; InTech: Croatia, 2012.
126. Faisal, M.; Khasim, S. *Adv. Mater. Res.* **2012,** *488–489,*557.
127. Micheli, D.; Apollo, C.; Pastore, R.; Marchetti, M. *Compos. Sci. Technol.* **2010,** *70,* 400.
128. Nanni, F.; Travaglia, P.; Valentini, M. *Compos. Sci. Technol.* **2009,** *69,* 485.
129. Wallace, G. G.; Spinks, G. M.; Kane-Maguire, L. A. P.; Teasdale, P. R. *Conductive Electroactive Polymers: Intelligent Polymer Systems* ;CRC Press: Boca Raton, FL, 2009.
130. MacDiarmid, A. G.; Chiang, J. C.; Halpern, M.; Huang, S. W.; Mu, S. L.; Somasiri, N. L.; Wu, W.; Yaniger, S. L. *Mol. Cryst. Liq. Cryst.* **1985,** *121,* 173.
131. Huang, S. W.; Humphrey, B. D.; MacDiarmid, A. G. *J. Chem. Soc. Faraday Trans. 1.* **1986,** *8,* 2385.
132. Sadia, A.; Akhtar, M. S.; Husain M, M. *Sci. Technol. Adv. Mater.* **2010,** *2,* 441.
133. Negi, W. S.; Adhyapak, P. V. *J. Macromol. Sci. Part C: Polym. Rev.***2002,** *42,* 35.
134. Freund, M. S.; Deore, B. A. *Self-Doped Conducting Polymers*; John Wiley & Sons Ltd: England, 2007.
135. Skotheim, T. A. *Handbook of Conducting Polymers*; 2nd edn: CRC Press, 1986.
136. Shirakawa, H. *Synth. Met.* **1995,** *3,* 69.
137. Heinze, J. *Synth. Met.* **1991,** *43,* 2805.
138. McCullough, R. D. *Adv. Mater.* **1998,** *10,* 93.
139. MacDiarmid, A. G. *Synth. Met.* **1997,** *27,* 84.
140. Wan, W. *Conducting Polymers with Micro or Nanometer Structure;* Springer: New York, 2008.
141. Cao, Y.; Andreatta, A.; Heeger, A. J.; Smith, P. *Polymer.* **1989,** *30,* 2305.
142. Armes, S. P.; Miller, J. F. *Synth. Met.* **1988,** *22,* 385.
143. Sapurina, I. *J. Stejskal, Polym. Int.* **2008,** *57,* 1295.
144. Mazerolles, L.; Folch, S.; Colomban, P. *Macromolecules.* **1999,** *32,* 8504.
145. Yuping, D.; Shunhua, L.;Hongtao, G. *Sci. Technol. Adv. Mater.* **2005,** *6,* 513.
146. Faisal, M.; Khasim, S. *J. Mater. Sci.: Mater Electron,* 2013 DOI 10.1007/s 10854-013-1080-y
147. Faisal, M.; Khasim, S. *Bull. Korean Chem. Soc.* **2013,** *34,* 99.
148. Chung, D. D. L. *Composite Materials: Science and Applications* ; Springer: London, 2003.
149. Oyharcabal, M.;Olinga, T.;Foulc, M. P.; Lacomme, S.; Gontier, Vigneras, V. *Compos. Sci. Technol.* **2013,** *74,* 107.
150. Olmedo, L.; Hourquebie, P.; Jousse, F. *Synth. Met.* **1995,** *69,* 205.

151. Moucka, R.; Mravcakova, M.;Vilcakova, J.; Omastova, M.; Saha, P. *Mater. Des.* **2011,** *32,* 2006.
152. Phang, S. W.; Hino, T.; Abdullah, M. H.; Kuramoto, N. *Mater. Chem. Phys.* **2007,** *104,* 327.
153. Klemperer, C. J. V.; Maharaj, D. *Compos. Struct.* **2009,** *91,* 467.
154. Makela, T.; Pienimaa, S.; Taka, T.; Jussila, S.; Isotalo, H. *Synth. Met.* **1997,** *85,* 1335.
155. Ohlan, A.; Singh, K.; Chandra, A.; Singh, V. N.; Dhawan, S. K. *J. Appl. Phys.* **2009,** *106,* 044305.
156. Koul, S.; Chandra, R.; Dhawan, S. K. *Polymer.* **2000,** *41,* 9305.
157. Phang, S. W.; Tadokoro, M.; Watanabe, J.; Kuramoto, N. *Curr. Appl. Phys.* **2008,** *8,* 391.
158. Wong, P. T. C.; Chambers, B.; Anderson, A. P.; Wright, P. V. *Electron. Lett.* **1992,** *28,* 1651.
159. Yu, X.; Shen, Z. *J. Magn. Magn. Mater.* **2009,** *321,* 2890.
160. Saini, P.; Choudhary, V.; Dhawan, S. K. *Polym. Adv. Technol.* **2009,** *20,* 355.
161. Sudha, J. D.; Sivakala, S.; Prasanth, R.; Reena, V. L.; Radhakrishnan Nair, P. *Compos. Sci. Technol.* **2009,** *69,* 358.
162. Chung, D. D. L. *Carbon.* **2001,** *39,* 279.
163. Roldughin, V. I.; Vysotskii, V. V. *Prog. Org. Coat.* **1999,** *39,* 81.
164. Strumpler, R.; Glatz-Reichenbach, J. *J. Electroceram.* **1999,** *3,* 329.
165. Kim, H. R.; Fujimori, K.; Kim, B. S.; Kim, I. S. *Compos. Sci. Technol.* **2012,** *72,* 1233.
166. Wang, L. L.; Tay, B. K.; See, K. Y.; Sun, Z.; Tan, L. K.; Lua, D. *Carbon* **2009,** *47,* 1905.
167. Markham, D. *Mater. Design.* **2000,** *21,* 45.
168. Wu, Z. P.; Li, M. M.; Hu, Y. Y.; Li, Y. S.; Wang, Z. X.; Yin, Y. H.; Chen, Y. S.; Zhou, Z. *Scripta Mater.* **2011,** *64,* 809.
169. Phang, S. W.; Daik, R.; Abdullah, M. H. *Thin Solid Films.* **2005,** *477,* 125.
170. Tong, X. C. *Advanced Materials and Design for Electromagnetic Interference Shielding;* CRC Press: Boca Raton, FL, 2009.
171. Nicolson, A. M.; Ross, G. F. *IEEE Trans. Instrum. Measur.* **1970,** *19,* 377.
172. Weir, W. B. *Proceedings of the IEEE.* **1974,** *62,* 33.
173. Dang, Z. M.; Yuan, J. K.; Zha, J. W.; Zhou, T.; Li, S. T.; Hu, G. H. *Prog. Mater. Sci.* **2012,** *57,* 660.
174. H. Ghasemi, U. Sundararaj, *Synth. Met.* **2012,***162,* 1177.
175. Shi, S.; Zhang, L.; Li, J. *J. Mater. Sci.* **2009,** *44,* 945.
176. Faisal, M. PhD Dissertation, Visvesvaraya Technological University, India, 2014

CHAPTER 6

A COMPREHENSIVE REVIEW OF MEMBRANE SCIENCE AND TECHNOLOGY AND RECENT TRENDS IN RESEARCH: A BROAD ENVIRONMENTAL PERSPECTIVE

SUKANCHAN PALIT[1*]

[1]*Department of Chemical Engineering, University of Petroleum and Energy Studies, Energy Acres, Post-Office-Bidholi via Premnagar, Dehradun 248007, Uttarakhand, India*

Corresponding author. E-mail: sukanchan68@gmail.com; sukanchan92@gmail.com

CONTENTS

ABSTRACT

Science and technology in today's world are moving at a rapid pace toward a newer visionary era. Environmental engineering science in a similar manner is witnessing drastic challenges. Ecological disbalance, environmental catastrophes, and stringent environmental restrictions have urged the scientific domain and the civil society to gear forward toward newer innovations and newer technologies. Scientific vision, scientific introspection, and deep scientific fortitude are today's torchbearers toward a newer visionary tomorrow. This treatise addresses and re-envisions the wide application of membrane science in environmental protection and industrial wastewater treatment. Reverse osmosis (RO), nanofiltration (NF), and ultrafiltration (UF) are the hallmarks of this treatise. The wide vision of mankind, the scientist's immense prowess, and the vast scientific rigor of environmental engineering science will lead a long and visionary way in the true realization of industrial wastewater treatment issues and the areas of membrane separation processes. Water science and technology are ushering in a new era of immense scientific understanding and vision. This treatise targets the vast environmental perspectives of the membrane applications. Scientific endeavor of this treatise is targeted toward membrane chemistry, structure, function, membrane properties, performance and engineering models, fouling and control, chemical process design, and the whole gamut of membrane applications. The author pointedly focuses on the immense potential of novel separation processes such as membrane science as well as the wide gamut of non-traditional environmental engineering techniques.

6.1 INTRODUCTION

Environmental engineering science and membrane separation processes are the visionary avenues of science. Technological vision and scientific objectives are the veritable pillars of scientific research pursuit. Global water challenges and drinking water treatment are witnessing new dimensions of scientific vision and deep scientific truth. Environmental catastrophes, destruction of ecological biodiversity, and grave concerns for environment have lead to a worldwide realization of environmental sustainability. Membrane science is today ushering in a new era of scientific vision and deep scientific understanding. The vision of applications of membrane science is far-reaching. The author deeply comprehends the varied success of membrane separation processes in environmental protection. Technological objectives and scientific motivation are the cornerstones of a greater visionary future in the field of membrane science. Chemical process engineering and environmental engineering science are two opposite sides of the visionary coin today. The challenge, the vision, and the immense potential of membrane separation phenomenon are widely opening the windows of innovation in environmental engineering science today. Technology of membrane separation phenomenon needs to be rebuilt as civilization moves from one paradigm toward another. This treatise also gives wide glimpses in the field of environmental process engineering and chemical process engineering as a whole. Application of NF and ultrafiltration (UF) in varied disciplines of science and engineering is deeply delineated in this treatise. The world of membrane science needs to be re-envisioned and re-envisaged as scientific research pursuit treads the newer path of scientific destiny.[1]

6.2   THE AIM AND OBJECTIVE OF THE STUDY

The major aim and objective of this study is to lucidly describe the vast applications of membrane science and the hurdles and barriers of fouling phenomenon. Technological vision is unimaginable as scientific research treads a visionary path toward zero-discharge norms. Green technology and green engineering are the visionary scientific challenges of the day. Scientific motivation and scientific introspection into the field of membrane science are the torchbearers toward a greater emancipation of green technology and environmental engineering science today. Technology and science are rapidly changing with more vision and targets toward eco-friendly innovations and techniques. Membrane science and other novel separation techniques in chemical process engineering are revolutionary in the path toward greater emancipation. Eco-friendly materials and nanomaterials are the ever-growing and challenging branches of scientific research pursuit today. The author in this treatise pointedly focuses on the vast and versatile applications of membrane technology and its interfaces with green engineering and green technology. The success and the wide vision of membrane science are changing the face of scientific research pursuit in the field of industrial wastewater treatment and drinking water treatment. Water purification is technologically crossing vast visionary frontiers of science. This treatise widely observes the technological and scientific challenges in the separation phenomenon of membranes. Nanomembranes are the latent yet visionary area of membrane science today. This treatise also delineates the fusion of membrane science and nanotechnology toward the furtherance of science and engineering. The holistic view of this study is to readdress and re-envision the world of challenges in the domain of environmental sustainability. Technology needs to be rebuilt and restructured as human scientific endeavor enters into a newer era.[1]

6.3 THE NEED AND THE RATIONALE OF THIS STUDY

The need and the rationale of this widely observed study is the worldwide water science and technology challenge. Global water crisis today stands in the midst of deep comprehension and scientific vision. Global water challenges are destroying the scientific landscape. The vast scientific rigor, the immense urge to excel, and the progress of science and technology are the pallbearers toward a greater vision of the global water initiatives today. Water purification and provision of pure drinking water are the need and the rationale of this well-observed study. Technology of water purification needs to be re-envisioned and rebuilt with the passage of scientific history and time. The author in this treatise also delineates the immense scientific potential, the wide challenges, and the immense scientific rigor in the field of non-traditional environmental engineering techniques such as advanced oxidation processes. The rationale of this comprehensive study targets mainly the wide world of water purification and the relevant domain of drinking water treatment. The vision, the challenge and the immense potential of membrane separation processes, as well as other novel separation processes are discussed in deep details in this chapter.[1]

6.4 WHAT IS MEMBRANE SCIENCE?

Membrane science and technology has a wide vision in today's world and is surpassing visionary frontiers. Global water crisis and drinking water crisis are the challenges of human civilization today. Technology needs to be re-envisioned and re-envisaged as scientific endeavor enters into a new era. Scientific truth, scientific cognizance, and the world of scientific challenges are today witnessing drastic challenges. Scientific history is in the path of a newer scientific rejuvenation. Water shortage, desalination, and water purification are the visionary areas of scientific endeavor today.[1–3]

The development of the Sourirajan–Loeb synthetic membrane in 1960 provided a valuable separation tool to the process industries and changed the face of frontiers of science and engineering. Filtration is defined as the separation of two or more components from a fluid stream based primarily on size differences. In scientific usage, it usually refers to the separation of solid immiscible particles from liquid and gaseous streams. Membrane filtration also involves the separation of dissolved solutes in liquid streams and for separation of gaseous mixtures.

6.5 SCIENTIFIC DOCTRINE OF MEMBRANE SCIENCE AND THE CLASSIFICATIONS OF MEMBRANE SEPERATION PROCESSES

Membrane science today is surpassing visionary frontiers. The technology of membrane science is far-reaching and visionary. Scientific doctrine, scientific truth, and deep scientific cognizance are the torchbearers toward a greater visionary future in the field of membrane science and the wide holistic domain of environmental engineering science. Fouling and bio-fouling are the critical issues facing membrane science today. Technology is changing at a rapid and unimaginable pace. The scientific truth and the wide domain of scientific forbearance will go a long and visionary way in unraveling the unknowns of membrane science and water technology in years to come. Membrane science is classified into different branches of scientific endeavor. The challenge of membrane separation phenomenon is truly beyond scientific imagination. Fouling stands as an impediment to the separation process (Table 6.1).[1]

6.6 WHAT DO YOU UNDERSTAND BY REVERSE OSMOSIS?

Reverse osmosis (RO) is one of the major membrane separation processes. Today's world of desalination science is widely dependent on RO technology. Global water crisis is at a state of tremendous turmoil. Technology stands at a state of major disaster as groundwater heavy metal contamination and water purification issues are rejuvenating the scientific journey in global water research and development initiatives. The challenge is immense and thought-provoking. Science has few answers to the desalination technology today. Yet RO and other membrane separation techniques are changing the wide scientific frontiers.[2,3]

RO is a water purification technology that uses a semipermeable membrane to remove ions, molecules, and larger particles from drinking water. In RO, an applied pressure is used to overcome osmotic pressure, a colligative property that is driven by chemical potential differences of

TABLE 6.1 Characteristics of membrane processes:

Process	Driving Force	Retentate	Permeate
Osmosis[1]	Chemical potential	Solutes/Water	Water
Dialysis	Concentration difference	Large molecules/Water	Small molecules/Water
Microfiltration	Pressure	Suspended particles/Water	Dissolved solutes/water
Ultrafiltration	Pressure	Large molecules/Water	Small molecules/Water
Nanofiltration	Pressure	Small molecules/divalent salts/dissociated acids/water	Monovalent ions/Undissociated acids/Water
Reverse Osmosis	Pressure	All solutes/water	Water
Electrodialysis	Voltage/Current	Nonionic solutes/water	Ionized solutes/Water
Pervaporation[1]	Pressure	Nonvolatile molecules/water	Volatile small molecules/Water

the solvent, a thermodynamic parameter. RO is today primarily used for drinking water treatment and water purification. RO can remove many types of dissolved solutes and suspended species from water, including bacteria, and is vastly used in both the industrial processes and the production of drinking water. Technology needs to be re-envisioned as scientific research pursuit enters into a newer frontier of scientific destiny and scientific vision. RO is most commonly known for its use in drinking water purification from seawater, removing the salt and other effluent materials from water molecules. Scientific vision and scientific profundity are changing the future directions of membrane separation processes. Science is today a visionary colossus with a definite aim of its own.[2,3]

6.7   WHAT DO YOU UNDERSTAND BY ULTRAFILTRATION AND NANOFILTRATION AND ITS VAST AND VARIED APPLICATIONS?

NF is a relatively recent membrane filtration process used most often with low total dissolved solids water such as surface water and fresh groundwater, with the purpose of softening (polyvalent cation removal) and removal of disinfection by-product precursors such as natural organic matter and synthetic organic matter.[2,3] NF is widely used in dairy industry and other food processing applications. Technology and science are veritably challenged today. The author in this treatise widely focuses on the success of application domain of membrane separation techniques.

Science and engineering in today's world are witnessing drastic challenges. The grave concern for environmental protection, the environmental disasters, and the loss of ecological diversity has plunged the scientific domain into the world of deep scientific introspection and scientific understanding. Membrane science and its vast and varied applications are a boon to scientific endeavor and a torchbearer toward a greater emancipation of environmental engineering science and global water initiatives today.

UF is a variety of membrane filtration in which forces like pressure or concentration gradients lead to a separation through a semipermeable membrane.[1–3] Suspended solids and solutes of high molecular weight are retained in the so-called retentate, while water and low molecular weight solutes pass through the membrane in the permeate (filtrate). Technological vision and deep scientific comprehension are the pallbearers toward a new world of scientific regeneration. Water science and technology are the true challenges of our times. UF is a type of membrane filtration. Industries such as chemical and pharmaceutical manufacturing, food and beverage processing, and wastewater treatment employ UF in order to recycle flow or add value to later products. UF's main attraction is its ability to purify, separate, and concentrate target macromolecules in continuous systems. Technologically, in today's world UF and NF are widening the vision of membrane science. Scientific objectives and definitive scientific vision are today emboldened as water science enters a newer realm. Also scientific girth and scientific progeny in the field of membrane science are in the path of newer visionary era.[1-3]

6.8 TECHNOLOGICAL VISION, SCIENTIFIC OBJECTIVES, AND THE WIDE WORLD OF MEMBRANE SCIENCE

Technological vision and scientific objectives are veritably opening a new era in the field of membrane science and technology. The scientific and the academic rigor are immense and versatile. The wide world of membrane science today stands in the midst of deep scientific comprehension and wide scientific regeneration. In a similar manner, water science and technology are today moving toward the culmination of scientific profundity and scientific vision. The challenge is immense and beyond scientific imagination and adroitness. Desalination science and drinking water treatment are the opposite sides of the visionary coin today. Global water initiatives are witnessing new challenges as science and engineering moves toward a newer era of scientific rejuvenation and scientific candor.

Groundwater heavy metal contamination stands as a major impediment to the progress of science. Developing countries and developed countries are faced with this global environmental crisis. Technology is retrogressive as heavy metal groundwater contamination issues ushers in an era of wide vision. The vision of science and engineering is entering into a new era as scientists and environmental engineers innovate and re-envision technologies. Today membrane science and desalination are immensely linked with groundwater remediation and global water initiatives. Membranes are the present generation and next generation environmental engineering tools. The success and the potential are beyond scientific imagination. Scientific rejuvenation and deep introspection are the torchbearers toward zero-discharge norms today.[1]

6.9 MEMBRANE FOULING AND THE WORLD OF CHALLENGES

Membrane fouling stands as a major impediment to the success of membrane separation phenomenon. Technological vision and scientific objectives are veritably bolstered as scientific advancements in membrane science enter a newer realm. Boundaries are surpassed as innovations in technology gains newer height. The challenge and vision of science are ever-growing as engineering science and applied science enter into a newer visionary eon. Membrane fouling and bio-fouling are changing the face of scientific research pursuit and deep scientific comprehension today. In this section, the authors deeply comprehend the success, the scientific imagination, and the scientific profundity behind the hurdles and barriers of membrane and bio-fouling.

6.10 GROUNDWATER CONTAMINATION AND ITS REMEDIATION

Heavy metal groundwater contamination is witnessing disastrous challenges. The success of science and engineering is faltering at each step of this vicious issue. A scientist's immense prowess, the scientific, and the academic rigor behind arsenic (As) groundwater remediation are challenged and moving from one paradigmatic shift over another. The utmost need of this century is the provision of clean drinking water. Global water issues are baffling and at the same time far-reaching. As groundwater contamination is revitalizing the world of scientific vision and scientific introspection. Desalination and membrane science are the ever-growing and challenging vistas of science today. Technology and science of desalination need to be rebuilt and reorganized as environmental protection reaches new heights. Desalination and water science are the imperatives of science today. Human civilization's wide progress and scientific rigor are changing the face of scientific research pursuit. As groundwater remediation technologies in South Asia need to be revamped and readdressed as science and engineering crosses newer frontiers. The immense success and the vast potential of membrane science are ushering in a new era of scientific vision and forbearance in the field of global water initiatives.

Groundwater remediation and drinking water problems are vexing issues of the present day human civilization. Water purification science and the world of desalination science are the visionary targets of present day scientific endeavor. Developed as well as developing countries throughout the world are faced with this tremendous problem. Technology has few answers to this monstrous water crisis issue. The intricacies of membrane science are still today in the latent phase. This branch of scientific research pursuit is opening new avenues of deep introspection and scientific cognizance in years to come.

6.11   ARSENIC GROUNDWATER REMEDIATION IN INDIA, BANGLADESH, AND SOUTH ASIA

As groundwater contamination in India, Bangladesh, and South Asia is witnessing devastating challenges. Human civilization is today in the state of unimaginable crisis. Science and technology have few answers to this worldwide vexing issue. Scientific and academic rigor is opening up new visionary avenues in the research trends in heavy metal groundwater remediation. In this treatise, the author reiterates and pointedly focuses on the application of novel separation techniques in groundwater remediation. The success of scientific advancements, scientific vision, and scientific profundity needs to be re-addressed as scientific research pursuit enters into a new era. The environmental situation in South Asia is vexing and totally disastrous. Mankind's deep distress, man's scientific rigor, and the world of scientific progress are changing the face of environmental engineering science today. The As groundwater contamination scenario in South Asia is the largest environmental disaster in our planet. The success of human civilization, the immense potential of environmental engineering, and the grave concerns of environmental disasters are the pallbearers of scientific regeneration and scientific vision today. Human scientific progress is so much retrogressive as regards As groundwater contamination today. Yet the scientific struggle and the scientific introspection are changing the landscape of environmental engineering science.

Some recent papers on As groundwater contamination in our planet are highlighted here in providing the readers the scientific insight behind its remediation.

Welch et al.[4] discussed with lucidity As in groundwater of the United States along with its occurrence and geochemistry. Concentrations of naturally occurring As in groundwater vary regionally due to combination of climate and geology. Although slightly less than half of 30,000 As analyses of groundwater in the United States were less than 1 µg/L, about 10% exceeded 10 µg/L. Technology and science have no answers to the intricacies and the vexing issue of As groundwater contamination. At a broader regional scale, As concentrations exceeding 10 µg/L appear to be more frequently observed in the Western United States than in the Eastern half. As release from iron oxide appears to be the most common cause of widespread As concentrations exceeding 10 µg/L in groundwater. This chapter gives a wide view of the As crisis in the planet's most developed country. The challenge and vision of remediation is immense and unimaginable.

Technology and engineering science need to be re-envisioned in the global sphere of As crisis.

Wang et al.[5] described with immense lucidity natural attenuation processes for remediation of As contaminated soils and groundwater. Natural attenuation of As contaminated soils and groundwater may be a cost-effective in situ remedial option. It veritably relies on the site intrinsic assimilative capacity and allows in-place cleanup. Sorption to solid phases is the principal mechanism immobilizing As in soils and removing it from groundwater. In this chapter, a wide view is presented on the development of conceptual and mathematical models to predict the fate and transport of As and to evaluate the site natural attenuation capacity.

Su et al.[6] deeply comprehended arsenate and arsenite removal by zerovalent iron and also discussed kinetics, redox transformation, and implications for in situ groundwater remediation. There has been a great interest in the in situ remediation of certain organic and inorganic contaminants in groundwater using zerovalent iron as a permeable reactive barrier medium. The zerovalent iron has been used to effectively destroy numerous chlorinated hydrocarbon compounds via reductive dehalogenation. Recent studies also present the concept that zerovalent iron effectively removes inorganic contaminants from aqueous solution. The removal mechanism appears to be reductive precipitation for these anions except for As(V) and As(III). Surface precipitation or adsorption appears to be the predominant removal mechanism for both As(V) and As(III) by zerovalent iron. The main objective of this study is to evaluate the effectiveness of four zerovalent metals for removing As(V) and AS(III) from water and examine the redox transformation of sorbed As on the surface of zerovalent iron. This chapter widely and effectively targets the success of As remediation techniques with a greater scientific understanding and greater vision toward water purification.

Safiuddin et al.[7] discussed deeply causes, effects, and remediation of groundwater As contamination in Bangladesh. The serious As contamination of groundwater in Bangladesh has come out recently as the largest environmental crisis. The citizens in 59 out of 64 districts in Bangladesh are suffering due to As contamination of drinking water. Most of the recognized stages of As poisoning have been identified in Bangladesh and the risk of As poisoning is increasing everyday dramatically. This chapter highlights the causes and mechanisms of As contamination of groundwater. The challenge of remediation is unimaginable. Technology and engineering have few answers.

6.12 ENVIRONMENTAL AND ENERGY SUSTAINABILITY, THE SCIENTIFIC RIGOR, AND THE SCIENTIFIC PROGRESS

Environmental and energy sustainability are the utmost need of the hour for human civilization's progress. The environmental catastrophes, the disasters of industrial wastewater treatment, and the loss of ecological diversity are truly leading a long way in the effective realization of sustainable development. The scientific vision and rigor in research and development initiatives in sustainability are challenging as well as far-reaching. Technology today is witnessing newer regeneration in such a challenging phase of science and engineering. Scientific progress today stands in the midst of vision and scientific profundity. Sustainable development is today the utmost backbone of technological vision and scientific motivation. Today human civilization is solely dependent on environmental and energy sustainability. The challenges of global water initiatives depend on sustainable infrastructural development. Scientific progress, scientific candor, and deep scientific vision are the pallbearers toward successful sustainability.[2,3]

Sustainability is the ability to continue a defined behavior indefinitely. For more detailed and practical detail, the behavior to be wished to continue indefinitely must be well defined. For example:

- Environmental sustainability is the ability to maintain rates of renewable resource harvest, pollution creation, and non-renewable resource depletion that can be continued indefinitely.
- Economic sustainability is the ability to support a defined level of economic production indefinitely.
- Social sustainability is the ability of a social system, such as a country, to function at a defined level of social well being indefinitely.

Brundtland Commission of the United Nations on March 20, 1987 defined sustainability as "sustainable development is development that meets the needs of the present without compromising the ability of future generations to meet their own needs."[2,3] This treatise focuses on environmental engineering tools that are sustainable. Here comes the immense importance of sustainability. Technology of environmental engineering science is indomitable today. Sustainable development and scientific rigor in the field of membrane science and desalination are linked by an unsevered umbilical cord. This question of sustainable development will stand weathering the winds of scientific rigor. Scientific destiny in the application of membranes is witnessing revolutionary challenges. The author in this treatise repeatedly focuses on the success, the immense potential, and the futuristic vision of membrane science and other novel separation processes as a whole. The author deeply treads the difficult path toward scientific future and scientific destiny of membrane science. Advancements of science in today's world are dependent upon the successful application of energy and environmental sustainability. Holistic infrastructural development, scientific and academic rigor, and the immense hurdles of novel separation processes are the torch-bearers toward a greater emancipation of membrane science today.[2,3]

6.13   RECENT SCIENTIFIC ENDEAVOR IN THE FIELD OF MEMBRANE SEPERATION PROCESSES

Scientific endeavor, deep scientific vision, and the immense scientific and academic rigor are all leading a long way in the true realization of environmental engineering tools today. Membrane separation science is today opening up new vistas of challenges and surpassing visionary frontiers. The research questions and research forays in environmental and energy sustainability are changing the face of scientific endeavor today. Water science and water technology are revolutionizing the domain of membrane separation phenomenon. Global water challenges and global water initiatives are witnessing wide barriers and needs to be re-addressed and re-envisaged as human civilization moves toward newer boundaries.

Nechifor et al.[8] discussed about nanostructured hybrid membrane polysulfone-carbon nanotubes for hemodialysis. The research endeavor of nano-functionalized membranes is one of the emerging trends for enhancing the performance of membrane separation processes. In this research treatise, the results concerning the synthesis of new polymeric-carbon nanotube composite membranes based on polysulfone with different types of nanotubes, single wall (SWNT) and double wall (DWNT) are shown with applications in medical field of advanced separations of heavy metal from blood or other physiological fluids.

Arnot et al.[9] deeply comprehended cross flow and dead-end microfiltration (MF) of oily-water emulsions and the widely observed area of mechanisms and modeling of flux decline. Technology and engineering science are challenged today. The veritable focus of this chapter is on the mechanisms and modeling of flux decline. Three distinct forms of modeling flux decline of cross flow filtration at constant transmembrane pressure were examined and deeply investigated. The experimental data used are that given in author's previous research work but the method of analysis is of wider interest. The data are for a dispersion of oil-in-water. Scientific objectives and deep motivation are in the path of immense interest as environmental engineering science and membrane science enters into a newer visionary realm. The discharge of crude oily wastewater into the sea and rivers has been under deep scrutiny in recent years. The challenge of environmental engineering science is ever-growing and widely observed as human civilization embarks on a newer visionary era. In addition to oily wastes from the petrochemical, metallurgical, and process industries, it should be remembered that the production of crude oil is accompanied by equal volume of water.

Sarkar et al.[10] discussed with deep and cogent insight the Donnan membrane principle and its opportunities for sustainable engineered processes and materials. The Donnan membrane principle can permit many engineered processes and materials to achieve better sustainability. Technology and science are at its scientific zenith as this treatise opens up new windows of innovation in membrane science and technology. The innovations of technology and science are beyond scientific imagination and scientific frontiers. The Donnan membrane principle is essentially a specific domain of the second law of thermodynamics dealing solely with completely ionized electrolytes. The conditions leading to the Donnan membrane equilibrium arise from the inability of certain ions to diffuse out from one phase (or region) to other in systems involving water or polar solvents. The authors decisively tread through the difficult intricacies of membrane science and technology. In accordance with the Donnan membrane principle, non-diffusible fixed charges in one phase in contact with water can be utilized to modulate the distribution of ions in both phases leading to efficient separation, product recovery, and a host of other innovative applications. The most popular application of the Donnan membrane principle is desalination science. Today, desalination science and membrane science are linked to water purification, industrial wastewater treatment, and other related issues. Technological vision and holistic scientific objectives are gaining new heights as the authors delineate the wide vision of Donnan principle in membrane science.

Yang et al.[11] discussed in details an electrocatalytic membrane reactor with self-cleaning function for industrial wastewater treatment. Industrial wastewater has become a global issue due to its high concentration of pollutants, especially refractory organic compounds. Membrane fouling is a worldwide difficult issue which needs to be solved with immediate effect. In order to solve the problem of fouling, the authors presented in this research work the design of a novel electrocatalytic membrane reactor with self-cleaning function. A tubular carbon membrane was employed as the conductive membrane substrate because of its significant mechanical strength, good chemical stability, and specific conductivity. This treatise moves in a visionary path of membrane technology. The authors widely observe the immense potential and success of membrane technology along with the latent scientific issue of membrane reactor.

Robeson[12] devised a correlation of separation factor versus permeability for polymeric membranes. The separation of gases utilizing polymeric membranes has emerged into an ever-growing area of scientific research pursuit. The technology is truly visionary as unit operation and chemical process engineering ushers in a new era. It has been recognized in the past decade that the separation factor for gas pairs varies inversely with the permeability of the more permeable gas of the specific pair. The authors rigorously discuss and redefine the world of polymeric membranes. Membrane separation of gases has emerged from a technical curiosity to intense research endeavor in today's world of science and technology. This treatise critically observes the technological vision of separation phenomenon of polymeric membranes.

Boussu et al.[13] rigorously discussed roughness and hydrophobicity studies of NF membranes using different methods of atomic force microscopy (AFM). Determination of surface roughness is vital support toward the furtherance of membrane technology. This technology is crucial in understanding of particle fouling in NF. Diverse technologies in membranes are addressed in this research work.

6.14   VISIONARY AND RELEVANT SCIENTIFIC ENDEAVOR IN THE DOMAIN OF MEMBRANE FOULING

Fouling stands as a major impediment to membrane separation phenomenon. The success of environmental engineering science is ever-growing as the world of scientific challenges and scientific vision trudges a weary path toward scientific destiny. The world of membrane fouling is challenging and is imperative to the research and development initiatives in membrane science and technology.

This section of this widely observed treatise redefines technological vision and scientific objectives in the field of membrane fouling. Global water challenges and global water science and technology endeavor are surging ahead in the pursuit of greater effectiveness and greater efficiency of membrane separation phenomenon. This is an endeavor of science which is latent yet challenging.

Mohammadi et al.[14] investigated deeply membrane fouling. Technology of membrane fouling and bio-fouling are today surpassing wide and visionary frontiers. Fouling of UF membranes in milk industries is mostly due to precipitation of micro-organisms, proteins, fats, and minerals on the membrane surfaces. Thus, chemical cleaning of the membranes is essential and vital to the furtherance of science. In this research work, results obtained from investigations on a polysulfon UF membrane fouled by precipitation of milk components have been presented in deep and instinctive details. Today technology is surpassing the wide boundaries of scientific instinct and scientific vision. Mankind's vision, the wide progress of science and technology and the furtherance of scientific candor and scientific forbearance will all lead a long way in the true emancipation of membrane science or environmental engineering science in future. Fouling is defined as existence and growth of micro-organisms and irreversible collection of materials on the membrane surface which results in flux decline. To overcome this process, a cleaning technique needs to be pursued. Cleaning usually performs in three forms: physical, chemical, and biological. Chemical methods are more relevant. The first step of chemical washing is finding appropriate materials as cleaning agents. This area needs to be re-envisioned. The choice of the best materials depends on feed composition and precipitated layers on the membrane surface and in most cases done by scientific intuition. The selected materials should be chemically stable, safe, cheap, and washable with water. The cleaning materials must also be able to dissolve most of the precipitated materials on the surface and remove them from the surface while not damaging the membrane surface. Some of these cleaning materials are acids, bases, enzymes, surfactants, disinfectants, and combined cleaning materials. The success of this research work is just visionary and crossing wide scientific imagination. Fouling of membranes are the areas of scientific research pursuit which needs to be re-envisioned and re-envisaged in this world of science and engineering. The authors of this treatise deeply delve into the intricacies of membrane fouling and the unknown and latent world of bio-fouling. In this chapter, results of investigations on the polysulfone membrane used in milk concentration are presented. In this widely observed treatise effects of different washing materials on the performance of the membrane recovery are investigated.

Rosenberger et al.[15] investigated the importance of liquid phase analyses to understand fouling in membrane assisted activated sludge processes in six different case studies of different European research groups. In combination with wastewater treatment by activated sludge process, micro or UF can be used either for direct sludge filtration (membrane bioreactor (MBR) concept) or as a polishing step after the settler (effluent filtration concept). In both applications, membrane fouling is a vital factor. Recent research activities pointedly focus toward the importance of soluble and colloidal material on membrane fouling in wastewater applications and water purification. The analytical methods applied by different research groups are presented and where possible compared. Measurements of organic material in activated sludge supernatant or wastewater treatment plant effluent have long be utilized in wastewater treatment engineering and science. The roots of scientific endeavor go back to the optimization of effluent quality, biofilm stability, or dewaterability of excess sludge. Research activities proposed correlations between the soluble and colloidal fraction of activated sludge and membrane performance in UF and MF applications. Results of six research activities are presented. In all six projects, micro or ultraporous membranes are used in different steps of advanced wastewater treatment concepts, either in effluent filtration or for direct sludge filtration. Technological forays into membrane science are being re-envisioned and re-envisaged in this well-observed research activity.

Chang et al.[16] discussed in wide and visionary details membrane fouling in MBRs for wastewater treatment. MBRs, in which membranes are applied to biological wastewater treatment for biomass separation, provide many advantages over conventional treatment. However, membrane fouling in MBRs restricts their widespread application because it reduces productivity and increases maintenance costs. Recent research and development has taken place to investigate, envision, and model fouling processes. In this treatise, fouling control techniques which have been included are low-flux operation, high shear slug flow aeration in submerged configuration, periodical air or permeate backflushing, intermittent suction operation, or addition of powdered activated carbon. The advantages of MBRs over conventional treatments are reviewed in details. They include reduced footprint and sludge production through maintaining a high biomass concentration in the bioreactor. Notwithstanding these prime advantages, the widespread application of the MBR process is constrained by membrane fouling. Membrane fouling is a serious problem affecting system performance. Fouling leads to permeate flux decline, making more frequent membrane cleaning and replacement necessary. This ushers in newer challenges in design and operation of MBRs.

Drews[17] reviewed membrane fouling in MBRs and its characterization. Despite the worldwide research activities in membrane fouling in MBRs, lots of questions remain unanswered. This author therefore aims at stepping back and critically re-evaluating fouling characterization methods and results. The authors deeply comprehend fouling analyses such as identifica-tion of foulants, characterization of foulants, and the use of model foulants. Recent findings include influence of soluble microbial products (SMP), influences on occurrences and properties of SMP, and the wide scientific world of hydrodynamics.

Howe and Clark[18] discussed with cogent insight fouling of MF and UF membranes by natural waters. Membrane filtration (MF and UF) has become an accepted process for water purification and drinking water treat-ment, but membrane fouling till today remains a vexing issue. The objective of this study was to systematically study the mechanisms and components in natural waters that contribute to fouling. Natural waters from five sources were filtered in a benchtop filtration system. A sequential filtration system was employed in most experiments. Technology and science of membrane separation processes are challenged today. This chapter critically over-views membrane fouling in MF and UF membranes. Membrane filtration has become an accepted process in industrial and domestic water treatment. Scientific vision and scientific adjudication are at its level best as regards newer innovations in environmental engineering. Howe and Clark reviews and redefines the wide world of membrane separation phenomenon.

Kimura et al.[19] discussed with deep and cogent insight irreversible membrane fouling during UF of surface water. In this research work, irreversible fouling caused by constituents in surface water was deeply investigated based on long term pilot plant study. Science of membranes and fouling is gaining new heights as technology and engineering crosses visionary frontiers. The membrane employed was a low pressure hydro-phobic UF membrane made of polysulfone and having a molecular cut-off of 750,000 Daltons. The main limitation of the use of membranes is its high energy consumption that can be mainly attributed to membrane fouling. This drawback and shortcoming of membrane science as a whole needs to be rebuilt with the passage of scientific history and time.

6.15 VISION OF SCIENCE, GLOBAL WATER SHORTAGES, AND RESEARCH AND DEVELOPMENT INITIATIVES IN WATER TECHNOLOGY

Global water shortages and environmental catastrophes are changing the wide frontiers of membrane science. Technology and engineering are under-going tremendous challenges. The success of human scientific endeavor needs to be re-envisioned and re-envisaged with the ever-growing steps of human life. Global research and development initiatives in water science and technology are witnessing a new dawn. This treatise definitely comprehends the wide vision of science and focuses on the success of application of membrane science in tackling global water crisis.

Research and development initiatives in water science and technology are wide and varied today. Technology and engineering science are emboldened today in every true spheres of scientific research pursuit today. Human mankind's immense and definite vision, the vast technological prowess, and the vast and varied fruits of engineering science will all lead a long way in the true realization of sustainability and environmental engineering techniques today. Water initiatives are vexing and in the similar manner far-reaching. The author in this treatise repeatedly points out the scientific intricacies and the scientific shortcomings in membrane science applications.

6.16   ENVIRONMENTAL PROTECTION AND THE SUCCESS OF MEMBRANE SCIENCE APPLICATIONS

Environmental protection and membrane science are two opposite sides of the visionary coin today. Membrane science applications are transforming the global water scenario. Membrane science applications are gaining new heights as global water shortage and drinking water issues stands today as vexing issues. A scientist's wide vision and the world of challenges in environmental engineering science are the torchbearers toward a greater emancipation of science and engineering today. Environmental protection and industrial wastewater treatment are the new dimensions of scientific endeavor today. Future of global water initiatives and the success of environmental science are crossing visionary boundaries.

6.17   DRINKING WATER PROBLEMS AND THE WIDE VISION OF MEMBRANE SCIENCE

Drinking water problems and water purification are the scientific enigmas today. These are the vexing areas of environmental engineering science today. Membrane science and novel separation processes are the only answers to this vicious global water issue. Non-traditional environmental engineering techniques are the other wide vistas of scientific endeavor. Technology and engineering science are moving at a rapid pace in our present day human civilization. The wide visionary area of membrane science today has an umbilical cord with global water initiatives. Desalination, industrial wastewater treatment, and drinking water treatment are the wide avenues of science and technology today. The author pointedly focuses on the wide visionary domain of novel separation processes such as membrane science and technology. The scientific challenges and the scientific profundity are immense as science and engineering steps into a newer eon.

6.18   RECENT TRENDS IN RESEARCH IN THE DOMAIN OF MEMBRANE SCIENCE

Science and technology are moving at a rapid pace in today's age of scientific regeneration and scientific vision. Membrane science and global water research and development initiatives are the two opposite sides of the visionary coin today. The barriers and hurdles of membrane separation processes are the wide world of membrane fouling. This area of scientific rigor needs to be re-envisioned and re-addressed with each step of science and engineering.

Van der Bruggen[20] delineated in details how a MF pretreatment affects the performance in NF. Today technology and engineering science of membrane science and NF are crossing vast and versatile visionary boundaries. The use of a well-chosen pretreatment system is a key element to avoid fouling in NF. Among the different possibilities for pretreatment systems, MF emerges as the most compatible with NF. As it allows the removal of components with relatively low molecular weight, NF is a process which is widely recognized as a process with numerous applications in drinking water production, process water recovery, and industrial wastewater recovery. This chapter gives a critical overview of the entire domain of MF and NF and its application domain for the furtherance of science.

Hajibabania et al.[21] deeply comprehended the effect of fouling on removal of trace organic compounds by NF. The aim of this chapter is to assess the impact of different organic-based fouling layers on the removal of a large range of trace organics. Technology and science of NF are widely delineated as membrane science ushers in a new era of scientific vision and scientific fortitude. Both model and real water samples were used to simulate fouling in NF under controlled environment. Membrane science is moving toward a wider futuristic trend in science and engineering.(Palit(2016),Palit(2015))[23,24]

Van der Bruggen et al.[22] described with deep and cogent insight in a review paper drawbacks of applying NF and how to avoid them. Scientific cognizance and scientific profundity are moving toward a newer futuristic dimension. This chapter identifies six challenges for NF where solutions are deeply scarce: (1) avoiding membrane fouling, (2) improving separation of solutes that can be achieved, (3) further treatment of concentrates, (4) chemical resistance phenomenon, (5) insufficient rejection of pollutants, and (6) the utmost need of modeling and simulation tools. The authors of this chapter greatly rejuvenate the success of NF phenomenon and the immense potential of membrane science. Scientific vision, scientific profundity, and deep scientific doctrine will all lead a long and visionary way in the true realization of membrane technology and water science in years to come (Palit(2015), Palit(2016)) [23,24]

6.19   CONCLUSION AND ENVIRONMENTAL PERSPECTIVES

Today, environmental engineering science is moving toward a newer visionary realm. Technology is gaining new heights. Perspectives of science and technology are crossing visionary boundaries today. Vision of science in the field of industrial wastewater treatment and membrane science is the utmost need of the hour. Global water crisis is the primary issue facing human civilization today. The other vexing issue which is an enigma today is the wide world of As and other heavy metal contamination of groundwater and its immediate remediation. The answers to global water technology and the vision of membrane science are being challenged today as human scientific research pursuit reaches a new path of regeneration. The present treatise pointedly focuses on the success of membrane separation phenomenon in environmental pollution control. The challenge and the vision of science are beyond scientific imagination today. Today, future environmental perspectives rely on zero-discharge norms and the wide world of traditional and non-traditional environmental engineering tools. Novel separation processes such as membrane science and non-traditional techniques such as advanced oxidation processes are the challenging and visionary areas of scientific research pursuit. Industrial pollution control and carbon sequestration are the wide avenues of science and engineering today. In this treatise, the author discussed lucidly the wide vision of membrane science and its vast and versatile applications. Technological motivation and vision are immense today. In such a crucial juncture of scientific history and time, protection of environment and the application of scientific tools assume immense importance. This treatise widely focuses on various branches of membrane science and the wide applications of membrane separation processes. Scientific endeavor, the vast and varied research output, and the academic rigor will lead a long way in the true realization of environmental engineering science.

ACKNOWLEDGMENTS

The author wishes to acknowledge the contribution of the chancellor, vice-chancellor, faculties, staff, and students of University of Petroleum and Energy Studies, Dehradun, India without whom this writing project would not have been completed. The author also wishes to respectfully acknowledge the contributions of Shri Subimal Palit, the author's late father and an eminent textile engineer from India who taught the author the rudiments of chemical engineering.

KEYWORDS

- **membrane**
- **ultrafiltration**
- **reverse osmosis**
- **nanofiltration**
- **vision**
- **water**
- **environment**

REFERENCES

1. Cheryan, M. *Ultrafiltration and Microfiltration Handbook;* Technomic Publishing Company Inc.: Lancaster, PA, 1998.
2. www.google.com
3. www.wikipedia.com
4. Welch, A. H.; Westjohn, D. B.; Helsel, D. R.; Wanty, R. B. Arsenic in Groundwater in the United States: Occurrence and Geochemistry. *Ground Water.* **2000,** *38* (4), 589–604.
5. Wang, S.; Mulligan, C. N. Natural Attenuation Processes for Remediation of Arsenic Contaminated Soils and Groundwater. *J. Hazard. Mater.* **2006,** *B138*, 459–470.
6. Su, C.; Puls, R. W. Arsenate and Arsenite Removal by Zerovalent Iron: Kinetics, Redox Transformation, and Implications for In situ Groundwater Remediation. *Environ. Sci. Technol.* **2001,** *35,* 1487–1492.
7. Safiuddin, M.; Masud Karim, M. In *Arsenic Contamination in Bangladesh: Causes, Effects and Remediation,* 1st IEB International Conference and 7th Annual Paper Meet, Chittagong, Bangladesh, Nov 2–3, 2001; Institution of Engineers: Bangladesh.
8. Nechifor, G.; Voicu, S. I.; Nechifor, A. C.; Garea, S. Nanostructured Hybrid Membrane Polysulfone- Carbon Nanotubes for Hemodialysis. *Desalination.* **2009,** *241,* 342–348.

9. Arnot, T. C.; Field, R. W.; Koltuniewicz, A. B. Cross Flow and Dead End Microfiltration of Oily-Water Emulsions, Part-II, Mechanisms and Modeling of Flux Decline. *J. Memb. Sci.* **2000**, *169*, 1–15.

10. Sarkar, S.; Sengupta, A. K.; Prakash, P. The Donnan Membrane Principle: Opportunities for Sustainable Engineered Processes and Materials. *Environ. Sci. Technol.* **2010**, *44*, 1161–1166.

11. Yang, Y.; Li, J.; Wang, H.; Song, X.; Wang, T.; He, B.; Liang, X.; Ngo, H. H. An Electrocatalytic Membrane Reactor with Self-Cleaning Function for Industrial Wastewater Treatment. *Angew. Chem. Int. Ed.* **2011**, *50*, 2148–2150.

12. Robeson, L. M. Correlation of Separation Factor Versus Permeability for Polymeric Membranes. *J. Memb. Sci.* **1991**, *62*, 165–185.

13. Boussu, K.; Van der Bruggen B.; Volodin, A.; Snauwaert, J.; Van Haesendonck. C.; Vandecasteele, C. Roughness and Hydrophobicity Studies of Nanofiltration Membranes Using Different Modes of AFM. *J. Colloid Interface Sci.* **2005**, *286*, 632–638.

14. Mohammadi, T.; Madaeni, S. S.; Moghadem, M. K. Investigation of Membrane Fouling. *Desalination.* **2002**, *153*, 155–160.

15. Rosenberger, S.; Evenblij, H.; te Poele, S.; Wintgens, T.; Laabs, C. The Importance of Liquid Phase Analyses to Understand Fouling in Membrane Assisted Activated Sludge Processes – Six Case Studies of Different European Research Groups. *J. Memb. Sci.* **2005**, *263*, 113–126.

16. Chang, I. S.; Clech, P. L.; Jefferson, B.; Judd, S. Membrane Fouling in Membrane Bioreactors for Wastewater Treatment. *J. Environ. Eng.* **2002**, *128*, 1018–1029.

17. Drews, A. Membrane Fouling in Membrane Bioreactors-Characterization, Contradictions, Cause and Cures. *J. Memb. Sci.* **2010**, *363*, 1–28.

18. Howe, K. J.; Clark, M. M. Fouling of Microfiltration and Ultrafiltration Membranes by Natural Waters. *Environ. Sci. Technol.* **2002**, *36*, 3571–3576.

19. Kimura, K.; Hane, Y.; Watanabe, Y.; Amy, G.; Ohkuma, N. Irreversible Membrane Fouling during Ultrafiltration of Surface Water, *Water Res.* **2004**, *38*, 3431–3441.

20. Van der Bruggen, B.; Segers, D.; Vandecasteele, C.; Braeken, L.; Volodin, A.; Van Haesendock, C. How a Microfiltration Pretreatment Affects the Performance in Nanofiltration. *Sep. Sci. Technol.* **2004**, *39*, 1443–459.

21. Hajibabania, S.; Verliefde, A.; Drewes, J. E.; Nghiem, L. D.; McDonald, J.; Khan, S.; Le-Clech, P. Effect of Fouling on Removal of Trace Compounds by Nanofiltration. *Drink.Water Eng. Sci.* **2011**, *4*, 71–82.

22. Van der Bruggen, B.; Manttari, M.; Nystrom, M. Drawbacks of Applying Nanofiltration and How to Avoid Them: A Review. *Sep. Purif. Technol.* **2008**, *63*, 251–263.

23. Palit, S. Filtration: Frontiers of the Engineering and Science of Nanofiltration – a Far-Reaching Review. In *CRC Concise Encyclopedia of Nanotechnology;* Ubaldo Ortiz-Mendez, Kharissova, O. V., Kharisov, B. I., Eds.; Taylor and Francis: Abingdon, UK, 2016; pp 205–214.

24. Palit, S. Advanced Oxidation Processes, Nanofiltration, and Application of Bubble Column Reactor. In *Nanomaterials for Environmental Protection;* Boris, I. K., Oxana, V. K., Rasika Dias, H. V., Eds.; Wiley: Hoboken, NJ, 2015; pp 207–215.

PART II

Chemometrics and
Chemoinformatics Approaches

CHAPTER 7

APPLICATION OF STATISTICAL APPROACHES TO OPTIMIZE THE PRODUCTIVITY OF BIODIESEL AND INVESTIGATE THE PHYSICOCHEMICAL PROPERTIES OF THE BIO/PETRO-DIESEL BLENDS

NOUR SH. EL-GENDY* and SAMIHA F. DERIASE

Department of Processes Design and Development, Egyptian Petroleum Research Institute, Nasr City 11727, Cairo, Egypt

Corresponding author. E-mail: nourepri@yahoo.com

CONTENTS

ABSTRACT

The environmentally friendly and renewable biodiesel fuel is steadily gaining attention and significance for use as an alternative or in blends with petro-diesel. For widespread commercialization of biodiesel, its production process should be cost-effective and optimized to produce high and qualified biodiesel yield. Blending biodiesel with petro-diesel would also overcome the high price of biodiesel and produce a fuel having the advantages of both bio- and petro- diesel. So, it is important to optimize both production and blending processes and evaluate the bio-petro-diesel blends to reach to a fuel that agrees well with the international fuel standards and characterized by lower greenhouse gas emissions, better lubricity, cetane number, flash point, and flow properties.

This chapter deals with the applications of different statistical techniques to optimize the biodiesel production process and predict mathematical models describing the changes of some basic fuel properties with the volumetric percentage of biodiesel in the bio-petro-diesel blends to reach for the optimum ratio of high quality bio-petro-diesel blend.

7.1 INTRODUCTION

The increased awareness of the negative impact of environmental pollution on human health and biodiversity comes with the depletion of the global reserves of high quality low sulfur content petroleum fuel and the frequent fluctuations in crude oil price. That makes the need for renewable and environment friendly fuels a mandatory aspect. Biodiesel as a non-toxic, biodegradable, environment friendly, renewable, and sustainable green fuel has become a field of great interest as an alternative and/or complementary for the conventional petro-diesel fuel. Transesterification reaction of triglycerides with a short chain alcohol in the presence of homogenous or heterogeneous catalyst to form mono-alkyl esters of long chain fatty acids (i.e., biodiesel) and glycerol as a byproduct is the most common process to produce biodiesel. There are different important factors affecting the transesterification reaction; feedstock, alcohol type and concentration, catalyst type and concentration, reaction time and temperature, and finally the mixing speed.

This chapter summarizes the application of different statistical techniques to investigate the significant and interactive effects of various parameters affecting the transesterification reaction and determine their optimal combination. Blending biodiesel with petro-diesel is a recommendable approach to overcome the problem of the high price of biodiesel and to produce a fuel blend having the advantages of both. This chapter also deals with the application of statistical techniques to establish mathematical correlations to describe the changes of some basic properties of the prepared fuel with the volumetric percentage of biodiesel in the bio-petro-diesel blends; that is, to optimize and evaluate the bio-petro-diesel blends to reach to a fuel blend with lower greenhouse gas emissions, better lubricity, cetane number, flash point, and flow properties.

7.2 BIODIESEL

Biodiesel is an alternative or complementary fuel for the conventional petro-diesel one.[1] It is defined according to the American Society for Testing Materials (ASTM 6751)[2] as "mono-alkyl esters of long chain fatty acids, that derived from renewable biolipids (i.e., fatty raw materials); vegetable oils or animal fats". The term "bio" is symbolic for its renewability and biological origin. The term "diesel" is symbolic for its similarity to diesel petro-diesel and its applicability in conventional diesel engines.[3]

Biodiesel has many advantages, such as higher combustion efficiency, flash point, lubricating efficiency, and cetane number than conventional petro-diesel. It is a clean, renewable, sustainable, biodegradable, and nontoxic fuel. It is characterized by better lubricity and thus enhances engine longevity. It has no sulfur and consequently does not emit SO_x on its combustion which is the precursor for acid rains. It has no aromatics and lacks metals and tiny particles of solid.[4-6] Biodiesel has approximately 95% energy content of petro-diesel and it has a positive net energy gain of approximately 3–4:1. Jairam et al.[7] and Farooq et al.[8] have reported that biodiesel returns about 90% more energy than that used in its production. It has similar physicochemical properties like those of petro-diesel fuel.[9] Thus, biodiesel is a realistic fuel alternative/complementary for diesel engines.[10] The standards for bio- and petro-diesel fuels are provided in Table 7.1; twenty-five parameters are listed in the ASTM-D6751[2], where mainly nine parameters are mainly applied to estimate the quality of biodiesel: flash point (FP), pour point (PP), cloud point (CP), kinematic viscosity at 40°C, density at 15°C, total acid number (TAN), water content, iodine number, and cetane number.[11]

Biodiesel can be used in any compression ignition engine without any required modifications. Its higher flash point makes it has a better and safer performance during its handling and storage. It has been reported that applying

TABLE 7.1 Standard Properties of Bio- and Petro-Diesel.

Fuel Standard Fuel Property	Petro-diesel ASTM D975[102]	Bio-diesel	
		ASTM D6751[2]	EN14214[96]
Kinematic viscosity cSt at 40°C	1.9–1	1.9–6.0	3.5–5
Density at 15.56°C			0.86–0.9
API gravity at 15.56°C	35–38		
Pour point °C	Depends on the location		
Cloud point °C	Depends on the location	Report*	−4
Total acid number mg KOH/g		<0.5	<0.5
Total S wt.%	<0.5	<0.05	<0.01
Water content ppm	<500	<500	<500
Flash point °C	>52	>130	>101
Calorific value MJ/kg	>44		32.9
Cetane number	>40	>47	>51
Iodine number mg I_2/100 g			<120
Calcium and Magnesium, combined ppm		<5	<5
Cold soak filterability s		<360	
Methanol content wt.%		<0.2	<0.2
Sulfated ash wt.%	<0.01	<0.02	<0.02
Copper strip corrosion	No.3	No. 3	Class 1
Carbon residue wt.%	<0.1	<0.05	
Free glycerin wt.%		<0.02	<0.02
Total glycerin wt.%		<0.24	<0.25
Phosphorus content wt.%		<0.001	<0.001
Distillation temperature** °C	282–338	<360	
Sodium and potassium, combined ppm		<5	<5
Oxidation stability hrs		>3	>8

*The cloud point of biodiesel is generally higher than petroleum based diesel fuel and should be taken into consideration when blending
**Atmospheric equivalent temperature, 90% recovered.

biodiesel leads to better engine performance and fuel consumption.[12-14] Moreover, its high oxygen content (approximately 10–12%) improves the combustion efficiency and reduces the greenhouse gas emissions; carbon mono- and di-oxides (CO and CO_2), soot, unburned hydrocarbons, and particulate matter (PM).[15] The CO_2 is continuously cycled and reused for growing plants instead of being released in the atmosphere, since the plants help in CO_2 fixation during the photosynthesis process. The National Renewable Energy Laboratory (NREL) and US-Department of Energy (US-DOE) reported that the overall life cycle emission of CO_2 from 100% biodiesel (i.e., B100) is approximately 78.45% lower than that of 100% petro-diesel (i.e., B0). The blend with 20% biodiesel (i.e., B20) would reduce the net CO_2 emissions by approximately 15.66%. The substitution of B0 by B100 in buses would reduce the life cycle consumption of petro-diesel by 95%. While applying B20 would drop the life cycle consumption of petro-diesel by approximately 19%. But biodiesel may cause a slight increase in NO_x approximately 10% more than petro-diesel, which can be positively influenced by delaying the injection timing in engines.[3,16] The suppression in soot formation is attributed to the biodiesel composition, since it is an oxygenated fuel having an oxygen mass fraction of approximately 10–12%.

One of the most important advantages of biofuels in general is that, politically, producing biofuels would make the nations totally or partially independent on other oil-producing countries. This would consequently save foreign currency. Besides the less global warming, it has additional benefits to the society such as rural revitalization and creation of new jobs opportunities.

7.3   BIODIESEL PRODUCTION TECHNIQUES

Biodiesel can be produced through pyrolysis or thermal cracking, microemulsification, dilution, and transesterification.[3,17]

Dilution of vegetable oil with petro-diesel can improve the viscosity, decrease the injector coking and carbon deposition, and enhance the engine performance. However, it is not recommendable for long-term usage in direct injection engine due to its lower volatility, higher viscosity, and reactivity of unsaturated hydrocarbons.[3] Thermal cracking or pyrolysis of vegetable oils in the presence of nitrogen and absence of air or oxygen breaks the chemical bonds, and procures alkanes, alkenes, alkadyns, aromatics, and carboxylic acid. The produced fuel has low viscosity and high cetane number, with appropriate amount of sulfur, water, deposits, and corrosion rate of copper. However, it has improper amount of ash, carbon residue, and cloud point, and the eco-friendly advantage of oxygenated oils vanishes. Catalytic cracking of triglycerides (TG) using Al_2O_3, MgO, SiO_2/Al_2O_3, NiMo/g Al_2O_3, $NiSiO_2$, zeolites composites, and others have been reported. However, a gasoline-like fuel is mainly formed than a diesel-like fuel.[18–20]

Micro-emulsions, using solvents; methanol, ethanol, or butanol and co-surfactant, decrease the viscosity of vegetable oils. But the produced fuel has low cetane number and low calorific value, and its combustion would be incomplete, which leads to the formation of carbon deposits.[3,17]

The transesterification of TG or the esterification of free fatty acids (FFAs) with low molecular weight alcohols, such as methanol, ethanol, propanol, etc., in the presence of catalyst is the best method for producing high quality biodiesel.[21] However, the common approach for biodiesel production is the methanolysis, which is the transesterification of vegetable oils or animal fats with methanol in the presence of catalyst, usually sodium hydroxide or potassium hydroxide. Methanol is the most widely used alcohol, due to its good polarity, small carbon chain, wide availability, low cost, and high activity.[15,22] Usually, the long chains fatty acids in the TG of vegetable oils are palmitic, stearic, oleic, linoleic, and linolenic.[23]

7.4 BIODIESEL GENERATIONS

Depending on the feedstock, the biodiesel can be classified into three categories: the first generation, where the biodiesel is derived from edible vegetable oils (e.g., soybean, palm, rapeseed, sunflower, corn, etc.); the second generation, which is one of the advanced biofuels and can be obtained from non-edible oils (e.g., castor, jatropha, neem, waste cooking oil, etc.) and animal fats (e.g., tallow and yellow grease);[24,25] and the third generation biofuels, which are obtained from alga.[26] Most of the biodiesel nowadays is produced from edible oils. However, this would cause many problems, such as the competition with the edible oil market, which could lead to the increase of the cost of edible oils and biodiesel. The overall production cost of biodiesel is estimated to be approximately 1.5 times higher than that for petro-diesel. That would come from the high cost of virgin vegetable oils, which contribute up to 75% of the total manufacturing cost.[27] The use of fresh vegetable oils will cause also the deforestation, since more and more forests would be felled for plantation of vegetable oil crops.[5] To overcome such drawbacks, there is a worldwide trend toward production of biodiesel from non-edible oils, such as the waste cooking oils (WCO) which is not suitable for human consumption as it contains some toxic compounds. Due to the lack of the process of planting, conserving, and harvesting, it leads to lowering of the total production cost of biodiesel and saving of farm-lands. Nowadays, biodiesel production from WCO is one of the most important aspects in biofuels, due to its low cost and wide availability.[27]

The WCO is much less expensive than virgin oils, approximately two- to three-fold cheaper than fresh vegetable oils.[28] The economic feasibility of biodiesel depends mainly on the low-cost feedstocks.[29] It has been estimated that the cost of raw materials represents approximately 60–95% of the total production cost.[30] Applying WCO instead of fresh oils as a feedstock will reduce the feedstock cost by 60–70%. The WCO is locally available wherever food is cooked or fired in oils. Frying or cooking oils can be derived from vegetable oils (such as coconut, corn, cottonseed, groundnut, linseed, olive, palm, palm kernel, rapeseed, sesame, soybean, sunflower, etc.) and animal fats/oils (such as butter, grease, fish oil, lard, and tallow). It has been reported that the worldwide annual production of WCO exceeds 25 million tons.[31] The annual production details of waste from different countries are recorded to be as follows: USA 10 MT, China 4.5 MT, EU 0.7–1 MT, Egypt >1 MT, Japan 0.45–0.57 MT, Malaysia 0.5 MT, Turkey 0.35 MT, Ireland 0.153 MT, Canada 0.12 MT, and Taiwan 0.07 MT.[11,32] These are in addition to the amounts that are annually discarded into water streams or dumped into land without any processing.[31,33] That causes a lot of problems; therefore, the use of the readily available waste edible oils as an efficient and cost efficient feedstock for biodiesel production would save the environment and solve many of the waste-management problems.[22,34] Many of the developed countries have set regulations that penalize the disposal of WCO into the waste drainage.[35] Even in developing countries like India, there is a great concern about using WCO for biodiesel production. It has been estimated that India can supplement about 41.14% of its total petro-diesel consumption, if WCO and other bio-wastes are used as raw materials for biodiesel production.[34] The FFA content in the WCO strongly affects the transesterification reaction and the type of catalyst to be used and it should be within a certain limit, otherwise it would cause a lot of operational problems and decrease the biodiesel yield. High FFAs and water contents would lead to soap formation and hydrolysis, respectively, and both reactions result in low biodiesel yield and high catalyst consumption.[36] Generally, the lower the cost of the feedstock is, the higher will be the FFA content.[23] Several techniques can be applied to reduce the high FFA content, such as acid esterification with methanol and sulfuric acid, neutralization with alkalis followed by soap separation by decanter, esterification of FFAs with ion exchange resins or extraction with polar liquids along with acid esterification, and distillation of FFAs. While, water content, usually eliminated by heating the WCO above 100°C, or alternatively, by vacuum distillation at 0.05 bar pressure. The suspended solids, food debris, phospholipids, and any other impurities can be removed by centrifugation or filtration.[23,37]

7.5 THE MAIN FACTORS AFFECTING THE TRANSESTERIFICATION REACTION

The parameters that significantly affect the rate of the transesterification reaction and the biodiesel yield are the reaction temperature and time, alcohol to oil molar ratio, agitation speed, the catalyst type (homogenous, heterogeneous, and enzymes), and its concentration and activity. Many researchers.[36,38,39] Ceclan et al.[40] reported other factors such as the WCO quality and the reactants' adding order too. Niju et al.[30] reported that, the surface area of the solid catalysts has a direct impact on the catalytic activity; thus, the higher the specific surface area, the higher will be the catalytic activity. The type of alcohol and its molar ratio have significant effects on the biodiesel yield and its characteristics.[41] The optimum values of the afore-mentioned factors mainly depend on the type of feedstock and its physico-chemical characteristics.[42] The cost of the applied catalyst affects the overall production cost as well. Thus, applying catalysts prepared from waste materials will decrease the cost and enhance the development of a sustainable biodiesel production process.[8,43]

7.6 CATALYSTS

Base catalysts, acid catalysts, and enzymes are the three categories of catalysts that can be used for biodiesel production.

Enzyme-catalyzed transesterification reaction weather by extra- or intra-cellular lipase is not yet industrially applied. It has the advantage of water tolerance, avoids FFAs' saponification, and takes place at low temperature and pressure, which in turn leads to lower energy consumption. But the enzymes are expensive and easily inhibited by methanol and the reaction requires long time and the produced glycerol is adsorbed on the enzyme surface, thus enzyme-catalyzed transesterification process seldom leads to completion.[21,44] Novozyme 435 is reported to be the most effective for methanolysis among lipases.[45] Homogenous base catalysts are alkaline liquids, such as sodium hydroxide, potassium hydroxide, and their methox-ides, while the homogenous acid catalysts are acidic liquids, such as sulfuric acid, hydrochloric acid, and sulfonic acids. One of the major drawbacks of homogenous catalysts, is that they cannot be recycled, regenerated, or reused, because it is consumed in the reaction and the separation of the unre-acted catalyst from the products is difficult and would require more equip-ment, which consequently increases the overall operation costs. The process is not environment friendly because a large amount of water is used in the purification of the product. This consequently produces huge amounts of toxic wastewater, which in turn leads to a new waste management problem.[7] The associated emulsification problem, due to the soap formation, especially with homogenous base-catalyzed reaction, would emulsify the biodiesel with glycerol, mostly if ethanol is used in the transesterification reaction.[46] Moreover, glycerol as a byproduct of this process is of very low quality, due to its high impurities content from unreacted catalyst and methanol, water, and produced soap. The purification of such glycerol through filtration, chemical additives, ion exchange resins, fractional vacuum distillation, bleaching, and deodorizing is costly especially for medium and small scale biodiesel producers.[47]

Heterogeneous catalyzed biodiesel production has been reported by many researchers as a milder, environment friendly, cost-efficient, and integrated process, which produces high quality biodiesel and glycerol. Heterogeneous solid catalysts overcome most of the drawbacks of the homogenous catalysts. It is more environment friendly. Solid catalysts are not consumed or dissolved in the reaction mixture, can be easily separated from the products, and can be regenerated and reused. No washing step for the product is required. It can be applied in continuous and batch processes. It is less corrosive, is easier to handle, generates less amount of toxic wastes, and prevents the undesirable saponification. Thus, heterogeneous catalysts eliminate the additional steps required by the liquid homogenous catalysts, which offers a reduction in the processing and overall costs.[25] Metal hydroxides, metal complexes, metal oxides, such as calcium oxide, magnesium oxide, zinc oxide, zirconium oxide, and supported catalysts have been reported in biodiesel production.[48] Effective factors on catalytic activity of solid catalysts are specific surface area, pore size, pore volume, and active site concentration on the catalyst surface. Type of precursor of active materials has also a significant effect on the catalytic activity of the supported catalysts. The most important factor for heterogeneous catalyst activity is the concentration of the active sites on the catalyst surface.[23] One of the major drawbacks of heterogeneous catalysts is that it is a three-phase system: catalyst, alcohol, and oil. It leads to diffusion limitations, which consequently lowers the reaction rate. The use of catalyst supports would improve the mass transfer limitations in the three-phase reaction. Alumina is preferred to other supports, such as silica and zinc oxide for transesterification reactions. Moreover, fixing metal oxides inside pores, catalyst supports would prevent active phases from sintering in the reaction medium. Optimization of the catalyst activity and the cost of its synthesis would promote the replacement of homogenous catalysts by the heterogeneous ones for commercial biodiesel production. Not only this, but the solid base catalysts would be more attractive than conventional alkaline metal alkoxides due to their economic and environmental benefits. Moreover, from the process point of view, on recovering the solid catalyst, there is no need for complex aqueous treatment or purification steps. Thus, this makes the process more simplified and produces higher biodiesel yield.[49] It can be applied in continuous catalytic biodiesel production process from low-grade WCO using slurry or fixed bed configuration reactors.[17,31,50] The advantages of heterogeneous acid catalysts are insensitivity to FFAs and water contents in the oil feedstock; they are recommended if low-grade oil is used, for the occurrence of simultaneous esterification and transesterification reactions and for the ease of the separation of the catalyst from the product. However, they have also, some drawbacks, such as high costs, complicated catalyst preparation, the requirement of high reaction temperature, large alcohol/oil molar ratio, long reaction time, energy intensive, and finally, the leaching of catalyst active sites, which would result in contamination of the products. Heterogeneous base catalysts solve most of the aforementioned drawbacks and offer relatively faster reaction rate than the acid-catalyzed ones. In addition, they are less energy intensive, require milder reaction conditions, offer ease of separation after reaction completion, ease of catalyst regeneration, reusability, and less corrosion, and finally are safer, cheaper, and environment friendly.[1] However, its drawbacks can be summarized in; the catalyst poisoning when exposed to air and leaching of the active sites that may results in contamination of the products.[31]

The use of homogenous alkaline catalysts is still the most common in industrial scale, because of its relatively cheap and active characteristics. The methoxides can give relatively higher yields (> 98%), in a short time (reaching to 30 min) at low molar concentration (0.5 mol%). The base-catalyzed transesterification reaction solved one of the major drawbacks of the acid-catalyzed reaction, that is, it can operate at atmospheric pressure and temperature of 60–70°C with an excess of methanol. But it still takes a long time; at least several hours to ensure that the catalyst completely transesterified the TG. So, one of the big challenges to decrease the cost of biodiesel is to decrease the time of the transesterification process with maximum yield production. However, due to the cheapness of NaOH and KOH, they are widely used in industrial biodiesel production. But it requires higher catalyst concentration 1–2 mol% and high alcohol/oil molar ratio (such as 6:1 or higher) instead of the stoichiometric 3:1 ratio to achieve high biodiesel yield.[51] However, to overcome the drawback of total high cost of biodiesel production, researchers have found that it is mandatory that future work should focus on the application of renewable, economic, and environment friendly heterogeneous catalysts in the biodiesel production from animal fats and waste oils as feedstock. The use of heterogeneous base CaO catalyst has started since 1980s as it can tolerate high concentration of FFAs content oils. However, calcium soap can also be produced as a side reaction, which would deactivate the catalyst. The low CaO-active species leaching and low methanol solubility (0.035%) promote its reusability, which further promote its future industrial applicability. However, its solubility in glycerol is very high, so it should be separated directly after the completion of the reaction. CaO should be thermally activated at approximately 700°C for 2–3 h before its application in the transesterification reaction to remove the unwanted $Ca(OH)_2$ and $CaCO_3$, as they have low to no transesterification activity, respectively. Although, the reaction time of CaO-catalyzed transesterification reaction is longer by two- to four-fold, relative to homogenous NaOH-catalyzed ones. However, the low cost of CaO, its preparation from readily available natural sources, and the simpler biodiesel production process make CaO a so promising candidate in biodiesel production.[1] The natural sources that can be used to prepare the basic heterogeneous catalysts can be categorized into shells, ashes, bones, and finally rocks and clay.[52] Organic wastes from food industries, domestic activities, seafood restaurants, butcheries, etc., cause a lot of waste management and environmental problems. Thus, the preparation of renewable, economic and environment friendly heterogeneous catalysts from such wastes (e.g. mollusks, crabs, egg, shrimp, snails... etc. shells and animal bones) and applying it in the biodiesel production from animal fats and waste oils as feedstock would make the process economic and fully eco-friendly. Thus, it would act as a four-fold solution for economy, environment, waste management, and energy problems.

7.7   PHYSICOCHEMICAL CHARACTERISTICS OF BIODIESEL

The physicochemical properties of the produced esters, that is, biodiesel vary depending on the types of oils used. The viscosity and oxidation stability of the produced biodiesel mainly depends on the type of oil feedstock. Oil feedstock riches in unsaturated fatty acids, for example, linoleic and linolenic acids, tends to give biodiesel with low oxidation stability. The cetane number depends on the proportion and location of the double bonds. The higher the iodine number, the higher is the number of double bonds.[53] The cetane number has a great effect on the engine performance. Low cetane number causes a delay in the engine start up and leads to noise with a smoke exhaust gas formation. The kinematic viscosity and the FFA content are the most important properties to determine the optimum catalyst type and concentration required for high biodiesel yield with good quality.[54] Low viscosity fuel does not provide enough lubricity, while high viscosity fuel forms deposits in the engine. The high viscosity biodiesel would deteriorate the atomization, evaporation, and air–fuel mixture formation characteristics. That leads to improper combustion and higher smoke emission. Not only this, but high viscosity biodiesel would generate some other operational problems, such as; difficulty in engine starting, unreliable ignition and deterioration in thermal efficiency. It creates problems in the atomization of the fuel spray and function of the fuel injectors.[55] The biodiesel stability includes; the oxidation, storage, and thermal stabilities. But, it is usually termed the oxidation stability. The lower the oxidation stability is, the higher the peroxide value, iodine number, and TAN are. Thus, increases the gum deposits, filter plugging, injector cocking, corrosion, damage to fuel delivery system, fusion of moving components, and hardening of elastomeric components.[56] Sulfur is known to provide diesel fuels with the required lubricity for engine performance, thus upon applying the stringent regulations toward ultra-low sulfur diesel oils, the lubricity would decrease. Thus, upon blending of 1–2% of biodiesel to conventional petro-diesel, this would retain the beneficial impact of restoring lubricity throughout an anti-wear action on the engine injection system. The fuel density is the main characteristic which determines the proper amount of fuel required in combustion system, pumps, and injectors to provide good combustion efficiency. The flash point of the fuel is related to its ignition, thus moderate value is required for safely handling and storage of the fuel. The TAN is a measure of the FFA and mineral acid content in the produced biodiesel. The high the TAN is the more the possibility of the occurrence of corrosion and engine deposits, especially in the fuel injectors. The increase of TAN during the storage of biodiesel indicates its degradation. The high the water content is, the lower the calorific value is and the higher the corrosion rate, microbial growth and biodiesel degradation and oxidation during the long-term storage are.[55] The cloud and pour point values known as the cold flow properties. The PP is often used to specify the applicability of fuel oil at cold temperature climates and it can be defined as the lowest temperature at which frozen oil can flow.

Blending biodiesel with petro-diesel is another recommendable way to overcome the problem of the high price of biodiesel and gain the advantages of both fuels.[22] The blended product is designated as Bxx (e.g., B5 is 5% biodiesel and 95% petro-diesel).

7.8 STATISTICAL APPROACHES IN BIODIESEL INDUSTRY

There are a lot of published reviews concerning the researches on the improvement in biodiesel processing and production, the types of feedstock, and catalysts especially, with the great and growing interest of biodiesel production from WCO, in the last two decades.[3,11,15,17,21,23,31–33,47,50,52,53,57] But to our knowledge there is no review summarizes the applications of statistical techniques on the optimization of the biodiesel production process.

This chapter gives a brief glance on the application of different statistical techniques to investigate the effects of various parameters affecting the transesterification reaction and determine their optimal combination. This chapter also summarizes the application of statistical techniques to correlate the changes in the fuel physicochemical properties with the bio-petro-diesel blends to reach to a fuel blend that retains the highest beneficial impact of both fuels.

7.8.1 WHAT IS THE MAIN PURPOSE OF RUNNING AN EXPERIMENT?

Typically, an experiment would be run for one or more of the following reasons;

- To determine the principal causes of variation in a measured response.
- To reach to the optimum conditions that give rise to a maximum or minimum response.
- To compare the responses achieved at different settings of controllable variables.
- To obtain a mathematical model in order to predict future responses.

7.8.2 SOURCES OF VARIATIONS

A source of variation is anything that can cause an observation (i.e., outcome) to be different from another observation.

Sources of variation are two types (Fig. 7.1):

1. Those that can be controlled and are of interest are called treatments or controllable variables or factors.
2. Those that are not of interest but are difficult to be controlled are nuisance (uncontrollable) factors.

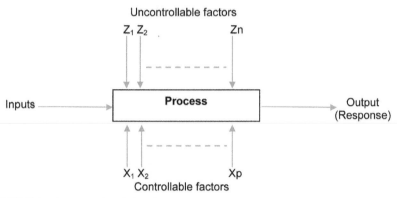

FIGURE 7.1 Sources of variation for a process.

For the transesterification process, the specific objectives of the experiments are;

- To determine the significant factors affecting the biodiesel yield.
- Analyze the effect of the determined factors on the process output.
- Predict a model equation relating the factors to the obtained biodiesel yield.
- Optimize the biodiesel production process.

Optimization study for transesterification process is crucial to assist researchers in development of mass production facilities in the future. Several researchers have focused on determining what yields of biodiesel can be obtained for a particular feedstock and catalyst used. Though this is necessary, it is equally important that the process be optimized and cost reduced to the minimum so as to make biodiesel competitive to petro-diesel, further sought to use statistical methods to determine the optimum parameters for the production of bio-diesel. Conventionally, the optimization study for biodiesel synthesis process is performed with the variation of one variable (factor) at a time and the response is a function of a single variable (one-factor-at-a-time OFAT) technique.

7.8.3 WHAT ABOUT OFAT?

The OFAT technique consists of selecting a starting point, or base line set of levels, for each factor, and then successively varying each factor over its range with the other factors held constant at the baseline level.

Actually, the OFAT technique is an inefficient and inadequate method of experimentation for identifying significant factors. OFAT requires more experiments than the factorial design to test the same number of factors. It cannot reveal any possible interactions between the factors and does not depict the complete effect of the variables on the process and miss optimal settings of factors. Moreover, OFAT technique is laborious, time consuming and exorbitant in cost. However, OFAT can be appropriate, whenever, the aim is to develop a functional relationship between a factor and the response, on a condition that the interactions between that factor and others are unimportant. In other words; when there is only one factor of interest or importance.[58]

7.8.4   DESIGN OF EXPERIMENTS (DOE) TECHNIQUE

Experiments often involve several factors, and usually the objective of the experiments is to determine the influence of these factors on the response.

The DOE or the experimental design, is a powerful statistics-based approach to design experiments in order to achieve a predictive knowledge of a complex, multi-variable process with the fewest acceptable trials. It enables designers to determine simultaneously the individual and interactive effects of many factors that could affect the output results in any design. Thus, DOE provides a full insight of interaction between design elements: it helps to pin point the sensitive parts and sensitive areas in designs that cause problems in the yield. So, designers would be then able to fix these problems to achieve robust performance, and produce higher yield designs prior going into production.[59] The essence of DOE, to plan informative experiments, to analyze the resulting data to get a good model, and from the model, creates meaningful maps of the system. Montgomery[60] has noted that, by using experimental design, engineers can determine which subset of the process variables has greatest influence on the process performance. The results of such experiments can lead to improved process yield, and reduced design, development time, and operation cost. Also Box and Wilson applied the idea of DOE to industrial experiments and developed the response surface methodology RSM.[58]

7.8.4.1   THE BENEFITS OF DOE

The DOE provides more information from fewer experiments, helps in predictions of future results and optimization of response variable. It reduces the number of experiment and saves time. Thus, it consequently reduces the operational cost. Moreover, it helps in choosing between alternatives and controlling the process. Finally, it studies the effect of multiple variables simultaneously.

7.8.4.2 A GLOSSARY FOR DOE TERMINOLOGY

- Coding factor levels: Transforming the scale of measurement for a factor so that the high value becomes (+1) and the low value becomes (−1).
- Design matrix: a matrix description of an experiment that is useful for constructing and analyzing experiments.
- Design of experiments: a plan for collecting the sample.
- Design space: a range of values over which factors are to be varied.
- Effect: how changing the settings of a factor change the response. The effect of a single factor is called a main effect.
- Experiment: a process of collecting sample data.
- Experimental unit: an object upon which the response variable is measured.
- Factorial design: two or more independent variables are manipulated in a single experiment, that is, experimental design with more than one independent variable.
- Factors: independent variables, or controllable variables or in other words, the input variables, that can be changed, that is, the variables that potentially affect the response.
- Interactions: that occurs when the effect of one factor on a response depends on the level of another factor(s).
- Level: an assumed value for a factor in an experiment, that is, number of different values a variable can assume.
- Model: a mathematical relationship which relates changes in a given response to changes in one or more factors.
- Nuisance variables: other variables which influence the response variable but are not of interest.
- Random error: error that occurs due to natural variation in the process (Random error is also called experimental error).
- Randomization: random order in which the runs of the experiment are to be performed.
- Replication: completely re-run experiment with the same input levels to obtain a more precise result. It is used to determine the impact of measurement error in the system.

- Response surface design: modeling and analysis of a process in which a response of interest is influenced by several variables and the objective is to optimize this response.
- Response variable: a dependent variable measured in experiment, that is, measured output of a process (objective function).
- Sample size: the overall number of experiments needed to be performed.
- Sample space: a set of experiments to be performed.
- Treatment or run: a particular combination of levels of all factors in an experiment.

7.8.4.3 CONSIDERATIONS WHEN DESIGNING AN EXPERIMENT

An experiment is a process or study that results in the collection of data. Experimental design is the process of planning a study to meet specified objectives. Thus, when designing an experiment there are some general considerations that should be followed, which can be summarized as follows.

- The objectives of the experiment should be clearly defined.
- Appropriate response and explanatory (i.e., controllable) variables must be determined and nuisance variables should be identified.
- Set of levels (i.e., the range of variability) for each explanatory variable (factor), that are to be controlled during the experiment, must be identified.
- Actual trials to take place must be selected.
- Performing the trials as planned.
- Statistical analysis of the experiment should be planned in detail to meet the objectives of the experiment and analyze the results.
- The experimental design should be economical.

7.8.4.4 THE MAJOR APPROACHES FOR DOE

7.8.4.4.1 Latin Square Designs

The Latin square design gets its name from the fact that it can be elucidated as a square matrix with Latin letters representing the factor's levels of the treatments. The term "Latin Square" was first used by Euler in 1782 Del Vecchio.[61] Latin square designs are used when the factors of interest have more than two levels and there are no interactions between factors (i.e., interested, mostly, in estimating main effects).

Actually, the idea of Latin square design is applicable for any factors $k > 3$; however, the technique is known with different names:

If $k = 3$: Latin square
If $k = 4$: Graeco–Latin square
If $k = 5$: Hyper–Graeco–Latin square

Although this technique is still applicable, but it does not give a particular name for $k > 5$ Cavazzuti.[58] A Latin square is a simple method of selecting a small set of trials in a situation where there are several factors and levels, without confounding the importance of the primary factor. It focuses on one particular factor referred to the primary factor (design factor, control factor, or treatment factor) the other factors are called the nuisance factors (or disturbance factors). An experiment with k controllable factors: X_1, X_2, ..., X_k, one of them X_k is of primary importance. The number of levels of each factor will be l_1, l_2, ..., l_k and the sample size is $N = l_1, l_2, ..., l_k$. Figure 7.2 graphically represents a Latin square design for three factors ($k = 3$) where X_1 and X_2 are nuisance factors, X_3 primary factor, $l_1 = l_2 = l_3 = l = 3$, that is, their levels are equal. The sample size is $N = l^2 = 3^2$.

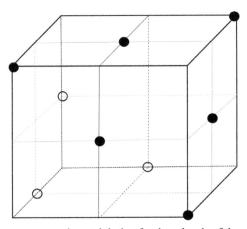

FIGURE 7.2   Latin square experimental design for three levels of three factors

7.8.4.4.2   *Full Factorial Design*

Experiments often involve several factors, and usually the objective of the experimenter is to determine the influence that these factors have on the response. The valuable approach to deal with several factors is to conduct a factorial experiment. In other words, this is an experimental strategy in which factors are varied together, instead of one factor at a time. It is worth to mention that in a factorial design, in each complete trial or replicate of an experiment, all possible combination of the factors' levels is investigated.

The major factors that affect the outcome of an experiment can be identified from; a few exploratory experiments, past experience or it can be based on a theory or hypothesis. Also, the levels at which these factors should be varied, must be determined considering a simple design, where, the total number of experiments (N) that are needed to be done is:

$$N = \prod_{i=1}^{k} l_i \tag{7.1}$$

where, k is the number of factors and the number of levels for the ith factor is l_i.

In a multivariable experiment, with $\{k\}$ factors and number of levels $\{l\}$ per factor, it demands $\{l^k\}$ number of measurements for complete understanding of the process, where the number of levels is the same for each factor. Figure 7.3 shows the graphical representations for 2^2, 2^3, and 3^3, full factorial design (FFD) not counting replications or center points. Advantage of FFD is that they make a very efficient use of the data and do not confound the effects of the factors, that is, produce efficient experiments where each

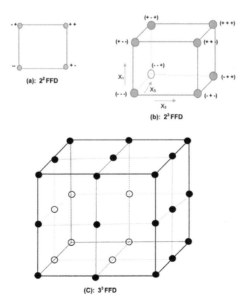

FIGURE 7.3   Geometric view of some examples for the full factorial experimental designs.

observation supplies information about all of the factors so that it is possible to evaluate the main and the interaction effects clearly. As an example Fig. 7.3b represents 2^3 FFD with three factors and two levels per factor, where the three factors are X_1, X_2, and X_3 and the arrows show the direction of increase of the factors. There are eight different ways of combining the two levels; high (+) and low (−) settings of each factor. These are shown at the corners of the diagram (cube).

7.8.4.4.3 Fractional Factorial Design

For large number of factors, and for number of levels per factor ≥ 2, a large number of experimental runs will be obtained on applying the FFD, since; one difficulty with FFD is that the number of combinations (i.e., sample size) increases exponentially with the number of factors manipulated and number of their levels. This will be too expensive and time consuming. It is important to obtain maximum realistic information with the minimum numbers of well-designed experiments. Since all researchers think about reducing the number of experiments to minimize the resources, that is, the equipment, materials, manpower, and time. However, the question is; "can the researcher reduce the number of experiments and yet get an adequate representation of the relationship between the outcome of the experiment and the variation of the factors?" the answer in general is "yes". Thus, through replacing the FFD with a fractional factorial design (FrFD).

In statistics, FrFD is an experimental design consists of a carefully chosen subset (i.e., fraction) of the experimental runs of a FFD, which helps to reduce the number of experiments.

FrFD is expressed using the following notation:

$$l^{(k-p)} \tag{7.2}$$

where l is the number of levels of each factor (k) indicates the number of factors investigated, and p describes the size of the fraction of the full factorial to be eliminated.

For example; $2^{(k-p)}$ design allows analyzing {k} factors of {2} levels per factor with only $2^{(k-p)}$ experiments. A design with p such generators is a {$1/l^p$} fraction of the fractional design.

$2^{(k-1)}$ design requires only {half} of the experiments.

$2^{(k-2)}$ design requires only {one quarter} of the experiments.

If there are three factors in the experiment (X_1, X_2, and X_3) and each has two levels; high (+) and low (−). The geometry of the experimental design for a full factorial experiment (2^3) requires eight runs (Fig. 7.3b), and a one half FFD of experiments $2^{(3-1)}$ requires four runs (Fig. 7.4), that is, in the half factorial design we would have to choose half the number of experiments corresponding to four of the eight corners of the cube.

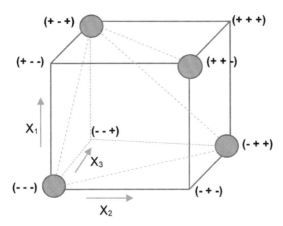

FIGURE 7.4   Geometric view of one-half factorial design 2^{3-1}.

7.8.4.4.4   Plackett–Burman Design

The Plackett–Burman design (PBD) is a very efficient screening design when only the main effects are of interest. It was early described by R. L. Plackett and J. P. Burman in 1946[58] to screen large numbers of factors in an efficient manner, that is, with the least number of observations necessary. For example, we can design and analyze an experiment with 127 factors and only 128 runs (observations) are still, able to estimate the main effects for each factor, and thus can identify quickly which ones are important to improve the process under study. The PBD has been developed for the construction of very economical designs with the run number a multiple of four (rather than a power of 2).[62]

7.8.4.4.5   Box–Behnken Design

The Box–Behnken design (BBD) was devised by George E. P. Box and Donald Behnken in 1960,[58] to achieve a design where each factor is placed at one of three equally spaced, that usually coded as (−1, 0, and +1) where, at least three levels for each factor are needed. Taking into consideration that;

the design should be sufficient to fit a quadratic model, and the ratio of the number of experimental points to the number of coefficients in the quadratic model should be reasonable. The geometry of BBD is shown in Figure 7.5.; the BBD does not need as many central points because points on the outside are closer to the middle. The BBD is desirable and safer, since there are more points in the middle range that are not as extreme as all of the factors.[63]

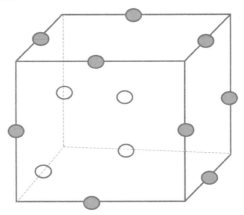

FIGURE 7.5   Geometric view of Box–Behnken design for three factors.

7.8.4.4.6   Central Composite Design

The Box–Wilson or central composite design, which commonly called the "central composite design (CCD)", is a design that contains an imbedded factorial or fractional factorial design with center points that is augmented with a group of "star points" which allows estimation of curvature. The addition of center points to the 2^k design based on the idea of replicating some of the runs in a factorial design. Runs at the center provide an estimate of error and allow the experimenter to distinguish between two possible models of first and second orders. The CCD is a very effective design for fitting a second-order response surface model.

Thus, the CCD has different structure from that of BBD, where in the former; a ball is used, in which all of the corner points lay on the surface. While in the BBD, the ball is located inside the box defined by a wire frame that is composed of the edges of the box. Also, the number of observations for BBD is lower than of CCD.[63] Based on the CCD matrix, the studied independent factors would vary within a defined range to reach the optimum condition for the response.[64] In order to reduce the effects of any uncontrolled factor on the response, the sequence of the experiments in the designed matrix can be randomized.[60]

There are three types of CCD; central composite circumscribed (CCC), central composite faced (CCF), and central composite inscribed (CCI), as pictured in Figure 7.6 One CCD consists of; cube points at the corners of a unit cube, star points along the axes or outside the cube and center point at the origin. The distance from the center of the design space to a star point is $\pm\,\alpha$ with $|\alpha| > 1$. The value of α depends on the number of experimental runs and the numbers of involved factors. For k factors, $2k$ star points and one central point are added to the 2^k full factorial, bringing the sample size for the CDD to $2^k + 2k + 1$, that is, CCD is a 2^k full factorial to which the central point and the star points are added. The CCC is the original form of the CCD. The CCC and CCI methods require five levels for each factor. While, the method CCF requires three levels for each factor. Both the CCC and CCI are rotatable designs, but CCF is not rotatable design.[57]

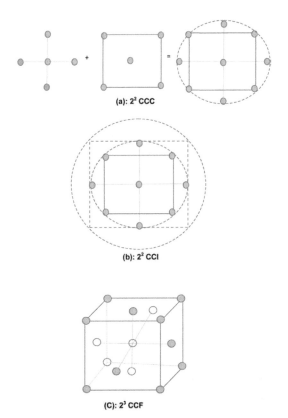

(a): 2^2 CCC

(b): 2^2 CCI

(C): 2^3 CCF

FIGURE 7.6 Examples for three types of CCD.

7.8.4.4.7 Determining α in CCD

To maintain rotatability, the value of α depends on the number of experimental runs in the factorial portion of the central composite design.

$$\alpha = [\text{number of factorial runs}]^{1/4} \qquad (7.3)$$

If the factorial is a full factorial, then

$$\alpha = [\ 2^k\ ]^{1/4} \qquad (7.4)$$

that is, the distance of the star points from the center point is given by $\alpha = 2^{k/4}$ for two factors, $\alpha = 1.414$

7.8.4.4.8 Taguchi Design

Dr. Taguchi's standardized version of DOE is popularly known as the Taguchi method or Taguchi approach and it was introduced in the USA in the early 1980s.[65] It is one of the most effective quality building tools used by engineers in all types of manufacturing activities Buasri et al.[59] Dr. Taguchi has developed his method which is based on "orthogonal array (OA)" experiments that serve as objective functions for optimization.[61]

The general steps involved in Taguchi method are as follows:

- Define the process objective.
- Determine the design factors affecting the process.
- Specify the number of levels that the factors should be varied at.
- Typically, the number of levels for all factors in the experimental design is chosen to be the same to aid the selection of the proper (OA).
- Create OA$_s$ for the factor design indicating the number and conditions for each experiment.
- Conduct the experiments to collect data.
- Perform a complete data analysis to determine the effect of different factors on the performance measurement.

Taguchi method is based on mixed levels of highly fractional factorial designs and other orthogonal arrays (OA) to perform the fewest number of experiments. It distinguishes between control variables, which are the factors that can be controlled, and noise variables, which are the factors that cannot be controlled except during experiments in the lab. Taguchi design provides information about the interaction between the controllable and noise variables.

The DOE using Taguchi approach can economically satisfy the needs of problem solving and process design optimization projects. By applying this technique, engineers, scientists, and researchers can significantly reduce the time required for experimental investigations. It helps in examining the effect of different process parameters on the mean and variance of performance characteristic, which determines the proper functioning of the process. It is very effective with nominal number of parameters (3–50) with few interactions between them and a very few significantly contributing ones.

7.8.4.4.9 *Optimal Design*

Optimal design is a good DOE technique used whenever the classical orthogonal methods may fail due to the presence of constraints on the design space. It is a response surface-oriented technique whose output depends on the RSM technique.[57] There are different optimal design methods: A—optimal, D—optimal, E—optimal, G—optimal, and I—optimal. The most popular design is the I—optimal which aims at the minimization of the normalized average, or integrated prediction variance. However, the D-optimal design (DOD) method is used for multi-factor experiments with both quantitative and qualitative factors, where, the factors can have a mixed number of levels. In this method optimization algorithms are usually applied to the search procedure. The procedure is stopped after a certain number of iterations, and the best solution found is taken as the optimal. The output is a set of samples spread through the whole design space. The researcher can apply the DOD when there is a limited budget and cannot run a completely replicated factorial design. Thus, the DOD reduces the cost of experimentations by allowing statistical models to be estimated with fewer numbers of experimental runs.

The choice of an experimental design depends on the objectives of the experiment and the number of factors to be investigated. Once the important factors have been determined and the screening experiments were performed, the next step is to produce a predictive model to detect interactions among the factors.

After the completion of an experimental plan, one must analyze the data to find out which factors influence the responses. This is done by fitting a polynomial model equation to the data.

7.8.5 REGRESSION MODELS

Regression is a simple statistical tool that can be used to model the dependence of a variable on one or more explanatory variables. Regression is a method by which a functional relationship in the real world would be described by a mathematical model, which can like all models, be used to explore or predict the relationship. This functional relationship may then be formally stated as an equation, with associated statistical values that describe how well this equation fits the data.

Considering the following regression model equations:

$$Y = \beta_o + \varepsilon \tag{7.5}$$

Equation (7.5) is a null model, where, the response Y can be predicted using no explanatory variables, that is, model contains, only, a constant (intercept) term.

$$Y = \beta_o + \beta_1 x + \varepsilon \tag{7.6}$$

The first-order model (eq 7.6) with one explanatory (i.e., independent) variable, is a simple linear regression model that contains an intercept and linear term for one predictor.

$$Y = \beta_0 + \sum_{i=1}^{k} \beta_i x_i + \varepsilon \tag{7.7}$$

While the first-order model with k independent variables (eq 7.7), is a multiple linear regression model (LRM) that contains an intercept term and linear terms for each predictor.

$$Y = \beta_0 + \sum_{i=1}^{k} \beta_i x_i + \sum_{i=1}^{k} \sum_{j=i}^{k} \beta_{ij} x_i x_j + \varepsilon \tag{7.8}$$

The interactions model (eq 7.8) contains an intercept term, linear terms, and all products of pairs of distinct predictors, but no squared terms are included.

$$Y = \beta_0 + \sum_{i=1}^{k} \beta_i x_i + \sum_{i=1}^{k} \beta_{ii} x_i^2 + \sum_{i=1}^{k} \sum_{j>i}^{k} \beta_{ij} x_i x_j + \varepsilon \tag{7.9}$$

The general form of the second-order regression model, or "quadratic polynomial model" with k explanatory variables is represented by (eq 7.9). It contains an intercept term, linear, and squared terms together with interactions terms. The quadratic models are almost always sufficient for industrial applications, these models can be typically used in response surface DOE's with suspected curvature and do not include the three-way interaction terms.

$$Y = \beta_0 + \sum_{i=1}^{k} \beta_i x_i + \sum_{i=1}^{k} \beta_{ii} x_i^2 + \sum_{i=1}^{k} \sum_{j=1}^{k} \beta_{ij} x_i x_j$$
$$+ \sum_{i=1}^{k} \sum_{p=1}^{k} \beta_{ijp} x_i x_j x_p + \varepsilon$$

$$i \neq j \qquad\qquad i \neq j \neq p$$

(7.10)

A full model (eq 7.10) could include many cross-product (or interaction) terms and involves squared x's, where, β_o is the overall mean response (i.e., the offset term), β_i is the main effect for factors ($i = 1,2,\dots k$, that is, the linear coefficients), β_{ii} is the quadratic coefficients, β_{ij} is the two-way interaction terms of all pairs of factors (i.e., the interaction coefficients), β_{ijp} is the three-way interaction terms of the three combinations (i^{th}, j^{th} and p^{th} factors), x_i, x_j, and x_p are the independent variables and ε is an error term. Planning the statistical analysis is an integral part of planning the experiment. The aim of the data analysis is to estimate numerical values of the model parameters. The least square methods are used to get the estimation of all β_s' regression coefficients. These values will indicate how factors influence the response. Such regression coefficients are easy to overview when plotted in Pareto chart.[60] For $k = 3$, the three terms with single x's are the main effects terms, also there are $k(k-1)/2$ two way interaction terms and 1 three-way interaction term which is often omitted, for simplicity.

The polynomial regression models are regression models that are second or higher order models, containing squared or higher powers of the predictor variable. In general, one could keep trying higher order polynomials, but that is not advised. It is theoretically possible to exactly fit any data set with N points with an $(N-1)^{th}$ degree polynomial. However, if this is really attempted, the researcher will get a widely oscillating function that does nothing but fit the observed data. Another drawback of higher-order models is that; they tend to become difficult to interpret. Due to all of the aforementioned drawbacks, it is very rare to use models with higher than second-order terms. Also the three-way interaction terms are not needed. However, in general these terms (three-way interaction terms) are not needed and most DOE software defaults to leaving them out of the model.

7.8.5.1   MODEL VALIDATION AND GOODNESS OF FIT

Once a regression model has been constructed, it is important to confirm the goodness of fit of the model and the statistical significance of the estimated parameters. It is important to select among the available regression models by using key statistical indicators. The goodness of fit measures is of two types: graphical and numerical. The residuals and prediction bounds are graphical measures, while the goodness of fit statistics and confidence bounds are numerical measures.

7.8.5.1.1 Residual Plot

This is simply the error, ε_i, in the model, plotted versus the observed (experimental) value of the independent variable. The error (residual) is calculated according the following equation:

$$\varepsilon_i = (y_{i\,exp} - y_{i\,cal})$$ (7.11)

Where y_{exp} is the observed (experimental) value of the response variable and y_{cal} is the fitted (predicted or calculated) value of the response variable, that is, the error (residual) is the difference between the observed and predicted value of the dependent variable. It is always very useful to examine residual plots of regressions to determine if there are any obvious trends, as the errors should be random (i.e., showing random distribution around zero). In other words, if the residuals appear to behave randomly, it suggests that the model fits the data well. However, if the residuals display a systematic pattern, it is a clear sign that the model fits the data poorly.

7.8.5.1.2 Prediction Bounds

The prediction bounds are often described as confidence bounds. They are calculated for the fitted function, that is, calculating confidence interval for a predicted response. They are displayed graphically. Each graph contains three curves: the fitted curve, the lower confidence bounds, and the upper confidence bounds. The width of the interval indicates how uncertain we are about the fitted curve.

7.8.5.1.3 Confidence Intervals

Statistician stresses on the importance of using confidence intervals (CI). It should be used to express the results of statistical tests because it conveys more information than the p-values alone. The CI sets the boundaries of a confidence interval, this conventionally set at 95%. In some studies, wider (e.g., 90%) or narrower (e.g., 99%) confidence intervals would be required.

A major indicator is the 95% confidence interval of each parameter. That is, there is a 95% probability that the actual values of the parameters are within the intervals. When the confidence interval is very large relative to the parameter, this suggests that the parameter may not be important in the correlation and perhaps should be set to zero.

7.8.5.1.4   Correlation Coefficient "R^2"

The R^2 is also called the square of the multiple correlation coefficients or the coefficient of multiple determinations

$$R^2 = 1 - \left(\frac{SSE}{SST} \right) \tag{7.12}$$

The R^2 ranges from zero to one. The higher the R^2 value is (i.e., closer to one), the stronger the indication of an existing relationship between the measured and predicted values. The R^2 measures how successful the fit is in explaining the variation of the data. For a good fit model, it is recommendable that R^2 should not be less than 0.8.

In general, one could keep trying higher order polynomials, but that is not advised. Even though adding additional terms will result in a higher R^2 and therefore a better fitting model there is always a danger upon over-fitting the model.

7.8.5.1.5 Adjusted R-Square "R^2_{adj}"

The adjusted coefficient of multiple determinations R^2_{adj} is an R^2 statistic adjusted for the number of parameters (k) in the equation and the number of data observations (N). It is a more conservative estimate of the percent of variance explained, especially when the sample size is small compared to the number of parameters (coefficients). The R^2_{adj} statistic is used to help determine the best fit. It is generally the best indicator of the fit quality when we add additional coefficients to the model. It can take on any value less than or equal one, with a value closer to one indicating a better fit. It is computed using the following formula:

$$R_{adj}^2 = 1 - \left(\frac{(N-1)}{(N-k)} (1 - R^2) \right) \tag{7.13}$$

The gap between the R^2 and R^2_{adj} tends to increase as non-significant independent variables are added to the regression model. As N increases, the difference between the R^2 and R^2_{adj} becomes less. The adjusted R-squared is always smaller than R-squared. It is possible that adjusted R-squared has a negative value, that occurs if the model is too complex for the sample size and/or the independent variable has too little predictive value, and some software just reports that the adjusted R-squared would be zero in that case.

7.8.5.1.6 Root Mean Squared Error "RMSE"

This statistic is also known as "the fit standard error" and "the standard error of the regression." It is a square root of the mean square error or the residual mean square. A "RMSE" value closer to zero indicates a better fit. Where "RMSE" indicates estimation errors, so with lower value of "RMSE", there are fewer prediction errors, that is, the optimal model should have the lowest "RMSE."

7.8.5.1.7 Sum of Squares due to Error "SSE"

The "SSE" statistic is also called the summed square of residuals. It is the least squares error of the fit, "SSE" statistically measures the total deviation of the response value from the fit. It is used to help to determine the best fit. The SSE with a value closer to zero indicating a better fit.

$$SSE = \sum_{i=1}^{N} (y_{i\,exp} - y_{i\,cal})^2 \tag{7.14}$$

7.8.5.1.8 The Variance

Often the variance is used to compare the goodness of fit for various models. The overall estimate of the model variance can be calculated by the following formula:

$$\delta^2 = \frac{\sum_{i=1}^{N} (y_{i\,exp} - y_{i\,cal})^2}{df} \tag{7.15}$$

where df $= (N-k)$, N is the data points and k is the number of model parameters.

7.8.5.2 ANALYSIS OF VARIANCE

Analysis of variance (ANOVA) is a set of statistical methods used mainly to compare the means of two or more samples (treatments). It can also be used in regression to establish the overall significance of the regression model being fitted and to establish the statistical significance of the model parameters (coefficients) at 95% confidence level. That is, the purpose of ANOVA is to determine whether the factor has a significant effect on the variable being measured or not. The computations for the ANOVA test are often summarized in a tabular form, known as the ANOVA table (Table 7.2).

TABLE 7.2 A Summary for the One-Way ANOVA Table to Compare Means of Several Groups.

Source of variations	Sum of squares (SS)	df	Mean square (MS) = SS/df	F	p-value
Treatment (between groups)	SST_r	I-1	$MST_r = SST_r/(I-1)$	MST_r/MSE	P
Error (within groups)	SSE	I(J-1)	$MSE = SSE/[I(J-1)]$		
Total	SST	IJ-1			

I is the number of treatments being compared and J is the sample size for the ith treatment.

The standard ANOVA table has six columns; the source of the variability, the sum of squares (SS) due to each source, the degrees of freedom (df) associated with each source, the mean squares (MS) for each source, which is the ratio (SS/df), the F-statistic, which is the ratio of the mean squares and finally the p-value (the associated significance probability).

There are some statistical calculations that are required for constructing the ANOVA table; the treatment sum of squares (SST_r), the sum of squares error (SSE), the total sum of squares (SST) which can be calculated using (eq 7.16), the mean square for treatment (between samples variations, MST_r), and finally the mean square error (within sample variation, MSE).

$$SST = \left(\sum_{i=1}^{N} y_i^2\right) - \frac{\left(\sum_{i=1}^{N} y_i\right)^2}{N} \tag{7.16}$$

The values for mean squares are determined by dividing the mean square by the corresponding degree of freedom:

$$\text{Mean Square Model} = SS_{model}/df_{model} \tag{7.17}$$

$$\text{Mean Square Error} = SS_{error}/df_{error} \qquad (7.18)$$

where the SS_{model} describes the variation within the fitted values of response variable (y). It is equal to the sum of the squared difference between each fitted value of y and the mean of y. While the SS_{error} describes the variation of experimental value of y from its fitted value.

For testing the overall model, the ANOVA table also included an F-statistic ("F-value" and "Prob(F)"). The F-test is a test to determine the overall significance of the regression model, and not just of one individual coefficient. The F value is the ratio of the mean regression sum of squares divided by the mean error sum of squares. For checking the significance of individual parameters, the t-tests is used. The t-statistic is computed by dividing the estimated value of the parameter by its standard error. It is a measure of the likelihood that the actual value of the parameter is not zero. The larger the absolute value of t, the less likely that the actual value of the parameter could be zero. The Prob(t) value (i.e., the p-value) is the probability of obtaining the estimated value of the regression coefficient. The smaller the value of the Prob(t), the more significant the coefficient. If Prob(t), for example, is 0.001 this indicates there is only one chance in 1000 that the parameter could be zero. If Prob(t) is 0.92 this indicates there is a 92% probability that the actual value of the parameter could be zero; this implies that the term of the regression equation containing the parameter can be eliminated without significantly affecting the accuracy of the regression. Most statisticians refer to statistically significant as $p < 0.05$ and statistically highly significant as $p < 0.001$ (less than one in thousand chance of being wrong).

The reason that ANOVA table splits into rows for model, error, and total is that to examine how much error the researcher would have when applying the predictive equation and to determine how much error has disappeared by using the predictive equation. There are different types of ANOVA; one-way ANOVA, two-way ANOVA and N-way ANOVA, that can reflect the different experimental designs and situations for which they have been developed.

7.8.6 BOX-PLOT

The box-plots are graphical tools to visualize key statistical measures. The box-plot, in its simplest form, presents five sample statistics: the minimum, the lower quartile, the median, the upper quartile and the maximum. In other words, box-plot displays the distribution of data around their medians. The following diagram represented in Fig. 7.7 shows a dot-plot of a sample of 20 observations together with a box-plot of the same data. The box of the plot is a rectangle which encloses the middle half of the sample, with an end at each quartile. The length of the box is thus the interquartile range of the sample. A line is drawn across the box at the sample median. The whiskers sprout from the two ends of the box until they reach the sample maximum and minimum.[66]

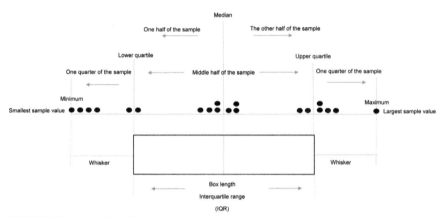

FIGURE 7.7 Box-plot of sample data.

Although box-plots can be drawn in any orientation, most statistical packages seem to produce them vertically by default, rather than horizontally. The diagrams (Figs. 7.8 and 7.9) show a variety of different box-plot shapes and how the shape of a box-plot encodes statistical measures. Figure 7.8 represents a simple Box–Whisker plot, while Fig. 7.9 represents the notched Box–Whisker plot. The whiskers extend to the minimum and maximum data values. There is a much more that can be read from a box-plot than might be surmised from the simplistic method of its construction, particularly when the box-plots of several samples are lined up alongside one another (parallel box-plots).

The widths of the boxes are proportional to the sample size. The notches (i.e., narrowing of the box around the median) are useful in offering a guide to significance of difference of medians. If the notches of two boxes do not overlap, this offers evidence of a statistically significant difference between the medians. The width of the notches is proportional to the interquartile range of the sample and inversely proportional to the square root of the size of the sample. The position of the box in its whiskers and the position of the line in the box also tell us whether the sample is symmetric or skewed,

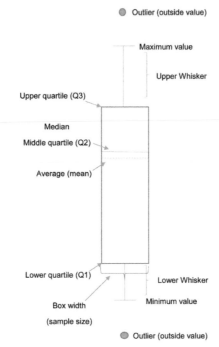

FIGURE 7.8   Simple Box–Whisker plot.

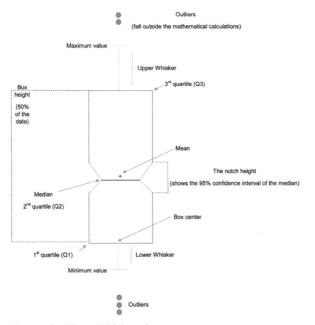

FIGURE 7.9   The notched Box–Whisker plot.

either to the right or left. Also if the median is not centered in the box, that is an indication of skewness. Outliers are data with values beyond the ends of the whiskers. Outlier in the data may be the result of a data entry error, a poor measurement or a change in the system that generated the data. The box portion of the box-whisker plot includes 50% of the data.[67]

7.8.7   PARETO CHARTS

A Pareto chart, also called a Pareto distribution diagram, was originally written by Italian economist "Vilfredo Pareto" in 1896.[58] It is a vertical bar graph in which values are plotted, in decreasing order of relative frequency from left to right. Pareto charts are extremely useful for analyzing what problems need attention, first because the taller bars on the chart, clearly illustrate which variables have the greatest cumulative effect on a given system. A statistical analysis provides estimates of how strongly each experimental factor affects process performance. Pareto chart is an interesting way to compare the relative importance of the estimated effects of the factors, as it allows one to detect the most important factor and interaction effects in a process or a design optimization study. In some cases, it displays the absolute values of the effects and draws a reference line on the chart so any effect that extends over this reference line is potentially important.[68]

7.8.8   OPTIMIZATION

Optimization involves substitution of an output value for the response variable and solving for the associated predictor variable values. However, the objective is to predict the response values for all possible combinations of factors within the experimental region and identify an optimal experimental point. Process optimization can be performed by the application of the design of experiment DOE and LINGO, contour plot or response surface methodology RSM.

7.8.8.1   LINGO

LINGO is a comprehensive tool designed to solve linear, nonlinear, quadratic, quadratically constrained, and integer optimization models. LINGO software is used to find out the optimum values of the factors for enhancing the process and achieve the optimum response variable (http://www.Lindo.com)[69].

7.8.8.2   CONTOUR PLOT

A contour plot is a graph that can be used to explore the potential relationship between three variables (i.e., the two independent variables and the response variable). It displays a 3-D relationship in two dimensions, with independent variables plotted on the x- and y-scales and response values represented by the contours. In other words, a contour plot is like a topographical map in which x-, y-, and z-values are plotted instead of longitude, latitude, and elevation. Where, the contour lines corresponding to different levels will not cross each other, while the contour lines of the same level may appear to intersect.[70]

7.8.8.3   RESPONSE SURFACE METHODOLOGY (RSM)

The RSM is a collection of mathematical and statistical techniques that are useful for modeling and analysis of a process in which a response of interest is influenced by several variables and the objective is to optimize this response in that complex process. Accordingly, RSM is a 3-D response surface plotted on the basis of the predicted model equation to investigate the interaction among the variables and to determine the optimum condition (range) of each factor. The response surface of the response variable is mapped out and the process is moved as close to the optimum as possible, taking into account all constraints. Suppose that the outputs are defects or yield, and the goal is to minimize defects and maximize the yield. If these optimal points are in the interior of the region in which the experiment is to be conducted, we need a mathematical model that can represent curvature so that it has a local optimum. Response surface models may have quadratic and possibly cubic terms to account for curvature. The simplest such model has the quadratic form represented by eq 7.9. The RSM helps on understanding the pattern in which the dependent variables are affected by the corresponding changes in the independent variables, improves the product of predicted property values, and predicts the interactive effects of two or more factors and the effects caused by the collective contributions of the measured response.[71]

The interactive response surface methodology is another tool for interactively investigating simultaneous one-dimensional contours of multidimensional response surface models, where, a sequence of plots is displayed, each showing a contour of the response surface against a single predictor, while all other predictors held fixed.

It is very important to study the effect of the transesterification reaction conditions on its rate and the biodiesel yield. The conventional experimental optimization has been widely used to study the effects of different parameters on transesterification reaction and to determine the optimal conditions to maximize the biodiesel yield.[72–75] However, this is not economical and time consuming. Thus, the statistical methods are more powerful, as they are used as tools to assess the effects of two or more independent factors on the dependent variables.[76–79]

The RSM usually accompanies factorial designs; the FFD, the FrFD, the CCD, the Taguchi design, the BBD or the DOD, which can be applied to optimize the transesterification reaction and study the effects of different parameters on the reaction rate and biodiesel yield. That reduces the number of experiments required to get sufficient data for a statistically relevant result.

Boy et al.[76] used RSM based on the CCD of experiments to optimize the transesterification reaction of WCO by a new catalyst system prepared from high basic boiler ash as an agricultural waste and mixed CaO prepared by calcination of mud crab shells and cockle shells at 900°C for 2 h, with an aim to decrease the reaction time of CaO-catalyzed reaction and increase the conversion efficiency for boiler ash-catalyzed reaction. The conversion reached approximately 99% with 3 wt.% boiler ash, reaction time of 30 min and mixed CaO of 5 wt.%. Ceclan et al.[40] studied the influence of different parameters (M:O molar ratio, KOH concentration and temperature, at constant mixing speed 600 rpm) on the biodiesel yield from the methanolysis of WCO by applying a full 2^3 factorial design, with a four replication experiments in the average experimental conditions (i.e., the center points). That yielded approximately 85 wt.% biodiesel at 12:1 M: O, 1 wt.% KOH and 65°C. Microbial lipase from *T. lanuginosus* was immobilized onto polyglutaraldehyde-activated olive pomace powder and was used to produce biodiesel using pomace oil and methanol. A five-level four-factor CCD consisting of 16 factorial points, 8 axial points, and 6 central points was employed through 30 experimental runs and RSM was used to optimize the biodiesel production parameters. The optimal conditions for the transesterification have been found to be reaction temperature of 40°C, alcohol:oil molar ratio of 5.3:1, biocatalyst concentration of 5.8% w/w and reaction time of 24 h. Biodiesel yield reached 93.73% by adding water 1% w/w in reaction medium and the predicted pomace oil methyl ester yield was 92.87%, under the optimal conditions. The properties of the produced methyl esters met the specified biodiesel standard (ASTMD6751).[80] RSM based on five-level, four-factor CCD has been used to optimize biodiesel production from *Jatropha curcas*-oil (JCO). A second-order polynomial model equation has been formulated and validated by ANOVA. A biodiesel of 98.3% was obtained with 11.1 M:O ratio using 1% w/w NaOH as catalyst concentration in 110 min reaction time at 55°C reaction temperature. ANOVA results revealed that the catalyst concentration, reaction time, and M:O molar ratio had a significant effect on *Jatropha curcas*-biodiesel (JCB).[81] Cunha Jr. et al.[29] reported the application of RSM based on 3^3 FFD to optimize the ethanolysis of a mixture of chicken and swine waste fats with KOH at 2000 rpm in a 10 L tubular glass reactor. Where the optimum condition; 30°C, 0.96 wt.% catalyst, 7:1 ethanol:fat produced approximately 83% fatty acid ethyl ester. Jazie et al.[47] reported the application of RSM to optimize the transesterification of rapeseed oil by CaO prepared by calcination of eggshells at 900°C for 2 h. That yielded 96% biodiesel at optimum conditions of 9:1 M:O molar ratio, 3 wt.% catalyst concentration and 60°C within 3 h reaction time. Koohikamali et al.[64] reported the application of CCD and RSM to optimize the FAME yield through the methanolysis of sunflower oil using a sodium methoxide catalyst. That led to approximately 100% conversion using the optimal conditions of; 25% M:O (w/w), 0.5 wt.% catalyst and 60 min reaction time at fixed reaction temperature (60°C), below the boiling point of methanol, to avoid the use of a pressure vessel and 300 rpm agitation speed with a little splashing to mix the oil and methanol phases. Aworanti et al.[82] reported the application of RSM based on rotatable central composite design (RCCD) at five levels and three factors to optimize the transesterification of WCO using methanol and CaO. That yielded approximately 94% biodiesel at optimum conditions of 9.14:1 M:O, 3.49 wt.% catalyst concentration, 60.49 min at constant reaction temperature (60°C), and mixing speed (300 rpm). Rezaei et al.[63] reported the application of RSM based on the BBD of experiments to optimize three important parameters; the calcination temperature for CaO preparation from mussel shell, catalyst concentration, and M:O molar ratio, for biodiesel production from soybean oil. That led to 99.68% purity and 94.17% biodiesel yield at 1050°C calcination temperature, 12 wt.% catalyst concentration, and 24:1 M:O molar ratio. Veličković et al.[83] applied RSM based on 3^3 FFD of experiments to optimize the ethanolysis of sunflower oil. That yielded 97.6% fatty acid ethyl ester at optimum conditions of 12:1 ethanol:oil molar ratio, 75°C, and 1.25% NaOH loading. Alhassan et al.[71] reported the application of RSM based on the widely used

RCCD of five-level-three-factor to optimize the transesterification of the low acid *Gossypium arboretum* oil with methanol and KOH, where a maximum biodiesel yield of approximately 95% was obtained at optimum conditions of 0.53 wt.% catalyst loading, 105 min, 60°C at constant M:O molar ratio (6:1), and mixing speed (550 rpm). El-Gendy et al.[54] reported the application of RSM based on face central composite face centered design (CCFCD) of experiments to statistically evaluate and optimize the conditions for maximum production of biodiesel from waste frying sunflower oil using CaO and to study the significance and interaction effects of M:O molar ratio, catalyst concentration, and reaction time on the yield. Based on an estimated quadratic model, the optimum operating conditions; 7.05:1 M:O, 8.21 CaO wt.%, and 1.5 h at 60°C reaction temperature and 300 rpm mixing rate yielded 93 wt.% biodiesel. El-Gendy et al.[84] employed RSM based on D-optimal design of experiments to study the significance and interactive effects of M:O molar ratio, catalyst loading, reaction time, and mixing rate on biodiesel yield, and to optimize the transesterification reaction of WCO with methanol using calcined waste mollusks and crabs shells at 800 and 700°C, respectively. At optimum operating conditions; 6:1 and 12:1 M:O molar ratio, 4.5 and 7.3 wt.% catalyst loading, 82 and 30 min reaction time, and 220 and 214 rpm; 96 and 98 wt.% biodiesel yields were obtained, respectively, at 60°C reaction temperature. The Taguchi L9 (3^4) orthogonal array was used to evaluate the factors (methanol/oil molar ratio, catalyst concentration, reaction temperature, and time) affecting the transesterification reaction of palm oil using CaO prepared from scallop waste shells (that prepared by calcination at 1000°C for 4 h). With the lowest number of experiments, the FAME conversion reached 95.44% at 9:1 M:O, 65°C, 10 wt.% catalyst loading within 3 h.[59] Taguchi experimental design was also used to optimize the transesterification of *Manilkara zapota* (L.) seed oil with methanol in the presence of KOH homogenous catalyst. That yielded 94.83% biodiesel yield at optimum conditions of; 6:1 M:O molar ratio, 50°C, 90 min, and 1 wt.% catalyst concentration.[65] Avramović et al.[85] reported the application of RSM to optimize the ethanolysis of sunflower oil using CaO, where the maximum fatty acid ethyl ester content (98%) was obtained at optimal conditions of; 14:1 M:O molar ratio, 17 wt.% catalyst concentration, 72°C, 440 min. Betiku et al.[86] reported the application of RSM based on RCCD to model and optimize the esterification of high FFA content (4.10 mg KOH/g) *Vitellaria paradoxa* oil, where the acid value was reduced to 1.19 mg KOH/g at optimum conditions; 3.3 O:M molar ratio, 0.15% (v/v) H_2SO_4, 60 min, and 45°C. Then the RSM based on the RCCD was also applied to optimize the transesterification of the pretreated oil using methanol and KOH, where approximately 99.94% biodiesel yield was obtained at 82°C, 2.62 O:M molar ratio, and 0.4% (w/v) catalyst concentration. Mahesh et al.[87] reported the preparation of KBr impregnated CaO catalyst by wet impregnation method, followed by calcination at 900°C. Then, the RSM using CCD of experiments was applied to study the effects of process variables on the biodiesel yield from WCO and optimize the process, where a maximum biodiesel yield of approximately 84% was obtained at 12:1 M:O ratio, 3 wt.% catalyst loading, and 1.8 h reaction time. The RSM based on a simplex centroid design of experiments was applied to optimize and study the effect of different ratios of oil feedstock (waste fish oil, palm oil, and WCO) on the percentage FAME and biodiesel yield, and the fuel properties; kinematic viscosity, oxidation stability, and cold flow properties of the produced biodiesel.[88] The highest biodiesel yield (90%) was obtained using oil feedstock mixture of 33.3 wt.% palm oil and 66.7 wt.% WCO; while the highest FAME% (98.5%) was obtained using 33.3 wt.% waste fish oil and 66.7 wt.% palm oil. The multi-objective optimization proved that single oil feedstock is more advantageous for biodiesel properties; while feedstock mixture of 42.1 wt.% waste fish oil and 57.9 wt.% WCO is preferable if the aim is maximum induction period and minimum completion of melt onset temperature, which represent the resistance time of biodiesel to oxidation and the temperature at which the FAME start to crystallize, respectively. El-Gendy et al.[39] reported the application of RSM based on RCCD of experiments with three levels (coded by −1, 0, and +1) and $\pm \alpha$ of ± 1.82116, to study the significance and interactive effects of five variables: M:O molar ratio, catalyst concentration, operating temperature, reaction time, and mixing rate on KOH catalyzed transesterification reaction of WCO to maximize the biodiesel yield. The advantages of applying the RCCD are that it permits the use of relatively few combinations of variables that cover a wide range of variables and determines and optimizes a complex response function. The total number of experiments was 26 runs. That included 11 factorial points and 10 axial points which allowed the estimation of all the main effects and the determination of all the quadratic terms, respectively, in addition to the five replicates at the center points which provided a check of the adequacy of the model prediction and assess the pure error. That design represented a big advantage since the number of tests was reduced in relation to the normal CCD of experiments, which gave 50 runs of experiments. Based on a multiple regression analysis, a second-order polynomial equation for biodiesel yield was obtained. Nearly, a complete conversion with approximately 99 wt.% biodiesel yield was obtained at optimum operating conditions of 7.54:1 M:O, 0.875 wt.% KOH concentration, 52.7°C, 1.17 h, and 266 rpm. El-Gendy et al.[45] have also employed the RSM based on D-optimal design of experiments to study the significance and interactive effects of the important parameters such as M:O molar ratio, catalyst loading, reaction time, and mixing rate at constant reaction temperature (60°C) on biodiesel yield throughout the transesterification of WCO with methanol using fluorapatite prepared from waste animal bones and Novozym 435. Quadratic model equations were obtained to maximize the response variable (i.e., the biodiesel yield). The prepared green catalyst yielded 96 wt.% biodiesel while Novozym yielded 62 wt.% at optimum operating conditions of 7.35:1 and 6:1 M:O molar ratio, 4.35 and 8.8 wt.% catalyst loading, 91 and 96 min reaction time, and 331 and 394 rpm, respectively. Stamenković et al.[89] applied RSM based on RCCD to optimize the methanolysis of hempseed oil, where approximately a complete conversion occurred at optimum conditions of 6.4:1 M:O molar ratio, 1.2% KOH catalyst concentration, and 43.4°C reaction temperature at constant reaction time and mixing rate of 30 min and 600 rpm, respectively. Yang et al.[90] applied RSM based on CCD to optimize *camelina* biodiesel production by an alkali-catalyzed transesterification process. The effects of temperature, time, molar ratio of methanol/oil, and catalyst concentration on the yield of both biodiesel and FAME were investigated. The *camelina* biodiesel product yield 97% and FAME 98.9% was achieved at the optimal reaction conditions of; 38.7°C reaction temperature, 40 min. reaction time, 7.7 molar ratios of M:O, and 1.5 wt.% catalyst concentrations. In another study, based on D-optimal design of experiments, El-Gendy et al.[91] conducted 20 runs of experiments for three levels of four independent variables—M:O molar ratio, catalyst concentration (wt.%), reaction time (min), and mixing rate (rpm)—to study their effects on the percent biodiesel yield from WCO at constant temperature 60°C, using natural CaO prepared by calcination of waste eggshells at 800°C and chemical CaO. Second-order quadratic model equations were obtained describing the interrelationships between dependent and independent variables to maximize the response variable (biodiesel yield). The three-dimensional response surface graphical diagrams of the obtained regression equations have been also plotted to determine the significant effect of the studied parameters and to study the interactive relationship between the independent variables and percent yield of biodiesel and determine the optimum conditions for maximum biodiesel production. The process optimization based on D-optimal design of experiments was found to be capable and reliable to optimize biodiesel production from WCO using natural and chemical CaO, where the activity of the produced green catalyst was comparable to that of chemical CaO, producing high yield of biodiesel ≈ 91 and 98%, respectively, at 8.57:1 M:O molar ratio, 3.99 wt.% catalyst concentration, 31 min reaction time, and 398.88 rpm mixing rate at 60°C. In order to optimize the transesterification of sunflower oil by CaO derived from palm kernel shell biochar, Kostic et al. [92] applied a RSM based on 3^3 FFD of experiments with five central points. That produced approximately 99.8% FAME within 4 h reaction time and 900 rpm mixing speed at optimum conditions of 65°C reaction temperature, 9:1 M:O molar ratio, and 3 wt.% catalyst concentration. The RSM based on three-level-four-factor D-optimal design was employed to optimize the transesterification of cotton-seed oil with methanol in the presence of *Pichia guilliermondii* lipase immobilized on hydrophobic magnetic particles, where approximately 89% biodiesel yield was obtained at optimum operating conditions of 4.715:1 M:O molar ratio, 38.76°C, 31.3% immobilized lipase, and 10.4% water content.[93] Onukwuli et al.[94] employed a two-level-four-factor CCD of experiments and RSM to optimize the transesterification of cotton seed oil with methanol in the presence of KOH basic homogenous catalyst. That yielded approximately 96% biodiesel at the following optimal conditions: 6:1 M:O molar ratio, 0.6 wt.% catalyst concentration, 55°C, and 60 min at constant mixing speed of 300 rpm. The RSM based on a Box–Behnken experimental design of four variables and three levels was employed to optimize the KOH-transesterification reaction of acid pretreated *Karanja* oil and to study the effect of methanol and ethanol on the biodiesel yield. The methanolysis yielded 91.05% biodiesel at optimum conditions of 10.44:1 M:O molar ratio, 1.22 wt.% catalyst concentration, and at reaction temperature and time of 66.8°C and 90.78 min, respectively. While, on applying the ethanolysis process, the separation of ethyl esters from glycerol were difficult; thus, a lower maximum biodiesel yield of 77.4% was obtained at 8.42:1 ethanol:oil molar ratio, 1.21 wt.% catalyst concentration, 61.3°C, and 120 min.[41]

7.9 CASE STUDIES

7.9.1 CASE STUDY No. 1

Application of Statistical Analysis to Study the Effect of the Type of WCO on the Biodiesel Yield and to Optimize a Heterogeneous Transesterification Reaction

In this study, the transesterification reactions were performed on a laboratory-scale-setup and CaO was used as a catalyst, for its advantageous; the elimination of neutralization step, lack of toxicity, the ability to withstand high temperatures, low cost, availability, ease of separation at the end of the process, recycling and reusability, application in batch and continuous processes, and its performance for biodiesel production is comparable to several homogenous catalysts. Different types of WCO were used; a mixture of different waste frying oil (WFO) samples that were collected from some local restaurants, waste frying sunflower oil (WFSFO), and waste frying corn oil (WFCO) that were collected from domestic wastes. A statistical design of experiments based on DOE strategy was performed to evaluate and investigate the biodiesel production process from those aforementioned types of WCO. The optimum values of the studied predictor variables were selected according to a response surface optimizer.[95]

7.9.1.1   EXPERIMENTAL DESIGN AND STATISTICAL ANALYSIS

To model and optimize the transesterification process using the pure chemical CaO, a series of controlled laboratory designed experiments were performed and a regression model defining the distribution of the response variable (percent biodiesel yield) in terms of four specified explanatory variables (factors) was predicted. These four variables were elucidated and investigated through DOE at three different levels (Table 7.3). The transesterification process of oil with methanol in the presence of heterogeneous catalyst is a three-phase reaction system. Accordingly, the reaction rate might be reduced due to the mass transfer resistance on the boundary between the oil and methanol phases. So, in this study, the mixing rate was set at 300 rpm to assure a proper contact between the three reactants (WCO, methanol, and CaO) without mixture turbulence. The physicochemical characteristics of the used WCO feedstocks are represented in Table 7.4. The total number of experimental runs was 15. The data analysis was performed to fit a regression model equation, specify the regression coefficients, recognize the significant model terms, and finally to determine the optimum values of the factors which would lead to a maximum response (i.e., percent yield of biodiesel).

TABLE 7.3   Factors and Levels of the DOE of Case Studies no. 1 and no. 3.

Factors	Levels		
	1	2	3
Methanol:Oil M:O molar ratio (X_1)	6:1	9:1	12:1
CaO concentration wt.% (X_2)	3	6	9
Reaction time min (X_3)	30	60	120
Type of WCO (X_4)	WFCO	WFSFO	WFO

An empirical regression model was employed for better understanding of the correlations between the explanatory factors and response.

TABLE 7.4 The Physicochemical Characteristics of Some Different Collected WCO Feedstock.

Parameters	Type of oil		
	WFO	**WFSFO**	**WFCO**
Density at 15.56°C, g/cm^3	0.9327	0.9219	0.9206
Viscosity at 40°C, cSt	64	36.6	33.3
Total acid number TAN, mg KOH/g oil	2.15	1.85	1.34
Iodine number mg I_2/100 g oil	111	121	127
Saponification value mg KOH/g oil	201	191	189
Saturated FFA	70.71	13.20	12.4
Unsaturated FFA	29.29	86.8	87.6

Thus based on DOE and obtained experimental data, a second-order-quadratic model (eq 7.19) was found to best estimate the response.

$$Y = 35.92 - 5.3327X_1 + 7.4808X_2 + 3.9312X_3 - 1.9945X_4 - 5.433X_1X_2$$
$$+0.357X_1X_3 - 3.0198X_1X_4 + 0.1748X_2X_3 + 15.119X_2X_4 - 0.3617X_3X_4 \tag{7.19}$$
$$+3.67X_1^2 - 5.864X_2^2 + 0.0028X_3^2 + 4.827X_4^2.$$

7.9.1.2   VALIDATION OF THE ELUCIDATED REGRESSION MODEL

The value of R^2 of the fitted model was found to be 0.9397 which substantiated that 93.79% of the total variation in response could be attributed to the experimental factors. Thus, that ensured the goodness of fit and confirmed the adequacy of the elucidated model. The "F" ratio (815.918) and "p" value (<0.0001) of the quadratic model indicated the high statistical significance of the model at 95% confidence interval. That was also reflected by the good agreement between experimental and predicted values of the response variable that ranged between 74.3 and 91% and between 75.9 and 93%, respectively. The plot of the predicted versus the experimental values of the response variable (Fig. 7.10a) with R^2 (0.9624) which is close to unity indicated that the model gave a good estimate of the response for the system within the studied range. In addition, investigation on residuals (i.e., the difference between the observed and predicted responses) to validate the adequacy of the model was performed. That was examined using the plot of residuals versus predicted response (Fig. 7.10b), where the residual distribution did not follow a certain trend or pattern with respect to the predicted values of response variable, which indicated that the predicted quadratic model adequately represents the biodiesel percent yield over the studied experimental range.

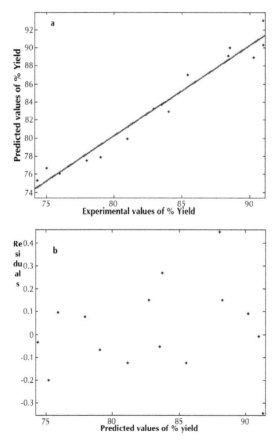

FIGURE 7.10   Model fit profile (a) and residual plot (b) of biodiesel yield for case study no. 1.

7.9.1.3   ANALYSIS OF VARIANCE

ANOVA was applied to establish the statistical significance of the model parameters or coefficients at 95% confidence level. The significance of each coefficient was determined by t- and p-values, where, the larger the magnitude of the t-values and the smaller the p-values, the more significant is the corresponding coefficient.

The obtained results implied that the variables with the largest effects, that is, highly statistically significant, were the linear effects of both M:O molar ratio (t-value–22.78 and p-value 0.001) and catalyst concentration wt.% (t-value 35.05 and p-value 0.001) and the interactive effect of catalyst loading and type of WCO feedstock (t-value 54.07 and p-value < 0.0001). While the linear effects of reaction time and type of WCO feedstock expressed a statistical significant effect on the biodiesel yield with t-values of –12.89 and –10.64 and p-values of 0.0393 and 0.0126, respectively. Also, the interactive effects of M:O with catalyst concentration and M:O with type of WCO expressed statistical significant effects on the biodiesel yield with t-values of 19.79 and –12.03 and p-values of 0.0210 and 0.03074, respectively. However, the interactive effect of M:O with reaction time showed a slight statistical significant effect on the response variable (t-value 0.06 and p-value 0.0645). But the interactive effects of catalyst loading with reaction time and that of type of WCO feedstock and reaction time expressed statistical non-significant effects on the biodiesel yield with t-values of 0.02 and –0.004 and p-values of 0.0817 and 0.872, respectively. The quadratic effects of both catalyst loading and type of WCO feedstock showed statistical significant effects on the biodiesel yield with t-values of –18.80 and 13.001 and p-values of 0.0231 and 0.0331, respectively. But, the quadratic effects of M:O and reaction time expressed statistical non-significant effects on the biodiesel yield with t-values of 2.17 and 0.01 and p-values of 0.198 and 0.674, respectively.

7.9.1.4 EFFECT OF USING DIFFERENT TYPES OF WCO FEEDSTOCK ON THE ACTIVITY OF CaO

The ANOVA was performed to investigate the effect of using different types of WCO feedstock on the percent biodiesel yield using pure chemical CaO. The obtained sum of squares SS (480.684), mean of squares (groups MSG 162.054 and errors MSE 13.048), degree of freedom (df 14), high F-statistic (12.42), and small p-value (0.0012) indicated that the change of the type of WCO feedstock have a statistically significant effect on the CaO activity.

The Box-plots (Fig. 7.11) graphically confirmed the above results and illustrated that the WFCO produced the highest biodiesel yield (>92%), followed by the other two types of WCO feedstock: WFSO ($\approx$ 89%) and WFO ($\approx$ 80%).The preference of WFCO and WFSFO as WCO feedstock than that of WFO might be attributed to their chemical constituents and physiochemical characteristics. Thus, it indicated the higher the percentage of unsaturated FFA and the lower the percentage of saturated FFA, the higher the biodiesel yield using a pure chemical CaO as a catalyst. Moreover, the lower the density, viscosity, TAN, Iodine number, and saponification value of the WCO feedstock, the higher the biodiesel yield using the heterogeneous pure chemical CaO as a catalyst. However, more research is recommended to investigate the effect of chemical composition of the feedstock on the transesterification process.

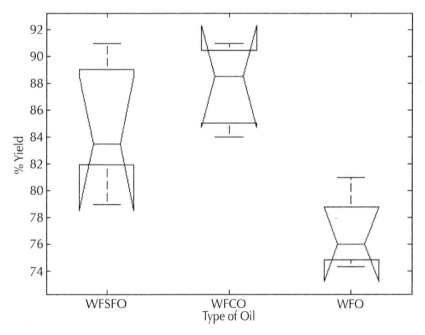

FIGURE 7.11 Box-plot for the effect of different types of WCO feedstock on the activity of pure chemical CaO and biodiesel yield.

7.9.1.5 OPTIMIZATION OF THE TRANSESTERIFICATION PROCESS

The numerical optimization was conducted through response optimizer available in MATLAB software (version 7.0.0, MathWorks, Inc., USA) to get the optimum values of the selected variables within the studied experimental range that would lead to a maximum biodiesel yield using the recommendable WFCO as a feedstock.

According to the response optimizer graph (Fig. 7.12), the excess M:O molar ratio (>10.55:1) resulted in a slight increase in the response. Since higher M:O is not favorable for product purification, and the separation and recovery of the excessive methanol will consume large amount of energy. With the increase of CaO concentration, the biodiesel yield increased, due to the increase of the total number of catalytic active sites. But excess catalyst concentration (> 6.65 wt.%) decreased the response. Since, the percent yield is affected by mass transfer between the reactants and catalyst. Consequently, at excess catalyst concentration, the reaction mixture becomes more viscous, which causes a decrease in the reaction rate and decreased biodiesel yield due to the mass transfer resistance. At longer reaction time (> 100.54 min) a slight increase the biodiesel yield was obtained.

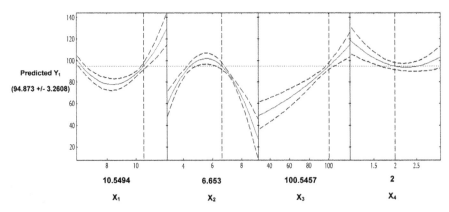

FIGURE 7.12 Response surface optimizer graph for transesterification of WFCO with methanol using pure chemical CaO.

Thus, from the economic point of view and according to the results obtained from the response optimizer graph (Fig. 7.12) the optimum M:O molar ratio, catalyst concentration, and reaction time, using WFCO as a feedstock were chosen to be 10.55:1, 6.65 wt.%, and 100.54 min, respectively, which yielded maximum predicted biodiesel yield of ≈94.87%. An experimental run was performed applying the obtained optimum conditions that yielded 95% biodiesel yield with physicochemical characteristics agreed well with the ASTM-D6751[2] and EN14214[96] standards (Table 7.5).

TABLE 7.5 The Physicochemical Characteristics of the Produced Biodiesel in the Illustrated Case Studies.

Parameters	Case studies							
	No. 1	No. 2	No. 3			No. 4	No. 5	No. 6
	WFCO-biodiesel	WFCO-biodiesel	WFO-biodiesel	WFSFO-biodiesel	WFCO-biodiesel	WFSFO-biodiesel	WFO-biodiesel	S. platensis-biodiesel
Density at 15.56°C, g/cm³	0.8902	0.8864	0.8898	0.8844	0.8854	0.8840	0.8884	0.8637
Viscosity at 40°C, cSt	5.68	4.83	5.22	4.63	4.73	4.65	5.69	12.4
Pour point, °C	-3	-3	3	-3	-3	-3	-9	-9
Cloud point, °C	0	0	0	0	0	0		-3
TAN, mg KOH/g oil	0.5	0.34	0.24	0.6	0.44	0.58	0.16	0.75
Sulfur content, wt.%	Nil	Nil	Nil	Nil	Nil	Nil	Nil	Nil
Water content, ppm	151	433	234	334	330	468		39
Flash point, °C	155	168	170	161	165	161	170	189
Calorific value, MJ/kg	39.51	40.01	40.08	39.37	39.51	39.40	37.54	45.63
Cetane number	44	43	49	43	44	43	42.29	70
Diesel index							40.29	67
Iodine number mg I_2/100 g oil	102	102	96	108	102	108	102	102

7.9.2 CASE STUDY No. 2

Optimization of the Transesterification Reaction of WFCO with Methanol Using Snails Shells-CaO

This study addressed environmental and economic aspects for the production of biodiesel, where, snails shells were used as a source for the heterogeneous CaO catalyst by simple calcination at 800°C and WFCO was used as TG feedstock.[97]

7.9.2.1 EXPERIMENTAL DESIGN

The specified four key independent variables with their three levels are represented in Table 7.6 to study their effects on the percent biodiesel yield at a constant reaction temperature (60°C). The 3^4 FFD of experiments led to 81 runs of experiments, which would have been time and energy consuming and consequently would have increased the final cost of the product. Thus, by applying the available functions in the statistical tool box in MATLAB software (version 7.0.0, MathWorks, Inc., USA) the required run design was selected to be 20 runs.[97]

TABLE 7.6 Factors and Levels of the DOE of Case Studies no. 2 and no. 4.

Factors	Levels		
	1	2	3
Methanol:Oil M:O, molar ratio (X_1)	6:1	9:1	12:1
CaO concentration, wt.% (X_2)	3	6	9
Reaction time, min (X_3)	30	60	120
Mixing rate, rpm (X_4)	200	300	400

7.9.2.2 STATISTICAL ANALYSIS

Once the important factors were identified and the screening experiments have been performed, the next step was to predict the model-detecting interaction among the design factors and estimating the response surface required to optimize the process, that is, to determine the local optimum independent variables with maximum biodiesel yield.

A quadratic regression model equation (eq 7.20) characterizing the influence of the considered variables on the process yield was predicted.

$$Y = 92.43 + 8.2541X_1 - 1.0585X_2 - 0.363X_3 - 0.1631X_4 + 0.2403X_1X_2$$
$$-0.0192X_1X_3 + 0.0058X_1X_4 - 0.0042X_2X_3 - 0.0130X_2X_4 + 0.0007X_3X_4 \quad (7.20)$$
$$-0.571X_1^2 + 0.2536X_2^2 + 0.0022X_3^2 + 0.0002X_4^2$$

7.9.2.3   VALIDATION OF THE PREDICTED MODEL EQUATION

The validity of the fitted model was evaluated by the determination coefficients; R^2 and R^2_{adj} which measure the reliability of the model and was found to be 0.9905 and 0.9829, respectively, which indicates that only 0.0095 of the total variations were not explained by the model. It consequently ensured the good adjustment of the predicted quadratic model equation (eq 7.20) to the obtained experimental data.

The statistical significance of the model equation was evaluated by F-test and p-values for ANOVA; where the F-value; 29.7623 and p-value; 0.0024 (<0.05) showed that the quadratic model equation (eq 7.20) was statistically significant at 95% confidence interval.

Confirmation of the adequacy of the regression model was reflected also by the good agreement between the experimental and predicted response variables, which ranged between 82.6 and 92% and 82.96 and 92.5%, respectively.

7.9.2.4   ANALYSIS OF VARIANCE

ANOVA was applied to establish the statistical significance of the model parameters at 95% confidence level. The linear effects of both M:O (t-value 22.78 and p-value < 0.0001) and catalyst concentration (t-value −5.05 and p-value 0.0001) showed the highest statistical significant effects on the biodiesel yield, followed by the statistical significant effects of their quadratic levels (t-value −2.17 and p-value 0.0231) and (t-value 1.8 and p-value 0.048), respectively, and their interactive effect (t-value 1.79 and p-value 0.0523), while the linear effects of both reaction time (t-value −0.89 and p-value 0.0393) and mixing rate (t-value -0.64 and p-value 0.0526) were statistically slightly significant.

7.9.2.5   RESPONSE SURFACE OPTIMIZATION

The empirical predicted quadratic model for response (i.e., percent biodiesel yield) in terms of process variables were plotted in 3D (Fig. 7.13).

Figure 7.13a represents the effect of varying M:O and catalyst concentration on percent biodiesel yield keeping the reaction time, and mixing rate constant at 60 min and 200 rpm, respectively. It was obvious that an increase in biodiesel yield occurred with the increase in M:O up to 7.5:1 with a slight increase in catalyst concentration up to 4 wt.%. However, the trend was reversed at higher M:O and catalyst loading, with maximum biodiesel yield at 6:1 M:O, and 3 wt.% catalyst concentration. Figure 7.13b illustrates the dependence of percent biodiesel yield on M:O molar ratio and reaction time at constant catalyst concentration and mixing rate of 3 wt.% and 200 rpm, respectively. At low M:O molar ratio the biodiesel yield increased with reaction time up to 80 min, then decreased with further increment of reaction time. The same trend was also observed at high level of M:O (12:1). While within M:O range of 7:1 up to 11:1, the percent biodiesel yield decreased over the reaction time range 30–120 min. However, the maximum biodiesel yield of ≈ 97% was obtained at 6:1 M:O and 60 min. The elliptical shape (Fig. 7.13c) indicates the strong interaction between varying the catalyst concentration and reaction time. It was obvious that low biodiesel yield occurred when the catalyst concentration ranged between 4 and 8 wt.% for reaction time of 50–70 min. High biodiesel yields were obtained at low and high reaction times (30 and 120 min) and at low and high levels of catalyst loading. Figure 7.13d represents the dependence of percent biodiesel yield on both M:O molar ratio and mixing rate at constant catalyst loading (3 wt.%) and reaction time (60 min). Where M:O showed a statistically negative influence on the response value, as with the increase of M:O and mixing rate, the biodiesel yield decreased, recording a minimum yield at 12:1 M:O and 400 rpm. While the maximum biodiesel yield 95–97% was obtained within 6:1–7.5:1 M:O and mixing rate of 200–250 rpm.

Based on the obtained results and on the viewpoint of design a transesterification process that is cost effective, with low preparation time, low alcohol and catalyst concentrations, and low energy consumption. The optimum transesterification conditions over the studied variables ranges were predicted to be 6:1 M:O molar ratio, 3 wt.% catalyst concentration, 60 min reaction time, 200 rpm mixing rate at 60°C reaction temperature, with theoretical maximum biodiesel yield of ≈ 96.76%. By applying the predicted optimum conditions an experimental biodiesel yield of ≈ 96% was obtained with good physicochemical characteristics that agreed well with the ASTM-D6751[2] and EN14214[96] standards (Table 7.5). The obtained biodiesel yield was comparable with that obtained using pure chemical CaO (≈ 95%) while it was higher than that obtained using Novozym435 (≈ 87%), applying the same predicted optimum conditions.

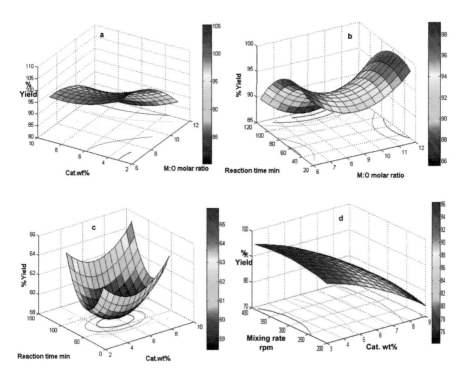

FIGURE 7.13   The response surface plots for biodiesel yield of case study no. 2.

7.9.3   CASE STUDY No. 3

Application of Statistical Analysis to Study the Effect of the Type of WCO on the Activity of Eggshells-CaO and to Maximize the Biodiesel Yield

As a waste recycling and renewable energy process, this study aimed to use the cheap and readily available eggshells as a source for the heterogeneous CaO catalyst by simple calcination at 800°C. Different types of WCO collected from different sources were used as TG feedstock to investigate the effect of feedstock type on the catalyst activity. Statistical analysis was applied to model and optimize the transesterification reaction to maximize the biodiesel yield.[98]

7.9.3.1   EXPERIMENTAL DESIGN AND STATISTICAL ANALYSIS

Once the important explanatory variables or factors were determined together with their levels (Table 7.3.), the next step was to perform a series of controlled laboratory designed experiments. On applying a (3^4) FFD of experiments, the total number of experimental runs was 81. But upon applying the D-optimal design of experiments, the number of experimental runs was reduced to 15. Once the designed matrix was illustrated, it was followed by predicting a regression model defining the distribution of the response variable (i.e., the biodiesel yield) in terms of the four explanatory variables. The model equation, which was found to best estimate the response, was the second-order-quadratic model.

Thus, based on DOE and experimental data, and by applying the regression analysis to determine the equation coefficients, the following predicted regression model equation (eq 7.21) was found to best estimate the response.

$$Y = 54.0717 + 2.884X_1 - 7.2601X_2 + 1.6297X_3 + 0.1929X_4 + 0.9666X_1X_2$$
$$-0.1185X_1X_3 + 0.07743X_1X_4 - 0.0307X_2X_3 - 0.2973X_2X_4 \qquad (7.21)$$
$$-0.0237X_3X_4 + 0.1253X_1^2 + 1.4547X_2^2 - 0.0026X_3^2 - 0.000273X_4^2$$

7.9.3.2   VALIDATION OF THE ELUCIDATED REGRESSION MODEL

To test the validation of the predicted model, the determination coefficient R^2 was evaluated to be 0.9872, which indicated that the model could explain 98.72% of the variability and only 1.28% of the total variations in eq 7.21 were not explained by the model. That ensured the goodness of fit and confirmed the adequacy of the predicted regression model. That was reflected also by the good agreement between experimental and predicted values of the response variable, which ranged between 70 and 92.1% biodiesel yield and 69.04 and 95.17% biodiesel yield, respectively.

Further, the validity of the model was checked with Fisher's test (F-test) and p-values. Their values were calculated to be 77.0855 and 0.0005, respectively. The recorded low p-value implied that the model was highly significant at 95% confidence interval.

The relationship between the predicted and actual response values was illustrated (Fig. 7.14a). The high correlation ($R^2 = 0.938$) between the predicted and experimental values indicated that they were in a reasonable agreement. That means the data fitted well with the predicted model and the model gives a convincing good estimate of response for the system within the studied experimental range.

The residual distribution is defined as the difference between the predicted (calculated) and actual (experimental) values of the response variable. Figure 7.14b represents a plot of the residual distribution versus the predicted ones. The quality of the fit was good, since the residual distribution did not follow a certain trend with respect to the predicted values of the response variable. Thus, it indicated that the quadratic model adequately represented the percent biodiesel yield over the studied experimental range.

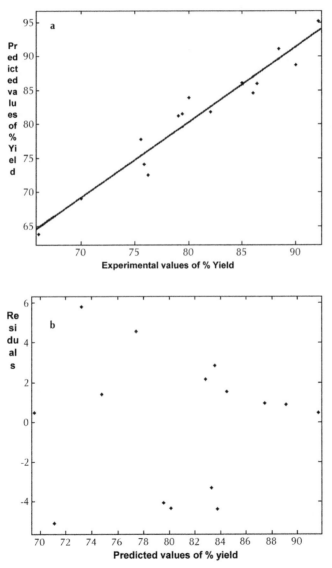

FIGURE 7.14   Model fit profile (a) and residual plot (b) of biodiesel yield for case study no. 3.

7.9.3.3 ANALYSIS OF VARIANCE

ANOVA was applied to establish the statistical significance of the model parameters at 95% confidence level by calculating t- and p-values. The results implied that the type of WCO feedstock has no statistical significant effect on the eggshells-CaO activity and thus the biodiesel yield (t-value 3.64 and p-value 0.0626). That contradicted with pure chemical CaO (case study no. 1), which adds to the favor of applying eggshells-CaO for biodiesel production. The variables with the largest statistical significant effects, were the linear effects of M:O molar ratio (t-value 23.278 and p-value 0.00274) and catalyst concentration (t-value 23.278 and p-value <0.0001) followed by and their quadratic effects (t-values 2.173 and 1.809 and p-values 0.0231 and 0.048, respectively). While the linear effect of reaction time expressed a statistical significant effect on biodiesel yield (t-value 11.99 and p-value 0.0393). The interactive effect of M:O with catalyst loading and that of M:O with reaction time showed statistical slight effects on the biodiesel yield (t-values 1.793 and −4.061 and p-values 0.0523 and 0.0545, respectively). However, all the other regression parameters (coefficients) expressed statistical no-significant effects on the biodiesel yield.

7.9.3.4 PARETO CHARTS

These are very useful in DOE, to easily visualize the main and interactive effects of all factors on the response variable. The Pareto chart (Fig. 7.15) illustrated that over the studied range; the catalyst concentration (wt.%) had

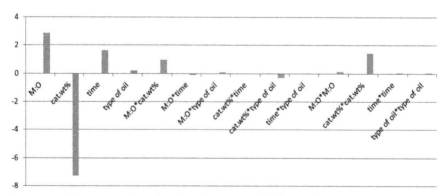

FIGURE 7.15 Pareto chart showing the effects of different independent variables on% biodiesel yield in case study no. 3.

a strong negative effect on the percent biodiesel yield. It means by increasing the amount of the catalyst loading, the yield would decrease. As the M:O molar ratio and reaction time had positive effects, the biodiesel yield would increase with the increase of M:O and the reaction time. But the type of WCO feedstock seemed to have no significant effect on the percent biodiesel yield. The interaction between M:O molar ratio and reaction time had a slight negative effect. That means the increment of the amount of M:O together with reaction time would reduce the biodiesel yield. The reverse would occur with the increment of M:O together catalyst concentration, since their interactive effect has a slightly positive effect on the biodiesel yield.

7.9.3.5 EFFECT OF USING DIFFERENT TYPES OF WCO FEEDSTOCK ON THE ACTIVITY OF EGGSHELLS-CaO

The statistical technique ANOVA, which compares the means of more than two groups, has been performed to investigate the effect of using different types of WCO feedstock on the activity of eggshells-CaO and biodiesel yield.

The obtained SS (763.337), MS (groups 24.8987 and errors 59.4617), degree of freedom (df 14), low F-statistic (0.42), and high p-value (0.6671; >0.05) indicated that the three types of WCO feedstock are statistically non-significant in their effects on the eggshells-CaO activity and consequently the biodiesel yield. The box-plots (Fig. 7.16) graphically confirmed the aforementioned results.

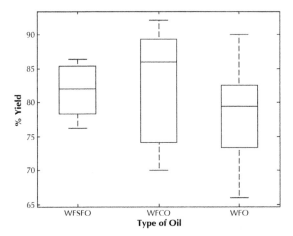

FIGURE 7.16 Box-plot for the effect of different types of WCO feedstock on the activity of eggshells-CaO and biodiesel yield.

7.9.3.6   OPTIMIZATION OF THE TRANSESTERIFICATION PROCESS

The optimum values of the selected independent variables were obtained by solving the regression equation (eq 7.21) using "LINGO 16.0" software for mathematical optimization (LINDO SYSTEMS INC., IL, USA). The maximum theoretical yield was predicted to be $\approx$ 90.22% at 9.15:1 M:O, 7.73 wt.% catalyst concentration, and 75 min reaction time, regardless of the WCO feedstock type. By applying the predicted optimum conditions, an experimental run was performed using the collected three types of WCO feedstock and biodiesel yields of $\approx$ 91, 90.5, and 90.4 wt.% were obtained using WFO, WFSFO and WFCO, respectively. The physicochemical characteristics of the produced biodiesel were evaluated and found to be agreed well with the ASTM-D6751[2] and EN14214[96] standards (Table 7.5).

It was obvious from obtained results that the physicochemical characterization of the produced biodiesel depends on the type of the WCO-feedstock and its characteristics. The WFO was characterized by the highest saturated FFA content, while the WFCO was characterized by the highest unsaturated FFA content (Table 7.4.). That might explain the obtained I_2 values for the produced biodiesels and the better density, viscosity, and cold flow properties (PP and CP) of the produced WFSFO and WFCO biodiesels than that of WFO-biodiesel (Table 7.5.). The I_2 value is a measure of unsaturation degree. The degree of unsaturation affects the fuel flow properties, cetane number, and oxidation stability. The higher the unsaturation is, the better the flow properties are, but the lower the oxidation stability and cetane number are. [22,54-56]

7.9.3   CASE STUDY No. 4

Application of Different Statistical Techniques to Optimize the Biodiesel Production from WFSFO Using Eggshells-CaO

The aim of this study was to model and optimize the transesterification of waste frying sunflower oil (WFSFO) collected from different domestic wastes with methanol using eggshells-CaO. That was performed by a series of controlled laboratory designed experiments and prediction of a regression model that defined the distribution of the response variable in terms of the defined explanatory variables. Data analysis was done to fit the regression model equation, specify the regression coefficients, recognize the significant model terms, and finally, determine the factors optimum levels that would lead to a maximum response (i.e., maximum percent biodiesel yield). Then the activity of eggshells-CaO at the predicted optimum transesterification conditions was compared with the most widely used immobilized enzyme; Novozym 435, and pure chemical CaO.[99]

7.9.3.1   EXPERIMENTAL DESIGN

The specified four key independent variables with their three levels are represented in Table 7.6 to study their effects on the percent biodiesel yield at constant reaction temperature (60°C). DOE was applied to reduce the number of experimental runs and obtain more information per experiment. The functions available in the statistical tool-box in MATLAB software package (version 7.0.0, MathWorks, Inc., USA) were used to generate the required run design, which was selected to be 20 runs for three levels of four independent variables.[99]

7.9.3.2   REGRESSION MODEL AND ITS VALIDATION

In order to better understand the correlations between the explanatory factors and response variable, a general linear interaction empirical regression model equation, which accounts for the explanatory factors and their interaction effects, was employed in this study.

Based on the DOE and the obtained experimental data, and by applying the regression analysis to determine the equation coefficients, the following predicted linear interaction regression model equation (eq 7.22) was found to best estimate the response.

$$Y = 40.157 + 4.1868X_1 + 2.8132X_2 - 0.7022X_3 + 0.2858X_4 - 0.1528X_1X_2$$
$$+0.0318X_1X_3 - 0.0189X_1X_4 + 0.0395X_2X_3 - 0.0175X_2X_4 + 0.0003X_3X_4 \quad (7.22)$$

The method of least squares was employed to ascertain the values of the model parameters, while the determination coefficient R^2 was evaluated to test the fit of the model. The obtained R^2 0.8959 indicated that the model could explain 89.58% of the variability and only 10.42% of the total variations in the model equation were not explained by the model. Thus, it confirmed and ensured the goodness of fit and the adequacy of the regression model. The model adequacy was also reflected by the good agreement between the experimental and predicted values of the response variable, which ranged between 73.9 and 90.0 and 71.24 and 92.56, respectively. Furthermore, the validity of the model was checked by the F-test and p-values, which were found to be 7.9771 and 0.0023, respectively, with an error of variance of ≈ 7.1811. The obtained low p-value (< 0.05), implied that the predicted model (eq 7.22) was significant at 95% confidence level.

7.9.3.3 ANALYSIS OF VARIANCE

ANOVA was applied to establish the statistical significance of the model parameters at 95% confidence level by calculating t- and p-values. The larger the magnitude of t-value and the lower the p-value, the more significant is the corresponding coefficient. It was noticed that the linear effect of M:O molar ratio was statistically highly significant (t-value 18.32 and p-value 0.0001), followed by the statistical significant linear effects of catalyst concentration and reaction time (t-values 7.09 and −2.38 and p-values 0.0018 and 0.0037, respectively). The linear effect of mixing rate was found to have a statistical possible-significant effect (t-value 0.88 and p-value 0.0261). All the interactive effects of the studied parameters, within the performed range expressed statistically non-significant effect on the biodiesel yield (p >0.05) except for the interactive effect of M:O with catalyst loading which might have expressed to some extent a statistical significant effect (t-value −0.62 and p-value 0.0225).

7.9.3.4 RESPONSE SURFACE OPTIMIZATION

The three-dimensional response surface was plotted on the basis of the predicted model (eq 7.22) to investigate the interaction among the independent variables and to determine the range of the optimum condition of each factor for maximum percent biodiesel yield. It was observed that at low catalyst loading, the biodiesel conversion increased with the increase in mixing rate. However, at low mixing rate, the increase of catalyst concentration had a slight positive effect on the biodiesel yield. Moreover, the overall increase in catalyst concentration (> 4 wt.%) and reaction time (> 40 min) decreased the biodiesel yield. The high mixing rate (350–400 rpm) at low reactants concentration (6:1 M:O and 3 wt.% eggshells-CaO) achieved high biodiesel conversion (> 95%) within a short reaction time (30–40 min).

Figure 7.17a represents the dependency of percent biodiesel yield on M:O molar ratio and catalyst concentration, at constant mixing rate (350 rpm), and reaction time (30 min). It was noticed that a recorded decrease of biodiesel yield occurred with the increase of both aforementioned variables. The maximum biodiesel yield (98.5%) was obtained at lower values of M:O (6:1) and catalyst concentration (3%). Figure 7.17b indicates that within the studied range, there was almost no interaction between the M:O molar ratio and reaction time. A maximum conversion of ≈95% occurred at low M:O (6:1) over a time range of 30–40 min and constant catalyst concentration (3%) and mixing rate (350 rpm). Figure 7.17c represents the influence of both M:O and mixing rate on the percent biodiesel yield at constant catalyst concentration (3 wt.%) and reaction time (30 min). It was noticed that at low values of mixing rate, the percent yield decreased all over the studied M:O molar ratio. But the percent yield increased at low value of M:O molar ratio and increased values of mixing rate. The minimum percent yield (< 75%) occurred at 6:1 M:O and 200 rpm, while the highest percent yield (>95%) obtained at the range of 6:1–7:1 M:O and 350–400 rpm.

The reaction mixture is a three-phase system. Thus, a decrease of FAME yield at higher levels of methanol might be due to the dilution effects on catalyst and reactants. With the increase of reaction time, the biodiesel yield increases to a certain limit, producing what is called the S-shape curve. That is due to the variation of the reaction order from zero-order with respect to oil concentration to first-order throughout the progress of the transesterification reaction.[99] At the early-stage it occurred at the boundary between oil and methanol, then at higher reaction time, due to the good miscibility in the reaction mixture, the rate changed to first-order kinetic. Moreover, the reaction rate might be reduced due to the mass transfer resistance on the boundary between the oil and methanol phase. Thus increase the mixing rate up to 350 rpm might have facilitated a proper contact between the reactants and consequently increased the biodiesel yield. But further increase in mixing rate (> 350) might have caused turbulence in the mixture, which consequently decreased the yield. Not only this, but the biodiesel yield in the three-phase system is mainly mass-transfer limited. Accordingly, the higher catalyst concentration (> 4%) might have caused the mixture to be more viscous, which might have consequently decreased the reaction rate and consequently the biodiesel yield due to the mass transfer resistance.

Briefly, the RSM optimization revealed that the maximum biodiesel yield (95–98%) can be obtained at 6:1 M:O, 3–4 wt.% eggshells-CaO, 30–40 min, 350–400 rpm at 60°C.

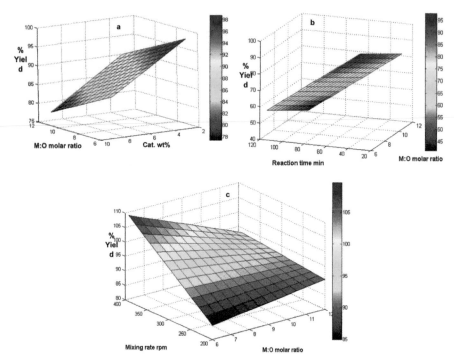

FIGURE 7.17   The response surface plots for biodiesel yield of case study no. 4.

7.9.3.5   LINGO OPTIMIZATION

The exact optimum values of the selected independent variables were obtained by solving the linear interactive regression equation (eq 7.22) using "LINGO 16.0" software for mathematical optimization (LINDO SYSTEMS INC., IL, USA). Where, the maximum theoretical conversion of ≈ 98.8% was predicted to occur at 6:1 M:O, 3% (w:w) catalyst concentration, 350 rpm, 30 min at 60°C.

Thus, in the viewpoint of preparation time energy consumption and cost-effective transesterification process with high biodiesel yield, an experimental run was performed applying the LINGO-recommended optimum parameters to produce biodiesel fuel form WFSFO using eggshells-CaO and compare its activity relative to that of pure chemical CaO and Novozym435. The eggshells-CaO expressed the highest activity relevant to the pure chemical CaO and Novozym435, producing biodiesel yields of ≈ 98, 85, and 50%, respectively. The produced high quality biodiesel fuel (Table 7.5.) recommends the application of eggshells-CaO on the industrial scale as readily available, cheap, and renewable catalyst for biodiesel production form WCO.

7.9.4   CASE STUDY No. 5

7.9.4.1   STATISTICAL OPTIMIZATION OF BIODIESEL PRODUCTION FROM WFO USING KOH AS AN ALKALINE HOMOGENOUS CATALYST

In this study the RSM based on CCD of experiments was used to optimize the biodiesel production from WFO collected from different local restaurants using methanol and KOH and to also investigate the influence of different process variables on the percent biodiesel yield. MATLAB software package (version 7.0.0, MathWorks, Inc., USA) was used for the design of experiments, regression, and graphical analysis of the data obtained. The used WFO was composed of 51% and 49% saturated and unsaturated FFA, respectively, with I_2 value of 114 mg I_2/100g oil, saponification value of 202 mg KOH/g oil, TAN of 1.15 mg KOH/g oil, viscosity 58.56 cSt, and density 0.9304 g/cm³.[100]

7.9.4.2   EXPERIMENTAL DESIGN

The experimental runs were carried out according to a 23 FFD for three identified design independent variables with two levels; low (−1) and high (+1) (Table 7.7), and the temperature was set at 60°C to be below the boiling point of methanol 65°C, to avoid any evaporation and the use of a pressure vessel. The total number of experimental runs was given by the simple formula $[24=2^k+2k+10]$ where k is the number of independent variables and eight factorial points from 14 full factorial CCD were augmented with 10 replicates at the center point to determine the experimental error.

TABLE 7.7   Factors and Levels of the DOE of Case Study no. 5.

Factors	Levels		
	−1	0	+1
Methanol:Oil M:O molar ratio (X_1)	6:1	7.5:1	9:1
CaO concentration wt.% (X_2)	0.6	0.9	1.2
Reaction time h (X_3)	1	2	3

7.9.4.3   THE STATISTICAL ANALYSIS AND MODEL VALIDATION

Once the experiments were performed, the next step was to predict the model detecting the interaction among the design factors. The regression model (eq 7.23) that characterized the influence of different considered variables on the process yield was obtained and its statistical significance was controlled by F-test.

$$Y = 113.78 + 11.4X_1 + 7.489X_2 - 1.933X_3 + 3.8X_1X_2 - 0.099X_1X_3$$
$$+ 3.79X_2X_3 + 0.76X_1^2 - 69.27X_2^2 + 4.21X_3^2 \quad (7.23)$$

The ANOVA for the response surface full quadratic model (eq 7.23) was done and indicated that the model is statistically significant with F-value of 99.3 and a very low probability p-value of 3.9941e-11. Furthermore, the R^2 was calculated to be 0.984 indicated that there was an excellent correlation between the independent variables and ≈98.46% of the response variability could be explained by the predicted model (eq 7.23), where the actual biodiesel yield ranged from 78.36 to 98.9% and its corresponding predicted values were 78.84 and 99.13%.

The ANOVA was also used to determine the significant effect of each factor on the response variable. The M:O and catalyst loading expressed statistical significant effect on the biodiesel yield (p <0.0001), while within the studied range, the reaction time expressed a statistical non-significant effect on the biodiesel yield ($p = 0.2708$).

7.9.4.4   RESPONSE SURFACE OPTIMIZATION

The RSM was used to understand the interactive relationship between the designed independent variables and the response one (percent biodiesel yield). The 3-D surface plot in Fig. 7.18a represents the graphical representation of the effect of varying M:O molar ratio and catalyst concentration wt.% at a fixed reaction time of 1 h on the percent yield of biodiesel. At low catalyst concentration (≤ 0.6 wt.%), the increase of M:O increased the biodiesel yield, recording its maximum value (≈99%) 9:1 M:O and 0.6 wt.% KOH. But the trend was reversed when increasing the catalyst up to 1.2 wt.% and decreasing M:O to 6:1, recording the minimum value of biodiesel yield (≈ 79%). Figure 7.18b represents the influence of M:O molar ratio and reaction time h on the process yield at constant catalyst concentration 0.6 wt.%. It was obvious that low M:O would lead to an incomplete conversion and low biodiesel yield and the increase of time (>1 h) has no significant effect on the yield. Thus, the higher methanol concentration resulted in a greater biodiesel conversion within a shorter time, where the maximum yield (≈ complete conversion) was obtained at 9:1 M:O and 1 h. Figure 7.18c represents the response plot for the interactive factors, reaction time h and catalyst wt.%, and the M:O molar ratio was kept constant at 9:1. The 3-D response surface plot indicated that the biodiesel yield decreased mainly with the increase in the catalyst loading while the effect of reaction time is non-significant, where the excess catalyst loading (> 0.6 wt.%) decreased the biodiesel yield within the studied time range (1–3 h), with maximum biodiesel yield (≈ complete conversion) at 0.6 wt.% and 1 h.

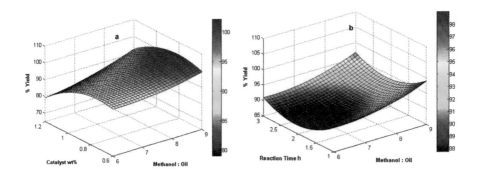

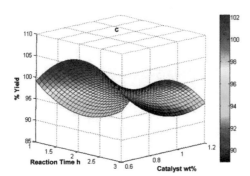

FIGURE 7.18 The response surface plots for biodiesel yield of case study no. 5.

7.9.4.5 CONTOUR PLOT

The graphical representation of the 2-D contour plots of the predicted model (eq 7.23) is illustrated in Fig. 7.19.

Figure 7.19a represents the effect of M:O molar ratio and catalyst loading wt.% at a fixed reaction time of 1 h, on the percent yield of biodiesel. It was clear that an increase in biodiesel yield occurred with the increase of M:O molar ratio. But, the trend was reversed when increasing the catalyst loading up to 1.2 wt.% and decreasing the M:O molar ratio, recording 80% yield within the range of 6:1 up to 6.5:1 M:O. While the maximum yield reached 98% within the range of 8.5–9:1 M:O molar ratio at 0.6 wt.% catalyst concentration and 1 h reaction time. Figure 7.19b represents the 2-D contour plot that illustrates the influence of M:O molar ratio and reaction time on the process yield at constant catalyst concentration of 0.6 wt.%. It was clear that a smaller M:O led to a low percent yield of biodiesel. It was also obvious that the increase in reaction time had no significant effect, while the higher methanol concentration resulted in a greater percent yield within shorter reaction time where the maximum percent yield of biodiesel

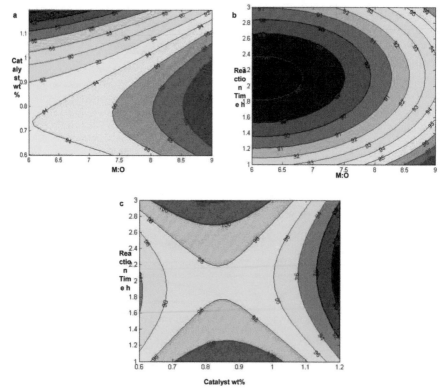

FIGURE 7.19   The contour plots for biodiesel yield of case study no. 5.

of 98% was recorded within the range of 8.75 up to 9:1 M:O molar ratio and reaction time of 1 h. The minimum value of percent biodiesel yield of 88% was obtained within the range of 6:1 up to 6.5:1 M:O and 0.6 wt.% catalyst load during 1.58 up to 2.4 h reaction time. Figure 7.19c illustrates the 2-D contour plot representing the interactive effect of reaction time and catalyst concentration, while M:O molar ratio was kept constant at 9:1 molar ratio. The contour plot indicated that percent yield decreased mainly with the excessive catalyst loading (>1 wt.%), while the reaction time expressed a non-significant effect. The maximum percent yield of 98% was obtained at catalyst concentration of 0.65 wt.% and reaction time of 1 h. However, the complete conversion (100%) has been obtained by increasing the catalyst concentration to 0.7 wt.%, within the same reaction time of 1 h. But a lower percent yield (≈90%) occurred at higher catalyst loading of 1.2 wt.% and higher reaction time of 1.6–2.6 h.

Thus, from the two applied statistical methods—RSM and contour plots—the optimum conditions for maximum biodiesel production from WFO using KOH and methanol was found to be 9:1 M:O, 0.6% catalyst concentration, 1 h at 60°C, and 300 rpm. The physicochemical characteristics of the produced biodiesel under the aforementioned optimum conditions (Table 7.5) were completely acceptable and met most of international standard specifications.[2,96]

7.9.5 CASE STUDY No. 6

7.9.5.1 ELUCIDATION OF MATHEMATICAL CORRELATIONS FOR THE CHANGES IN THE PHYSICOCHEMICAL PROPERTIES OF BIO-PETRO-DIESEL BLENDS

The blend level (i.e., the percentage of biodiesel in the bio-petro-diesel mixture) determines many important characteristics of the blended fuel. The physicochemical characteristics of the produced blends should fit with the international fuel standards.[2,96,102,103] A lower blend level of biodiesel may reduce the expected benefits, such as fuel lubricity and tail pipe emission. A higher-than-specified level of biodiesel may exceed the engine manufacturer's recommended limitations, compromising the engine performance. Biodiesel is usually sold at a higher price than petro-diesel fuel; thus, the price of the fuel is dependent on the blend level.[101]

In this study, regression analysis was applied to describe the changes of some basic fuel properties with the volumetric percentage of the biodiesel in the bio-ptero-diesel blends. Two biodiesel feedstocks—biodiesel prepared by the transesterification of WFO with methanol in the presence of KOH[100] and another biodiesel prepared by the transesterification of lipid extract from *Spirulina platensis* using methanol and NaOH[101] were used to prepare different bio-petro-diesel blends. Table 7.8 summarizes the FAME composition of both feedstocks for blending. They were mainly composed of ≈ 51 and 90% saturated FAME, respectively, and all of their physicochemical characteristics (Table 7.5) were completely acceptable and met most of the international standard specifications,[2,96] except for the viscosity of the *S. platensis* biodiesel. Thus, they can be ranked as a realistic fuel and as an alternative to petro-diesel fuel. The preparation of different bio-petro-diesel blends is recommended to lower the viscosity of biodiesel and improve the lubricity of petro-diesel. It is also recommended to lower the sulfur content in the petro-diesel fuel to decrease the SO_x emissions. Moreover, it is better to improve the flash point for better handling and safety during transportation and storage and other physicochemical characteristics (calorific value, aniline point, diesel index, cetane number, and distillation temperature) to

TABLE 7.8   A Summary of the FAME Composition of WFO and S. platensis Biodiesel.

Fame	wt.%	
	WFO biodiesel	**S. platensis biodiesel**
Myristic acid methyl ester (C14:0)	10.63	3.56
Palmitic acid methyl ester (C16:0)	20.38	14.87
Palmitoleic acid methyl ester (C16:1)		0.98
Stearic acid methyl ester (C18:0)	18.09	28.11
Oleic acid methyl ester (C18:1)	21.93	5.39
Linoleic acid methyl ester (C18:2); ω6	26.40	1.12
Linolenic acid methyl ester (C18:3)	1.02	
Arachidic acid methyl ester (C20:0)	1.55	32.43
Eicosenoic acid methyl ester (C20:1)		1.94
Eicosadienoic acid methyl ester (C20:2); ω6		0.99
Behenic acid methyl ester (C22:0)		6.65
Lignoceric acid methyl ester (C24:0)		3.96
Saturated FAME	50.65	89.58
Unsaturated FAME	21.93	8.31
Polyunsaturated FAME (PUFA)	27.42	2.11

enhance the combustion properties and ignition performance which consequently will reduce the greenhouse emissions (e.g., CO and CO_2). The lower
water content and the higher C-chain length in *S. platensis* biodiesel might
explain its higher calorific value.[55] The higher cetane number of *S. platensis*
biodiesel might be attributed to the FAME composition as the increase in
saturated FAME content and its chain length positively enhance the cetane
number of biodiesel.[97]

Some important properties—the density, kinematic viscosity, total acid
number, total sulfur content, flash point, calorific value, cetane number,
and diesel index—have been studied to predict the regression equations
describing their changes with the percentage of biodiesel in some bio-petro-
diesel blends.[100, 101]

This study suggests that the empirical n^{th} order polynomial (eq 7.24) is
the best one correlating the fuel properties to the blend composition.

$$P = b_o + \sum_{k=1}^{n} b_k x^k \qquad (7.24)$$

where P is the fuel property, b_0, b_1, b_2,... B_n are correlation coefficients; x is
the volumetric percentage of biodiesel in bio-petro-diesel blends (B_0, B_2, B_5,
B_{10},... B_{100}).

Table 7.9 represents some of the studied fuel properties together with
their corresponding polynomials with the measurements of goodness of fit,
correlation coefficient (R^2), adjusted correlation coefficient (R^2_{adj}), root mean
square error (RMSE), and sum of square error (SSE). The recorded high R^2
and R^2_{adj} of these polynomials (≥ 0.9235 and ≥ 0.8088, respectively), with
the recorded low values of RSME and SSE, suggested that the studied fuel
properties can be calculated as polynomial functions of biodiesel volumetric
percentage in the bio-petro-diesel blends. The recorded low R^2_{adj} of the *S.
platensis* biodiesel calorific value (0.6615) is attributed to the low predictive value of coefficient b_4 ($-2.846e-6$). The order of the polynomial seemed
to depend on the biodiesel feedstock type. It was found that for bio-petro-
diesel blends prepared from the WFO-biodiesel, the density, TAN, and CN
followed the fourth-order polynomial. But the kinematic viscosity, sulfur
content, FP, CV, and DI were found to follow third-order polynomial. For
the bio-petro-diesel blends prepared from the *S. platensis* biodiesel, it was
found that the density and kinematic viscosity followed the first-order polynomial. The TAN was represented by second-order polynomial. While, the
total sulfur content, FP, and DI followed the third-order polynomial and the
CN followed the fourth-order polynomial.

TABLE 7.9 The Empirical n^{th} Order Polynomial Correlating Some Fuel Properties to the Bio-Petro-Diesel Composition.

Property	Coefficients					Order of Polynomial	R^2	R^2_{adj}	RMSE	SSE
	b_0	b_1	b_2	b_3	b_4					
WFO-Biodiesel Case										
Density at 15.56°C, g/cm³	0.8371	2.414e-3	-6.181e-5	6.41e-7	-2.154e-9	4th	0.9901	0.9803	2.308e-3	2.13e-5
Viscosity at 40°C, cSt	3.55	0.03489	-4.717e-4	3.357e-6		3rd	0.9986	0.9977	0.03464	0.006
TAN, mg KOH/g oil	0.4078	-8.05e-3	1.91e-4	-2.203e-6	8.511e-9	4th	0.9964	0.9928	7.319e-3	2.143e-4
Sulfur content, wt.%	0.8455	-0.02014	5.039e-4	-3.862e-6		3rd	0.9931	0.9889	0.02645	3.498e-3
FP, °C	87.56	1.589	-0.0342	2.649e-4		3rd	0.9965	0.9944	1.903	18.1
CV, MJ/kg	45.01	-0.0335	-7.787e-4	3.631e-6		3rd	0.9948	0.9917	0.2547	0.3242
CN	61.43	-0.7731	0.01893	-2.056e-4	7.447e-7	4th	0.982	0.964	1.257	6.32
DI	59	-0.6092	0.0101	-5.895e-5		3rd	0.9775	0.964	1.257	7.895
S. platensis Biodiesel case										
Density at 15.56°C, g/cm³	0.8423	0.0002139				1st	0.997	0.9962	0.0005078	1.032e-6
Viscosity at 40°C, cSt	3.522	0.08873				1st	0.9961	0.9951	0.2403	0.2311
TAN, mg KOH/g oil	0.02621	0.003494	3.742e-5			2nd	0.9995	0.9992	0.007907	0.0001876
Sulfur content, wt.%	0.1344	-0.01331	0.0004423	-3.226e-6		3rd	0.9828	0.9571	0.01098	0.0002412
FP, °C	80.01	4.308	-0.1611	0.001289		3rd	0.9967	0.9918	3.559	25.34
CV, MJ/kg	45.37	0.09246	-0.01046	0.0003802	-2.846e-6	4th	0.9323	0.6615	0.06129	0.003756
CN	50.94	4.1	-0.3642	0.01181	-8.558e-5	4th	0.9658	0.829	3.033	9.197
DI	50.3	2.402	-0.0995	0.000772		3rd	0.9235	0.8088	3.011	18.14

ANNEX 1   Abbreviations

Two-dimensional	2-D
Three-dimensional	3-D
American Society for Testing and Materials	ASTM
Box–Behnken Design	BBD
Calorific Value	CV
Central Composite Design	CCD
Central Composite Face Centered Design	CCFCD
Central Composite Faced	CCF
Central Composite Inscribed	CCI
Cetane Number	CN
Cloud Point	CP
Composite Circumscribed	CCC
Confidence Intervals	CI
Degrees of Freedom	df
Design of Experiments	DOE
Diesel Index	DI
D-Optimal Design	DOD
Fatty Acid Methyl Ester	FAME
Flash Point	FP
Fractional Factorial Design	FrFD
Free Fatty Acids	FFAs
Full Factorial Design	FFD
Jatropha curcas-Biodiesel	JCB
Jatropha curcas-Oil	JCO
Interquartile Range	IQR
Linear Regression Model	LRM
Mean of Squares Errors	MSE
Mean of Squares Groups	MSG
Mean Square for Treatment	MSTr
Mean Squares	MS
Methanol:Oil	M:O
Multiple Linear Regression	MLR
National Renewable Energy Laboratory	NREL
Nitrogen Oxides	NOx
One-Factor-At-A-Time	OFAT
Orthogonal Arrays	OA
Particulate Matter	PM
Plackett–Burman design	PBD
Pour Point	PP

ANNEX 1 *(Continued)*

Response Surface Methodology	RSM
Root Mean Squared Error	RMSE
Rotatable Central Composite Design	RCCD
Sulfur Oxides	SOx
Sum of Squares	SS
Sum of Squares Error	SSE
Total Acid Number	TAN
Total Sum of Squares	SST
Treatment Sum of Squares	SStr
Triglycerides	TG
US-Department of Energy	US-DOE
Waste Cooking Oil	WCO
Waste Frying Oil	WFO
Waste Frying Corn Oil	WFCO
Waste Frying Sunflower Oil	WFSFO
Waste Vegetable Oil	WVO

ANNEX 2   Some Statistical Software Packages

The statistical software packages are specialized computer programs for analysis in statistics and econometrics. The following table gives information about a number of statistical software packages (https://en.wikipedia.org/wiki/Comparison_ of statistical _packages).

Product	Developer	Latest version	Interface	Written in
Maple	Maplesoft	March 2, 2016	CLI / GUI	
Mathematica	Wolfram Research	September 28, 2016	CLI / GUI	
MATLAB	Math Works	New releases twice/ year	CLI /GUI	C++, Java
Minitab	Minitab Inc.	February 18, 2014	CLI / GUI	
SAS	SAS Institute	July, 2013	CLI / GUI	
Statistica	Dell Software	September, 2015	GUI	C++
SPSS	IBM	March 3, 2015	CLI / GUI	Java
NCSS	NCSS, LLC	February, 2015	GUI	
SOCR	UCLA and university of Michigan	May 10, 2015	GUI	Java
Design of Expert®10	State-Ease, Inc.	2016		
NCSS	NCSS, LLC Dr. Jerry Hintze	2007		
NLREG	Phillip H. Sherrod	July 27, 2009	GUI	C++

KEYWORDS

- **statistical optimization**
- **transesterification**
- **biolipids**
- **catalyst types**
- **bio-petro-diesel blend**
- **fuel properties**

REFERENCES

1. Nur Syazwani, O.; Rashid, U.; Taufiq, Y. H. Low-Cost Solid Catalyst Derived from Waste *Cyrtopleura Costata* (Angle Wing Shell) for Biodiesel Production Using Microalgae Oil. *Energ. Convers. Manage.* **2015,** *101,* 749–756.
2. ASTM Standard D6751- 15ce1. *Standard Specification for Biodiesel Fuel Blend Stock (B100) for Middle Distillate Fuels;* American Society for Testing and Materials: West Conshohocken, PA, 2008.
3. Tan, Y. H.; Abdullah, M. O.; Nolasco-Hipolito, C. The Potential of Waste Cooking Oil-Based Biodiesel Using Heterogeneous Catalyst Derive from Various Calcined Eggshells Coupled with an Emulsification Technique: A Review on the Emission Reduction and Engine Performance. *Renew. Sust. Energ. Rev.* **2015,** *47,* 589–603.
4. Wilson, K.; Lee, A. F. Rational Design of Heterogeneous Catalysts for Biodiesel Synthesis. *Catal. Sci. Tech.* **2012,** *2,* 884–897.
5. Ferrero, G. O.; Almeida, M. F.; Alvim-Ferraz, M. C. M.; Dias, J. Glycerol-Enriched Heterogeneous Catalyst for Biodiesel Production from Soybean Oil and Waste Frying Oil. *Energ. Convers. Manage.* **2015,** *89,* 665–671.
6. Corro, G.; Sanchez, N.; Pal, U.; Banuelos, F. Biodiesel Production from Waste Frying Oil Using Waste Animal Bone and Solar Heat. *Waste Manage.* **2016,** *47,* 105–113.
7. Jairam, S.; Kolar, P.; Sharma-Shivappa, R.; Osborne, J. A.; Davis, J. P. KI-impregnated Oyster Shell as a Solid Catalyst for Soybean Oil ransesterification. *Bioresour. Technol.* **2012,** *104,* 329–335.
8. Farooq, M.; Ramli, A.; Naeem, A. Biodiesel Production from Low FFA Waste Cooking Oil Using Heterogeneous Catalyst Derived from Chicken Bones. *Renew. Energ.* **2015,** *76,* 362–368.
9. Li, M.; Zheng, Y.; Chen, Y.; Zhu, X. Biodiesel Production from Waste Cooking Oil Using a Heterogeneous Catalyst from Pyrolyzed Rice Husk. *Bioresour. Technol.* **2014;** *154,* 345–348.
10. Vashist, D.; Ahmad, M. Statistical Analysis of Diesel Engine Performance for Castor and Jatropha Biodiesel-Blended Fuel. *Int. J. Automot. Mech. Eng.* **2014,** *10,* 2155–1169.

11. Diya'uddeen, B. H.; Abdul Aziz, A. R.; Daud, W. M. A. W.; Chakrabarti, M. H. Performance Evaluation of Biodiesel from Used Domestic Waste Oils: A Review. *Process. Saf. Environ.* **2012,** *90,* 164–179.

12. Rahim, R.; Mamat, R.; Taib, M. Y.; Abdullah, A. A. Influence of Fuel Temperature on a Diesel Engine Performance Operating with Biodiesel Blended. *J. Mech. Eng. Sci.* **2012,** *2,* 226–236.

13. Abbaszadeh, A.; Ghobadian, B.; Najafi, G.; Yusaf, T. An Experimental Investigation of the Effective Parameters on Wet Washing of Biodiesel Purification. *Int. J. Automot. Mech. Eng.* **2014,** *9,* 1525–1537.

14. Kumaran, P.; Gopinathan, M.; Kantharrajan, S. Combustion Characteristics of Improved Biodiesel in Diffusion Burner. *Int. J. Automot. Mech. Eng.* **2014,** *10,* 2112–2121.

15. Gorji, A.; Ghanei, R. A Review on Catalytic Biodiesel Production. *J. Biodivers. Environ. Sci.* **2014,** *5* (4), 48–59.

16. Liaquata, A. M.; Masjukia, H. H.; Kalama, M. A.; Varmana, M.; Hazrata, M. A; Shahabuddin, M.; Mofijur, M. Application of Blend Fuels in a Diesel Engine. *Energy Procedia.* **2012,** *14,* 1124–1133.

17. Boro, J.; Deka, D.; Thakur, A. J. A Review on Solid Oxide Derived from Waste Shells as Catalyst for Biodiesel Production. *Renew. Sust. Energ. Rev.* **2012,** *16,* 904–910.

18. Xu, J.; Jiang, J.; Zhang, T.; Dai, W. Biofuel Production from Catalytic Cracking of Triglyceride Materials Followed by an Esterification Reaction in a Scale-p Reactor. *Energ. Fuel.* **2013,** *27* (1), 255–261.

19. Vu, X. H. Hierarchical Zeolite Catalysts for Cracking of Triglyceride-Rich Biomass to Hydrocarbons: Synthesis, Characterization and Catalytic Performance. PhD Dissertation, University of Rostock, Germany, 2014.

20. Vu, X. H.; Schneider, M.; Bentrup, U.; Dang, T. T.; Phan, B. M. Q.; Nguyen, D. A.; Armbruster, U.; Martin A. Hierarchical ZSM-5 Materials for an Enhanced Formation of Gasoline-Range Hydrocarbons and Light Olefins in Catalytic Cracking of Triglyceride-Rich Biomass. *Ind. Eng. Chem. Res.* **2015,** *54,* 1773–1782.

21. Kiakalaieh, A. T.; Amin, N. A.; Mazaheri, H. A Review on Novel Processes of Biodiesel Production from Waste Cooking Oil. *Appl. Energ.* **2013,** *104,* 683–710.

22. Ullah Z., Bustam M. A., Man Z. "Biodiesel Production from Waste Cooking Oil by Acidic Ionic Liquid as a Catalyst". *Renew. Energy.* **2015,** *77,* 521–526.

23. Diamantopoulos, N.; Panagiotaras, D.; Nikolopoulos, D. Comprehensive Review on Biodiesel Production Using Solid Acid Heterogeneous Catalysts. *J. Thermodyn. Catal.* **2015,** *5,* 143. doi:10.4172/2157-7544.1000143

24. Alam, F.; Date, A.; Rasjidin, R.; Mobin, S. M.; Moria, H.; Baqui, A. Biofuel from Algae—is it a Viable Alternative? *Process. Eng.* **2012,** *49,* 221–227.

25. Su, F.; Guo, Y. Advancements in Solid Acid Catalysts for Biodiesel Production. *Green Chem.* **2014,** *16,* 2934–2957.

26. Brennan, L.; Owende P. Advanced Biofuels and Bioproducts. *Biofuels from Microalgae:* Towards Meeting Advanced Fuel Standards. In: Lee J. W., Eds; Springer: New York, NY, 2013; pp 553–599.

27. Singh, S.; Patel, A. Mono Lacunary Phosphotungstate Anchored to MCM-41 as a Recyclable Catalyst for Biodiesel Production via Transesterification of Waste Cooking. *Fuel.* **2015,** *159,* 720–727.

28. Mazubert, A.; Poux, M.; Aubin, J. Intensified Processes for FAME Production from Waste Cooking Oil: A Technological Review. *Chem. Eng. J.* **2013,** *233,* 201–223.

29. Cunha Jr., A.; Feddern, V.; De Prá, M.C.; Higarashi, M. M.; de Abreu, P. G.; Coldebella, A. Synthesis anCharacterization of Ethylic Biodiesel from Animal fat Wastes. *Fuel.* **2013,** *105,* 228–234.

30. Niju, S.; Begum, K. M. M. S.; Anantharaman, N. Modification of Egg Shell and its Application in Biodiesel Production. *J. Saudi Chem. Soc.* **2014,** *18,* 702–706.

31. Said, N. H.; Ani, F. N.; Said, M. F. M. Review of the Production of Biodiesel from Waste Cooking Oil Using Solid Catalysts. *J. Mech. Eng. Sci.* **2015,** *8,* 1302–1311.

32. Hindryawati, N.; Maniam, G. P. Novel Utilization of Waste Marine Sponge (*Demospongiae*) as a Catalyst in Ultrasound-Assisted Transesterification of Waste Cooking Oil. *Ultrason. Sonochem.* **2015,** *22,* 454–462.

33. Maddikeri, G. L.; Pandit, A. B.; Gogate, P. R. Intensification Approaches for Biodiesel Synthesis from Waste Cooking Oil: A Review. *Ind. Eng. Chem. Res.* **2012,** *51,* 14610–14628.

34. Nantha Gopal, K.; Pal, A.; Sharma, S.; Samanchi, C.; Sathyanarayanan, K.; Elango, T. Investigation of Emissions and Combustion Characteristics of a CI Engine Fueled with Waste Cooking Oil Methyl Ester and Diesel Blends. *Alex. Eng. J.* **2014,** *53,* 281–287

35. Prafulla, D. P.; Veera Gnaneswar, G.; Harvind, K. R.; Tapaswy, M.; Shuguang, D. Biodiesel Production from Waste Cooking Oil Using Sulfuric Acid and Microwave Irradiation Processes. *J. Environ. Prot.* 2012, *3,* 107–113.

36. Gnanaprakasam, A.; Sivakumar, V. M.; Surendhar, A.; Thirumarimurugan, M.; Kannadasan, T. Recent Strategy of Biodiesel Production from Waste Cooking Oil and Process Influencing Parameters: A Review. *J. Energ.* **2013,** Volume (2013), Article ID 926392, 10 pages. http://dx.doi.org/10.1155/2013/926392

37. Santacesaria, E.; Vicente, G. M.; Di Serio, M.; Tesser, R. Main Technologies in Biodiesel Production: State of the Art and Future Challenges. *Catal. Today.* **2012,** *195,* 2–13.

38. Noshadi, I.; Amin, N. A. S.; Parnas, R. S. Continuous Production of Biodiesel from Waste Cooking Oil in a Reactive Distillation Column Catalyzed by Solid Heteropolyacid: Optimization Using Response Surface Methodology (RSM). *Fuel.* **2012,** *9* (1), 156–164.

39. El-Gendy, N. Sh.; El-Gharabawy, A. S. A.; Abu Amr, S. S.; Ashour, F. H. Response Surface Optimization of an Alkaline Transesterification of Waste Cooking Oil. *Int. J. ChemTech Res.* **2015,** *8* (8), 385–398.

40. Ceclan, R. E.; Pop, A.; Ceclan, M. Biodiesel from Waste Vegetable Oils. *Chem. Eng. Trans.* **2012,** *29,* 1177–1182.

41. Verma, P.; Sharma, M. P. Comparative Analysis of Effect of Methanol and Ethanol on *Karanja* Biodiesel Production and its Optimization. *Fuel.* **2016,** *180,* 164–174.

42. Carlini, M.; Castellucci, S.; Cocchi, S. A Pilot-Scale Study of Waste Vegetable Oil Transesterification with Alkaline and Acidic Catalysts. *Energy Proc.* **2014,** *45,* 198–206.

43. Zhang, H.; Ding, J.; Zhao, Z. Microwave Assisted Esterification of Acidified Oil from Waste Cooking Oil by CERP/PES Catalytic Membrane for Biodiesel Production. *Bioresour. Technol.* **2012,** *123,* 72–77.

44. Balasubramaniam, B.; Perumal, A. S.; Jayaraman, J.; Mani, J.; Ramanujam, P. Comparative Analysis for the Production of Fatty Acid Alkyl Esterase using Whole Cell Biocatalyst and Purified Enzyme from *Rhizopus Oryzae* on Waste Cooking Oil (sunflower oil). *Waste Manage.* **2012,** *32,* 1539–1547.

45. El-Gendy, N.Sh.; Hamdy, A.; Abu Amr, S.S. Application of D-optimal design and RSM to Optimize the Transesterification of Waste Cooking Oil Using a Biocatalyst Derived

from Waste Animal Bones and Novozym 43. *Energ. Sources Part A.* **2015**, *37* (11), 1233–1251.

46. Atadashi, I. M.; Aroua, M. K.; Aziz, A. R. A.; Sulaiman, N. M. N. The Effects of Catalysts in Biodiesel Production: A Review. *J. Ind. Eng. Chem.* **2013**, *19,* 14–26.

47. Jazie, A. A.; Pramanik, H.; Sinha, A. S. K. Egg Shell as Eco-Friendly Catalyst for Transesterification of Rapeseed Oil: Optimization for Biodiesel Production. *Int. J. Green Econ.* **2013**, *2* (1), 27–32.

48. Borges, M. E.; Díaz L. Recent Developments on Heterogeneous Catalysts for Biodiesel Production by Oil Esterification and Transesterification Reactions: A Review. *Renew. Sust. Energ. Rev.* **2012**, *16,* 2839–28.

49. Farooq, M.; Ramli, A.; Subbarao, D. Biodiesel Production from Waste Cooking Oil Using Bifunctional Heterogeneous Solid Catalysts. *J. Clean. Prod.* **2013**, *59,* 131–140.

50. Borges, M. E.; Díaz, L. Catalytic Packed-Bed Reactor Configuration for Biodiesel Production using Waste Oil as Feedstock. *Bioenergy Res.* **2013**, *6,* 222–228.

51. Atapour, M.; Kariminia, H.R.; Moslehabadi, P.M. Optimization of Biodiesel Production by Alkalicatalyzed TYransesterification of Used Frying Oil. *Process. Saf. Environ.* **2014**, *92* (2), 179–185.

52. Nurfitri, I.; Maniam, G. P.; Hindryawati, N.; Yusoff, M. M.; Ganesan, S. Potential of Feedstock and Catalysts from Waste in Biodiesel Preparation: A Review. *Energ. Convers. Manage.* **2013**, *74,* 395–402.

53. Pullen, J.; Saeed, K. An Overview of Biodiesel Oxidation Stability. *Renew.Sust. Energ. Rev.* **2012**, *16,* 5924–5950.

54. El-Gendy, N.Sh.; Abu Amr, S. S.; Abdul Aziz, H. The Optimization of Biodiesel Production from Waste Frying Sunflower Oil using a Heterogeneous Catalyst. *Energ. Sources Part A.* **2014**, *36* (15), 1615–1625.

55. Margaretha, Y. Y.; Prastyo, H. S.; Ayucitra, A.; Ismadji, S. Calcium Oxide from P*omacea* sp. Shell as a Catalyst for Biodiesel Production. *Int. J. Energy Environ.* **2012**, *3,* 33. http://www.journal-ijeee.com/content/3/1/33.

56. Sorate, K. A.; Bhale, P. V. Biodiesel Properties and Automotive System Compatibility Issues. *Renew. Sust. Energ. Rev.* **2015**, *41,* 777–798.

57. Thanh, L. T.; Okitsu, K.; Van Boi, L.; Maeda, Y. Catalytic Technologies for Biodiesel Fuel Production and Utilization of Glycerol: A Review. *Catal.* **2012**, *2,* 191–222.

58. Cavazzuti, M. Optimization Methods: From Theory to Design. Springer: Verlag Berlin, Heidelberg, 2013. DOI:10.1007/978-3-642-31187-12

59. Buasri, A.; Worawanitchaphong, P.; Trongyong, S.; Loryuenyong, V. Utilization of Scallop Waste Shell for Biodiesel Production from Palm Oil-Optimization Using Taguchi Method. *APCBEE Proc.* **2014**, *8,* 216–221.

60. Montgomery, D. C. Design and Analysis of Experiments, 8th ed.; John Wiley and Sons, Inc.: New York, 2013.

61. Del vecchio, R. *Understanding Design of Experiments: A Primer for Technologists,* Carl Hanser Publisher, Munich, Hanser/Gardner Publications, Inc.: Cincinnati, Ohio, United States, 2015.

62. Paintsil, A. Optimization of the Transesterification Stage of Biodiesel Production Using Statistical Methods. Electronic Thesis and Dissertation Repository, Paper 1693, University of Western Ontario, Canada, 2013.

63. Rezaei, R.; Mohadesi, M.; Moradi, G. R. Optimization of Biodiesel Production Using Waste Mussel Shell Catalyst. *Fuel.* **2013**, *109,* 534–541.

64. Koohikamali, S.; Tan, C. P.; Ling, T. C. Optimization of Sunflower Oil Transesterification Process Using Sodium Methoxide. *Scientific World J.* 2012, Volume 2012, Article ID 475027, 8 pages. doi:10.1100/2012/475027.

65. Sathish Kumar, R.; Sureshkumar, K.; Velraj, R. Optimization of Biodiesel Production from *Manilkara zapota* (L.) Seed Oil Using Taguchi Method. *Fuel.* **2015,** *140,* 90–96.

66. Spitzer, M.; Wildenhain, J.; Rappsilber, J.; Tyers, M. BoxPlot R. A Web Tool for Generation of Box Plots. *Nat. Methods.* **2014,** *11,* 121–122.

67. BoxPlotR: a web-tool for generation of box plots. http://boxplot.bio.ed.ac.uk.

68. Antony, J. *Design of Experiments for Engineers and Scientists. 2nd ed.*; Elsevier Science and Technology Books: London, 2014.

69. LINDO Software for Integer Programming, Linear Programming, Nonlinear Programming, Stochastic Programming, Global Optimization. http://www.Lindo.com.

70. Anderson, J. M.; Whitcomb, J. P. *DOE Simplified: Practical Tools for Effective Experimentation. 3rd ed.;* CRC Press, Taylor & Francis Group: New York, NY, 2015.

71. Alhassan, Y.; Kumar, N.; Bugaje, I. M.; Mishra, C. Optimization of Gossypium Arboreum Seed Oil Biodiesel Production by Central Composite Rotatable Model of Response Surface Methodology and Evaluation of its Fuel Properties. *J. Pet. Technol. Altern. Fuels.* **2014,** *5* (1), 1–12.

72. Liu, C. C.; Lu, W. C.; Liu, T. J. Transesterification of Soybean Oil Using CsF/Cao Catalysts. *Energy Fuel.* **2012,** *26,* 5400–5407.

73. Wang, B.; Li, S.; Tian, S.; Feng, R.; Meng, Y. A New Solid Base Catalyst for the Transesterification of Rapeseed Oil to Biodiesel with Methanol. *Fuel.* **2013,** *104,* 698–703.

74. Wu, H.; Zhang, J.; Wei, Q.; Zheng, J.; Zhang J. Transesterification of Soybean Oil to Biodiesel Using Zeolite Supported CaO as Strong Base Catalysts. *Fuel process. Technol.* **2013,** *109,* 13–18.

75. Kaur, M.; Ali, A. Ethanolysis of Waste Cottonseed Oil over Lithium Impregnated Calcium Oxide: Kinetics and Reusability Studies. *Renew. Energ.* **2014,** *63,* 272–279.

76. Boey, P. L.; Ganesan, S.; Maniam, G. P.; Khairuddean, M. Catalysts Derived from Waste Sources in the Production of Biodiesel using Waste Cooking Oil. *Catal. Today.* **2012,** *190* (1), 117–121.

77. Dias, J. M.; Alvim-Ferraz, M. C. M.; Almeida, M. F.; Diaz, J. D. M.; Polo, M. S.; Utrilla, J. R. Selection of Heterogeneous Catalysts for Biodiesel Production from Animal. *fat. Fuel.* **2012,** *94,* 418–425.

78. Liao, C. C.; Chung, T. W. Optimization of Process Conditions Using Response Surface Methodology for the Microwave Assisted Transesterification of Jatropha Oil with KOH Impregnated CaO as Catalyst. *Chem. Eng. Res. Des.* **2013,** *91,* 2457–2464.

79. Mahdavi, V.; Monajemi, A. Optimization of Operational Conditions for Biodiesel Pproduction from Cottonseed Oil on CaO-MgO/Al_2O_3 Solid Bas Catalysts. *J. Taiwan Inst. Chem. Eng.* **2014,** *45,* 2286–2292.

80. Yücel, Y. Optimization of Biocatalytic Biodiesel Production from Pomace Oil Using Response Surface Methodology. *Fuel Process. Technol.* **2012,** *99,* 97–102.

81. Goyal, P.; Sharma, M. P.; Jain, S. Optimization of Transesterification of *Jatropha Curcas* Oil to Biodiesel Using Response Surface Methodology and its Adulteration with Kerosene. *J. Mater. Environ. Sci.* **2013,** *4* (2), 277–284.

82. Aworanti, O. A.; Agarry, S. E.; Ajani, A. O. Statistical Optimization of Process Variables for Production from Waste Cooking Oil using Heterogeneous Base Catalyst. *Br. Biotechnol. J.* **2013,** *3* (2), 116–132.

83. Veličković, A. V.; Stamenković, O. S.; Todorović, Z. B.; Veljković, V. B. Application of the Full Factorial Design to Optimization of Base-Catalyzed Sunflower Oil Ethanolysis. Fuel. **2013,** *104,* 433–442.

84. El-Gendy, N. Sh.; Hamdy, A.; Abu Amr, S. S. An Investigation of Biodiesel Production from Wastes of Seafood Restaurants. *Int. J. Biomater.* **2014,** Volume 2014, Article ID 609624. http://dx.doi.org/10.1155/2014/609624

85. Avramović, J. M.; Veličković, A. V.; Stamenković, O. S.; Rajković, K. M.; Milić, P. S.; Veljkovic V. B. Optimization of Sunflower Oil Ethanolysis Catalyzed by Calcium Oxide: RSM versus ANN-GA. *Energy Convers. Manage.* **2015,** *105,* 1149–1156.

86. Betiku, E.; Okunsolawo, S. S.; Ajala, S. O.; Odedele, O S. Performance Evaluation of Artificial Neural Network Coupled with Generic Algorithm and Response Surface Methodology in Modeling and Optimization of Biodiesel Production Process Parameters from Shea Tree (*Vitellaria Paradoxa*) Nut Butter. *Renew. Energ.* **2015,** *76,* 408–417.

87. Mahesh, S. E.; Ramanathan, A.; Begum, K. M. M. S.; Narayanan, A. Biodiesel Production from Waste Cooking Oil Using KBr Impregnated CaO as Catalyst. *Energ. Convers. Manage.* **2015,** *91,* 442–450.

88. de Almeida, V. F.; Garcia-Moreno, P. J.; Guadix, A.; Guadix, E. M. Biodiesel Production from Mixtures of Waste Fish Oil, Palm Oil and Waste Oil: Optimization of Fuel Properties. *Fuel Process. Technol.* **2015,** *133,* 152–160.

89. Stamenković, O. S.; Veličković, A. V.; Kostić, M. D.; Joković, N. M.; Rajković, K. M.; Milić, P. S.; Veljković, V. B. Optimization of KOH-catalyzed methanolysis of hempseed oil. *Energy Convers. Manage.* **2015,** *103,* 235–243.

90. Yang, J.; Corscadden, K.; He, Q. S.; Caldwell, C. The Optimization of Alkali-Catalyzed Biodiesel Production from *Camelina Sativa* Oil Using a Response Surface methodology. *J. Bioprocess. Biotech.* **2015,** *5,* 235. doi:10.4172/2155–9821.1000235

91. El-Gendy, N. Sh.; Ali, A. A.; Abu Amr, S. S.; Abdul Aziz, H.; Mohamed, A. S. Application of D-optimal Design and RSM to Optimize the Transesterification of Waste Cooking Oil Using Natural and Chemical Heterogeneous Catalyst. *Energ. Sources Part A.* **2016,** *38* (13), 1852–1866.

92. Kostic, M. D.; Bazargan, A.; Stamenkovic, O. S.; Veljkovic, V. B.; McKay, G. Optimization and Kinetics of Sunflower Oil Methanolysis Catalyzed by Calcium Oxide-Based Catalyst Derived from Palm Kernel Shell biochar. *Fuel.* **2016,** *163,* 304–313.

93. Li, Y. X.; Dong, B. X. Optimization of Lipase Catalyzed Transesterification of Cotton Seed Oil for Biodiesel Production Using Response Surface Methodology. *Braz. Arch. Biol. Technol.* **2016,** *59,* e16150357. http://dx.doi.org/10.1590/1678-4324–2016150357.

94. Onukwuli, D. O.; Emembolu, L. N.; Ude, C. N.; Aliozo, S. O.; Menkiti, M. C. Optimization of Biodiesel Production from Refined Cotton Seed Oil and its Characterization. *Egypt. J. Petrol.* **2017,** *26(1),* 103-110.

95. El-Gendy, N. Sh.; Deriase, S. F. Statistical Optimization of Bio-Diesel Production from Different Types of Waste Cooking Oils Using Basic Heterogeneous Catalyst. *IJCBS.* **2013,** *4, 79–88.*

96. JUS EN 14214. Automotive fuels. *Fatty Acid Methyl Esters (FAME) for Diesel Engines-Requirements and Test Methods*; Standardization Institute: Belgrade, Serbia, 2004.

97. El-Gendy, N. H.; Deriase, S. F., Hamdy A. The Optimization of Biodiesel Production from Waste Frying Corn Oil Using Snails Shells as a Catalyst. Energ. Sources Part A. **2014,** 36 (6), 623–637.

98. El-Gendy, N.Sh.; Deriase S.F. Waste Eggshells for Production of Biodiesel from Different Types of Waste Cooking Oil as Waste Recycling and a Renewable Energy. *Process. Energ. Sources Part A*. **2015,** *37* (10), 1114–1124.

99. El-Gendy, N. Sh.; Deriase, S. F.; Hamdy, A.; Abdallah, R. I. Statistical Optimization of Biodiesel Production from Sunflower Waste Cooking Oil Using Basic Heterogeneous Biocatalyst Prepared from Eggshells. *Egypt. J. Petrol.* **2015,** *24,* 37–48.

100. El-Gendy, N. Sh.; Deriase, S. F.; Osman, D. I. The Optimization of Bio-Diesel Production from Waste Frying Oil Using Response Surface Methodology and the Investigation of Correlations for Changes in Basic Properties of Bio-Petro-Diesel Blends. *Energ. Sources Part A*. **2014,** *36* (5), 457–470.

101. Mostafa, S. S. M.; El-Gendy, N. Sh. Evaluation of Fuel Properties for Microalgae *Spirulina Platensis* Bio-diesel and Its Blends with Egyptian Petro-diesel. *Arabian J. Chem.* **2017,** 10, S2040–S2050.

102. ASTM Standard D975s. *Standard Specification for Diesel Fuel Oils. Annual Book of ASTM Standards. Petroleum Products and Lubricants (I–III),* Vols. 05.01–05.03; American Society for Testing and Materials: West Conshohocken, PA, 1991.

103. ASTM Standard D7467. *Standard Specification for Diesel Fuel Oil, Biodiesel Blend (B6 to B20). In: Book of Standards* Volume: 05.04; American Society for Testing and Materials: West Conshohocken, PA, 2015.

CHAPTER 8

SMART MOBILE DEVICE EMERGING TECHNOLOGIES: AN ENABLER TO HEALTH MONITORING SYSTEM

HERU SUSANTO[1,2*]

[1]*Department of Information Management, College of Management, Tunghai University, Taichung, Taiwan*

[2]*The Indonesian Institute of Sciences, Jakarta, Indonesia*

Corresponding author. E-mail: heru.susanto@lipi.go.id

CONTENTS

ABSTRACT

The adoption of smart mobile device (SMD) emerging technologies in every dimension of life has been phenomenal. Ever-more sophisticated mobile technology has fundamentally altered the ways in which people communicate and conduct business. SMD has become indispensable for individuals nowadays, as it is considered as a common necessity. With the fast pace of technology advancement, SMD is no longer used only for "call and text" but is now is also able to perform other functions through a series of high-end applications. With the popularity and functions offered, SMD has seen an increase of demand. SMDs have been used in various areas such as engineering, education, healthcare, business, production, and logistics. The increasing presence of SMD in health institutions (mobile health), government, and NGO, initiatives to improve the information and communication technology (ICT) supported it. Mobile health makes it feasible for patients to collect and share relevant data at any time, not just when they happen to visit a clinic, allowing more rapid convergence to optimal treatment. Mobile health apps can contribute to a rapid, learning health system. The recent proliferation of wireless and mobile technologies provides the opportunity to connect information in the real world via wearable sensors and, when coupled with fixed sensors embedded in the environment, to produce continuous streams of data on an individual's biology, psychology (attitudes, cognitions, and emotions), behavior, and daily environment. These data have the potential to yield new insights into the factors that lead to disease. They also have the potential to be analyzed and used in "real time" to prompt changes in behaviors or environmental exposures that can reduce health risks or optimize health outcomes. The main goal of this study is to introduce a promising future research direction, which may shape the future of mobile health. This study reveals healthcare expectations of mobile health services and the way these services can enhance them through their business process re-engineering, which actively increases efficiency in various healthcare processes.

8.1   INTRODUCTION

Smart mobile device (SMD) has become indispensable for individuals nowadays as it is considered as a common necessity. With the fast pace of technology advancement, SMD is no longer used only for "call and text" but now is also able to perform other functions through a series of high-end applications, distinctive operating systems, fast connectivity, and features that create a compact SMD. The adoption of SMD technologies in every dimension of life has been phenomenal. Ever-more sophisticated mobile technology has fundamentally altered the ways in which people communicate and conduct business. "Smart devices are characterized by the ability to execute multiple, possibly concurrent, applications, supporting different degrees of mobility and customization and by supporting intermittent remote service access and operating according to local resource constraints".

With the popularity and functions offered, SMDs have seen an increase of demand. SMDs have been used in various areas such as education, healthcare, business, production, and logistics. The increasing presence of SMD in health institutions (Government and NGO) initiatives to improve the information and communication technology (ICT) supported it. In short, an SMD is an all-in-one compact mobile gadget, an ideal substitution for all of these bulky devices where it provides convenience, equipped with advanced application programming interfaces (APIs) for running third-party applications to have integration with the phone's operating system (OS) and hardware. The disruptive power of these new technologies and the accompanying waves of innovation they have sparked are transforming the healthcare industry, propelling stakeholders to reassess and repurpose how they provide services.

The healthcare and life sciences industry is recognized as one of the top three fields (along with consumer products and the financial services industry). Digital technologies, including ubiquitous mobile devices, can play a key role in transforming healthcare into a more-efficient, patient-centered system of care in which individuals have instant, on-demand access to their medical records and powerful clinical decision support tools that enhance them to actively participate in their treatment plans. As applications become more user-friendly and technologies expand their reach to remote populations, they will be seamlessly integral to the provision and financing of health services. The future of mobile communication device utilization in healthcare industry mirrors the bigger challenges facing this industry: to accommodate intense demand for value from purchasers and consumers, new ways of doing business that leverage technologies effectively are requisite. Mobile health is a key element of the industry's response to the market's quest for value.

The main goal of this chapter is to introduce a promising research direction which may shape the future of mobile health. This study highlights and reveals the SMD role in applying mobile health to conduct re-engineering processes of patient disease monitoring tools and the way these services can enhance healthcare through their re-engineering processes that actively increase efficiency in various healthcare processes.

The rest of this chapter is organized as follows. Related works on critical resources and ICT application emerging technology are discussed in Section 8.2. SMD features, trends, and technology development are discussed in Sections 8.3 and 8.4. Section 8.5 contains the business process reengineering (BPR) through mobile health and discusses BPR in the perspective of engineering technology. Section 8.6 discusses patient enhancement as an impact of re-engineering processes through mobile health. Finally, Sections 8.7 and 8.8 present our conclusions and recommendation.

8.2 LITERATURE REVIEW

Healthcare system is the system or program that is made available to the public and invested by government, private companies, or both. ICT is the use of computer, networking, and other physical devices such as mobile phones, infrastructure, and procedures to create, process, store, secure, and exchange all forms of electronic data. Health information technology (HIT) is an area that involves with creation, design, development, and maintenance of information system for the healthcare industry. In addition, health information systems (HIS) are predicted to improve efficiency, to reduce error, and to lower cost, while also providing for better consumer care and service; for instance, the management of patient care through secure and sharing of health information.

8.2.1 ICT IN HEALTH SYSTEM

HIT systems improve the quality, patient safety and reduce the cost of healthcare. HIT can be introduced in many forms such as electronic health record (EHR), electronic medical record (EMR), and clinical decision support system (CDSS). HIM is the main focus on managing medical records and since it became electronic, the overlaps between informatics grew. Moreover, the most frequently used HIT is the EMR, but it has been replaced by EHR that shows more extra information about the patient. In addition, personal health record (PHR) became more appeal as it kept patient health record in private. In other side, health information exchange (HIE) also gained interest as health information of the patient can be exchanged with other HIE within a region.

HIT play a major role in enhancing the quality of health system especially in terms of healthcare. There are many aspects where information technology can improve for example in documentation, ability in accessing any crucial information, and increase the communication availability. With improvements in information technology, decision making can be easily made of each party.

8.2.2 USAGE OF ICT IN HEALTH SYSTEM

Disobedience is one of the problems among patients with serious mental disorders. It creates challenges for the mental health professionals. To overcome this problem, the use of ICT has increased in order to prompt the patients, such as using text message and e-mail by stating the purpose of it being sent.

The usage of ICT has increased in the healthcare organizations, which is similar to what happened in other companies that rely in a well-developed ICT infrastructure. ICT infrastructures may include the use of web, databases, and network infrastructures. Additionally, HIS should ensure efficiency and security of information flows and, efficiency and proximity of health system. With availablity of IT in the health system, there is strong opposition toward the adoption of e-health systems called EMRs. EMRs allow the patient data information system from a paper record to electronic format in the form of files which is easier and enables more effective management. To add on, EMR is one type of the HITs. The functions of EMR is to keep and gather the patient's past medical records, to inform any medical care and to publish any results they conduct in diagnostics testing. EMR helps to reduce cost, improve healthcare service quality, and increase productivity among physicians.

EHR provides the patients a summary of their recent visit, medications, drug allergies, appointments, payments, and some medical forms through the internet. All of the patient information is kept secure in their database. Electroencephalograph is a new health technology that will monitor an individual's brain waves from home; this is to aid diagnosis of a disease through telemedicine. This technology is hoped to be the next generation device for communication in the developments of brain science and medical area, as this is useful in investigating human mental condition and health diagnosis.

A clinical document consists of patient's records, notes, discharge summaries, and doctor's referral letters. Natural language processing (NLP) is a successful device which helps to reduce cost in medical records and improve quality of the patient's health and the accuracy of the documents. This software is one of the successful programs to keep information and provide solution space for annotating and organizing the documents into the database for the ease of health professionals to analyze. The first unsupervised approach was produced which is known as Prefix Span; this is used for medical concept extraction. Second, C-Value and its extraction, NC-Value, were produced for statistics and linguistics information. Lastly, Text Rank algorithm is used for the document summary task.

The technology use for dental implant is CAD/CAM technology in the process, as it is automatic and precise. Dental implant helps to replace a damaged tooth with titanium.

Healthcare professionals consume extra time and resources when making decision for their patients as the decision made by healthcare professionals will be put into HIT systems. Thus, with these capabilities, it helps doctors and other medical experts to manage their patients with ease. These systems are called CDSS. CDSS is defined as "software applications that integrate patient data with a knowledge-base and an inference mechanism to produce patient specific output in the form of care recommendations, assessments, alerts, and reminders to actively support practitioners in clinical decision making". Hence, CDSS can make decision based on the health situation. Additionally, a decision support system (DSS) will help surgeons and doctors to schedule their patients. The main components of DSS are database, user interface, and DSS software system. To integrate DSS and HIS, an update service calls a web service in AIDA for a request to integrate data warehouse with data in HIS and shared the database to update the DSS database.

8.2.3 HEALTH FACILITIES OR COMMODITIES IN ICT

Healthcare organizations require doctors or nurse or hospital staffs to comprehend that their future in HIS, the need to construct supportable well-being frameworks, is irrefutably attached to great execution and administration, detailed and useful information for medications need to be easily understood to enable consumers to use their medications safely and effectively since some studies show that multimedia education can be helpful than the usual care. In addition, the programs use different kinds of style to issue the information such as words, diagrams, pictures, together with audio, animation, and video. Then, the finalized information can be accessed through DVD, CD-ROM, or the internet, and there is insufficient proof to replace written education or education by the health professionals with multimedia education. Therefore, it must be used together with usual care provided by health providers.

8.2.4 ADVANTAGES AND DISADVANTAGES OF ICT IN HEALTH SYSTEM USAGE

However, there is also the disadvantage of HIT such as, EHR where elder patients find it difficult to use this technology, especially in the consumer health information technology (CHIT) as this make them uncomfortable, less efficient, and uncontrollable, which they have to go online to use this systems. In future, the population of elderly will be increased as the standard of living increases. Physicians found it hard to use EMR as it makes them lose control of their workflow, provides too much irrelevant information about the patient, and causes distraction when the physicians interact with the computer which will lower their productivity and will not improve efficiency, as has been proven.

In addition, the data must be recovered from paper-based records that frequently do not have the desired features of electronic frameworks to have an inseparable tie to the patient's well-being. Exchanging the substance of a paper record to an electronic patient record (EPR) framework will not suffice later on for increases to the framework if the framework is to be utilized for patient-driven purposes, since doctor-driven frameworks do not gather all the information expected to look at patients. However, information technology in health system could be risky with the patient's information due to scammers in online world.

8.3 SMD USABILITY

The SMDs were intentionally used as enterprise devices and were considered expensive for most consumers. Now, a SMD is a necessity in today's day and age. In the 1990s the first smartphone device created was the IBM Simon in 1993. The Simon was made to act as a multipurpose type of phone that combines voice and data services into one package. This device has characteristics of a mobile phone, a PDA, and even a fax machine. In addition it has its own touch screen that could be used to dial phone numbers. In 1996, the Palm Pilot was invented; though it was not considered a smartphone, it does deserve some credit in popularizing the use of mobile data by enterprise users.

With every successful business these days, there must be a presence of technology which has been developing rapidly to an extent it has contributed in all sectors. SMD is a new device that should be functioned pre-eminently by using communications technology available. Explained that the influence of innovation drastically changes the way a business works. The usage of SMD has spread widely to the entire world in a short period of time. Ten years ago, the technology was only used by professionals, but currently it is used by the public for research, social networking, and educational purposes. The high demand for SMD is the result of the decreasing rate of sales of common mobile phone devices with common features. SMD does not take a long time to surpass computing through desktop, notebook, and net book.

The main advantage of using SMD is that it is fast and efficient. Based on the latest technologies on SMD, the largest shipment of the devices is Samsung, while Android dominates the most efficient operating system of the SMD. Consumers are perceived to be dependent on their SMD. Having been highly engaged with SMD means that the consumers not only have personal knowledge about their requirement but also have the personal experience about how to satisfy their needs. Consumers' behavior for future purchase is influenced by their past experience as they depend on SMD. The consumers' high dependency on SMD is positively correlated with consumers' future purchase behavior. Moreover, consumers who have prior experience with a SMD learn quickly from their experience. Whether the experience of depending on smartphones is positive or negative, experienced consumers will quickly adjust their subsequent evaluations in purchasing their next SMD.

To help patient and healthcare institutions improve the socio-economic situation, Generation Y needed to harness their skills toward enhancing social and economic policies. To ensure the plan is feasible, there was a development on technology to enable the youth or Generation Y to become more productive, creative, and innovative. Generation Y have read news, health information, and research through smartphone. This would lead to an increasing knowledge and are more social activeness that "SMD can move into classrooms worldwide, we may be witnessing the start of healthcare revolution."

8.4 TECHNOLOGY DEVELOPMENT OF EMERGING TECHNOLOGY

SMDs are upgraded with WAP, GPRS, and UMTS technologies to deliver next generation services, that is, 4G network which is a lightning speed internet connection for mobile devices—the fourth generation of wireless mobile communication technology, set to take over the current 3G network we currently use. With the availability of fast internet connection, providers foresee this as an opportunity to cater users with the ability of accessing their accounts even online transaction. 4G network offers a speed of 20 MB per second, faster than home broadband connection but slower than the current home fibre optic offerings which provide speed of up to 100 MB.

However, another advantage of SMD is the availability of thousands of small applications—called apps. A wide variety of apps can be downloaded from online stores, allowing one to customize and personalize their phones to the way they like. Some of the apps can also enhance the use of SMDs, making them more efficient to the users. Mobile apps can make travelling easier with the maps apps that guide user to discover new places, learn language, check flight's departure time, trigger wake-up call, and keep in touch with the home country news, family, and friends through all of the social media such as sharing the current location, photos, skype, or email facility. Apps such as calendar with reminders also help some to remember important tasks and keep one well-organized of their time, and also the QWERTY keyboards make handling these handheld devices easy and simple. The size of the apps is commonly small; it is indicated that each SMD could have at least 20 apps in them. Though SMD market size is approximately 2 million, it is stated that potentially 40 million market sizes of mobile apps are available. The new business segmentation is created here; mobile apps providers have to be proactive to provide a range of mobile apps to grab the market demand.

The healthcare organizations require doctors or nurses or hospital staffs to comprehend that their future in HIS, the need to construct supportable well-being frameworks, is irrefutably attached to great execution and administration. The healthcare organization required scholastics and specialists to quit offering the most recent administration designs and to help the framework to contextualize its decisions over details and controllers to enhance their insight into such details, to take educated choices and to contribute in the advancement of those territories where there is an aggregate deficit in execution administration, for example, group administrations.

Part of the expansions in general well-being consumption has been coordinated for the utilization of steady innovations, planning to upgrade social insurance procurement. The utilization of medicinal services has advanced with the fast improvement of information technology and advances in social insurance innovation in parallel with current concerns emerging over patients' security and obviously how to cure patients effectively. The important thing in any information system is the accuracy of the data being provided. Data input is derived from the patients and other available sources which contain all of the patient's medical record. The information will then be processed and the output which comprises all of the patients' medical records.

Another dimension is the workflow and communication that define the procedure that assures that patient care tasks are done effectively. Next two dimensions are internal organizational features such as policies, processes, culture, and external rules and regulations. Both of these dimensions provide many features of the preceding dimensions. Last but not least of the eight-dimensional model is the measurement and monitoring. These dimensions measure and evaluate both known and unknown effects of the usage and implemented HIT.

HIT is the combination of information systems, computer science, and healthcare. As a result of advancement in technology, it is gaining attention globally. With the available of HIT systems, patients can monitor either in clinical setting or from outside especially home. Patient monitoring systems applied sensor network technology for collecting physiological data of a patient who is suffering from different diseases such as diabetes. For instance, Jog Falls, a diabetes management system using sensor devices for collecting physiological and activity data, monitors patient's physical activities such as food intake. Furthermore, a system of monitoring such as Type 1 diabetic patient using mobile phone for a diet management system and web-based medical diagnosis that is used to predict patient's condition. Additionally, HIT is beneficial to the patients and the providers as it helps to improve the patient's health as well as cost saving as they can access the information easily.

The usage of electronic HIT is to store health information such as EHRs, claims data, registries, and payment system. Primary care providers in North Carolina have used health information to help better asthma care and look for their performance on a number of key metrics, decrease hospital admission rates, and emergency room admissions. One of the evidence developments of HIT is that it can be used to organize investigation including comparing the success of observable studies. In addition, Cancer Care Outcomes Research and Surveillance Consortium (CanCORS) project uses many information that requires IT such as demographic, contact and medical information to investigate information on lung and colorectal cancers in America. Another evidence of development of HIT is the use of claims data, which shows that it enables to look for nearly 17,000 patients over a year to discover relative risk of heart attack among patients taking both drugs and started to warn physicians through the results.

EVIDENT Program is the one of the theory-based intervention programs. This program is useful for giving information specifically under healthy nutrition and exercise activities about the patients that they record daily. They can access their information through their own mobile online applications in the smartphones, and so some comparison on their everyday activities and food intake, which makes progress monitoring easier for the patients. Second, social-cognitive theory is produced to monitor obesity treatment by the research group in online or traditional face-to-face method. This program will monitor their nutrition and exercise activities as they will be provided counseling. Lastly, this mobile intervention program helps to prevent an unhealthy behavior among the teenagers, such as smoking. This program provides information which can aid to develop social strategies and provide useful activities for the adolescent.

Patients with severe mental health problems are more likely to disobey their medical treatment such as forgetting their appointment and when to take the medicines. This can cause them not getting the medicines at the right time which can lead to poor health condition and being hospitalized. Several strategies have been introduced to help the patients with their medications. One of the strategies is prompting, which helps to remind the patients to follow their treatments by using telephone calls, sending letters, and personal visit by the hospital's staff. Recently, another type of prompting has been introduced which is information and technology-based prompts. Laboratory information systems (LIS) are a complicated machine that functions in the information systems, integrated clinical information systems, and Electronic Laboratory Records (EMRs). LIS creates a link between analyzing in the laboratory, medical technologies and clinical providers. This can help to monitor and improve the quality of healthcare; therefore, this reduces any human error that could be made. In addition, LIS should provide computerized reports as it can support a high performance laboratory and automatic laboratory results. However, designing a LIS is a challenge as LIS must have the ability to communicate across the technology platforms. This means that new laboratory IT systems will be incompatible with the present laboratory hardware. Furthermore, poor performance can happen when the cost of administration is high and use of information technology is limited.

EHR is accepted due to chronic conditions, as before it was rejected by most of the patients. This facility was introduced in other countries, as this was proved to improve patients' health and effectiveness of treatment. With this proved, there is an increase shown in terms of investment. EHR is different from EMR; EMR provide each patient's information such as drug allergies, drug-to-drug interactions, and past treatment in one hospital. EMR also consists of sensitive personal information such as sexually transmitted disease, abortions, emotional problems, and instances of physical abuse. Other than that, basic information of the patient are also included, such as, height, weight, and blood pressures. EMR is only used within the hospital. However, EHRs provide a wide view for the patients' healthcare. PHRs help EMR to keep the patient information data and to help the patients by sharing the information which other hospitals may have a better cure for the patients. Furthermore, EHR elaborates the usage of these systems by taking care and giving more safety to the patient's health.

The patient's information privacy in EMRs still remains a concern to them and the hospital staff. Nurses have an important role for collecting the patient's information and keep them private, and they can secure the record-keeping in effectively and efficiently. However, some of the nurses are only familiar in paper-based medical records and have no intentions to keep the information private and secure. These issues can be improved by giving adequate training to all nurses, covering ethics, information security procedures, and IT skills. Nurses also have to know the importance of using EMRs and should know what information should be kept private. In physicians' case, they have difficulty in using EMR, as most physicians have practice in paper-based medical and now they have to change into EMR environment. This difficulty occurs among the elder physicians; they have been using paper-based medical records for years and they do not have any technical skills. In addition, EMR gives them a hard time to all the patient medical needs and they hoped EMR help them to work effectively and efficiently; however that did not happen. NVivo9 was introduced to improve the ability to access information and they can make better decisions for the patients' care within a short time accurately. This will improve the patients' health and increase healthy population. Lastly, physicians can improve time-management with other staff through the EMR messaging capability.

Electroencephalograph is cheap, light with approximately 100 g, easy to set up and has a rechargeable battery. However, the battery only gives 1–2 h of consecutive measurements, and the electrodes are too small due to the electric potential. Other problems occur when it is easy to setup because only a few electrodes used in the central part of the head, and this can cause the life and right side are more advantageous. Second problem occur when they have to use a simplified electroencephalograph. This simplified electroencephalograph is used for medical and brain science area of research. It will give a person caution about his/her health conditions. One example of simplified electroencephalograph is the 3B Band which is produced by the B-Bridge International and this type of headband is known in the world.

This is a headband which contains NeuroSky chips. 3B Band uses Bluetooth to connect to a computer; therefore, the usage of wire is not needed, and a person who uses this headband will not feel bound. However, these headbands have an implication that the shape of earlobe and sweat status can lead to a wrong measurement of the brain wave.

NLP is an unsupervised approach to manage a clinical document, and there is three types have been introduced. TextRank was the first unsupervised approached being familiarized by Mihalcea and Tarau, this is to aid staff to summarize text. Second types of unsupervised approach is the C-Value and NC-Value that introduced by Frantzi et al., this C-value helps to produce a unit-hood score depends on the length of the phrase were entered. NC-Value helps to enhance the accuracy and quality of the term data entered. And lastly, frequent sequence mining or known as PrefixSpan help to cooperate long text in little time needed, and to get use this data only a minimal training is needed.

CAD/CAM technology helps the production of dental implants easy and it can be done in-house if there resources are available to a collaborating partner. This technology needed a massive investment for Research and Development (R&D), as this step needs a huge amount of money for the special kits and accessories to make a successful dental implants, however, this will give them a long-term benefits. The benefits are saving times and money and meets the international standards in terms of technology of producing dental implants.As health consumers demand for exact and evidence based information to be delivered in a way that is easy to understand about their health and its proper treatments, health professionals want to save consultation time and improve the medication compliance. Consultation is usually presented verbally alongside with written materials. Consequently, high chances that the patients will forget about the information that was delivered to them. Therefore, multimedia educational programs give more benefits as it is convenient because it can be accessed anywhere by the individuals and their families which is cost saving rather than having consultation with the doctor. Additionally, multimedia programs also allow individuals to alter the information according to their need.

In ICU, healthcare system is needed as it gives relevant and accurate information on time. However, these high technologies give some limitations such as high cost of maintaining the machines, delayed products being sent, and wrong delivery date. RFID is one of enforcement technology which monitored a manufacturing in a critical healthcare such as stents. The act of passing the duty of care for a patient to another nurses is called nursing handover. During this time, there is a possibility of getting error especially when the important medical information is not shared efficiently, accurately and in a timely manner thus may result in adverse events (AEs). Therefore, Information Technology is used to support the process in order to decrease the potential risk such as miscommunication, misunderstanding, and the omission of crucial information. Furthermore, poor handover might cause delays for the patients' treatments. Consequently, an accurate handover of important information is crucial to continuity and safety of care for hospitalized patients. In any field, particularly in the well-being framework field, one of the numerous focal points of data innovations is to offer the specialists, some assistance with nursing, healing facility staff and patients to pay in contact with one another inside and out structure the association. A standout amongst the most critical points of interest of data advances is the formation of one exhaustive asset, which is upgraded and utilized by specialists, medical attendant or doctor's facility staff.

One of the challenges for retrieving the health information is the mismatch between a consumer's terms and professionals vocabularies used in medical literature. To overcome this, a system called MeshMed has been introduced where it combined different functional search into one. It has two new search components known as term browser and tree browser which provide unique information about the search topic. In addition, Mesh vocabulary is downloaded from the National Library of Medicine (NLM) in Extensible Markup Language (XML) format to support both browsers. Both browsers provide quick access to the information needed which make it more efficient and beneficial to find the definitions as well as synonyms for medical terms. The advantage and disadvantage of any system in Health System can only be known when the users such as doctors, surgeons and patients, accept and use it. Therefore, there are studies that facilitate the adoption of use of HIT system. Technology Acceptance Model (TAM) is one of the popular theories for studying the perception and factors to the acceptance of a new technology in this case the acceptance of technology in healthcare system. Davis (1986) designed TAM for modeling user acceptance of Information Systems. Moreover, by promoting its acceptance, it will increase the use of IT. TAM focuses on the users' behavioral intentions towards accepting a new technology specifically self-diagnosis system for reducing cost and improving the quality in healthcare system. Not only that, TAM was re engineer to TAM2 and TAM3. TAM2 focused on identifying sources of usefulness and moderating variables and TAM3 centered on interventions that can affect the acceptance and use of IT in a healthcare system. A new model developed from TAM2 called Information and Communication Technology Acceptance Model (ICTAM). ICTAM is for predicting and showing consumer's health information and services usage behavior on the internet. An intervention program has been introduced into the online applications and this has been beneficial to the patients. Moreover, interventions have the intention to provide a prevention or treatment to the patients who need them, which this is depending on the theory and model. An example that when researchers such as the clinicians uses electronic health information, they can easily look for patients health information anytime and anywhere, in other words, electronic health information is very convenient and they can observe their quality in terms of their healthcare services which may vary with the other healthcare center. Different policies may give different outcomes like in designing formula and payment procedure. Therefore, HIT such as EHR or EMR enables efficient early investigations and this may also be one of the advantages of HIT.

Healthcare has invested a large amount of money for Information Technology (IT). The utilization of IT and Information System (IS) has taken different directions due to the demands and needs in various sectors of government and public organizations. The technological advancement in the form of electronic patient s records, clinical applications, and health management information system (HMIS) has encouraged many healthcare organizations to use it. However, this system can be hard to implement in both public organizations and healthcare organizations. Discussed and believed overseen care focuses on decreasing conveyance costs and enhancing medicinal services financing through strict use administration, money related impetuses to doctors and restricted access to suppliers. Right now, oversaw care exists as the overwhelming financing and conveyance framework; as anyone might expect, access, expense, and quality predicaments are imperative; procedures and arrangements, in this manner, must be received to address these issues. A methodical examination of these systems is valuable for administrators and experts endeavoring to enhance care quality under healthcare organization.

The acceptance of using HIT is low as some believe that it is not very useful and has less benefit. HIT needs to focus on changing the care services, improve patient results, created to support the needs the hospitals. In order for HIT to be successful is that its focal point must be on the usage or assumption rather that the effect of the people's health. Therefore, if HIT is being applied in a right way, it may be successful in terms of supporting and expand clinician and patient efforts to intensify the population's health and welfare. New generation in information technology helps in improving the quality of healthcare services for example in handling the documentations especially the patient's information as well as their confidential prescription.

Decision making in each organizations also plays a major role in having the information technology where it can fasten the decision making amongst the end-user not only the administration management, but it also include the medical expert. The functions in IT involves interchange any content of information which leads continuation of healthcare management, between departments in the organizations, helps the end user to decode any scientific documentation into practice, treatment procedures and healthcare system's security.

Administration control frameworks as an apparatus for top supervisors to control doctors and to distribute top-down assets. Few associations have grown more cooperative methodologies, in which the administration control framework gives a chance to encourage dialog in the middle of administration and doctors. Targets and assets are not just allotted top-down, but rather will be fairly arranged from the base up and taking into account a common "sharing of psyches," a typical comprehension of issues and a mutual need setting where showing clinics were and still are late movers, as colleges opposed the presentation of any execution estimation framework. EHRs can lead to dangerous matters due to internet access. Scammers can easily access any patient's record history as well as able to change any content or information within the records itself. Security in information technology department plays a crucial part to secure any confidential report.

8.5   ENABLER OF HEALTHCARE BPR

Mobile health as ICT emerging technology brings new paradigm and business processes of healthcare institution. In line with that, mobile health contributes to rising the patient enhancement, particularly in monitoring of their disease and symptom. Here, ICT contributes efficiency BPR to healthcare organizations which provide a high-transact organizational outcome. The information processing competences certify by modern ICT. For example, enterprise software provides a probable data support to healthcare organization and also work cells with the opportunity being able to do their work efficiently and productively. Moreover, broadband networks provide a convenient way that allows employee teams to get access among each other through network wherever they are. Managers are able to keep track on the performance of their employees through internet based networks which enable everything follow in their plan with all the decision and rules that is being implemented[10.]

Properly implementing information technology in BPR of healthcare institutions can have several advantages to the companies. Turnaround time can be reduced by using information technology rather than manual approaches which will be more time consuming. Less chance of fraud and corruption will occur and also more quantity of work such as reports can be done in less time. It can produce good quality of work results, services and products and in a team, a quick communication can be formed. Beside in a team, faster communication can also formed with the customer and other stakeholders with the help of information technology in BPR.

Khalil[2] states that there are four ways in which ICT contributes in BPR, which include: managing BPR project, provide technological vision, developing process-oriented and process management. Chan[3] states an important aspect of ICT are covered, including the roles of IT; as the initiator, facilitator and enabler of a business process, the logical, physical and conceptual views of ICT, a review on the impact of ICT on a design, specific business processes and management. Attaran[4] argues that the way a business is coordinated and developed depend on its investment in ICT. The increasing affordability of ICT in relation to its function may have influenced the business to proceed with the investment. ICT achieves the reengineering goals in three ways: (1) information sharing and enhanced communication between those involved in an organization from top-level management to labor workers, (2) improving the performance of business processes, and (3) acts as a framework for planning, optimizing and risk management.

ICT is also required to achieve the proper understanding and evaluation of an organization as a whole, its requirements and system purpose. The mere introduction or modification of ICT in a business organization initiates BPR because it is crucial to recognize the impact of ICT implementation in BPR during the designing stage. They also claim that ICT plays an important role to coordinate the changes in the business processes. Radman[5] illustrates how the scheme of the design relies on the ICT infrastructure; it acts as a guideline to ensure that the chosen ICT infrastructure is suited to the BPR project/processes. It ensures that the results of BPR processes remain positive in any typical healthcare organization; if ignored it may cause a downfall to the project. The BPR team plays a significant role in the redesigning process as they are the ones responsible on ensuring that the IT chosen for the process is compatible with the new system.

Krishnankutty[6] gives a brief explanation of how ICT is adapted to BPR. It also concludes that the success of a BPR project revolves around the relationship between BPR and ICT. Asgarkhani and Patterson[7] state that ICT is an important component that can enhance the effectiveness and efficiency of an organization in many aspects. They also mention how the business systems are running presently. ICT creates a better relationship between the healthcare organization, the customers, and the suppliers. They summarize that ICT is needed to correct the errors in the old processes due to the fragmentation of old processes and poor system flow. Moreover, the internet makes it possible to centralize and decentralize which allow users to take control of the processes. Najjar et al.[8] emphasize the impact of ICT on the business and customers' networks and its capability in business operations. When healthcare organization can maximize the ICT usage, effective and efficient implementation of BPR can be achieved.

8.5.1 ROLE OF ICT

The term of BPR is in order to achieve high efficiency and productivity in an organization, there must be a major change or modification in how the business is process. In addition, it is also stated that the implementation of information technology in an organization will improve work quality.[9] Trkman (2010) states that TTF theory which stands for Task Technology Fit shows that information technology will help to improve an individual performance, depends on how the user uses to perform the task. Moreover, according to Huscroft (2013), TTF theory suggests that when it is accompanied with preferable ICT skills, it will increase productivity outcomes. Here, the organizations that are adopting information technology in improving their BPR would be a likely to experience the following benefits:

8.5.2 WORKERS EFFICIENCY

It leads to an increase in workers efficiency where manual work is done automatically where the processes become much faster, convenient and less time-consuming.

8.5.3 EASY COMMUNICATION

ICT may lead to easy communication as it breaks down communication barriers within the organization, both internal and external. In other words, information technology can connect people throughout the world. Due to broken barriers, it is useful in analyzing the company's strengths and weaknesses and market structure and opportunities.

8.5.4 BUSINESS ASPECTS

The opportunity of ICT provides utilization of newer and better technology to develop a strategic vision and to help improve the business process before it is designed. It supports the organization's mission and reduces costs (Joseph Sungau, 2012).

8.5.5 CUSTOMER SERVICE AND INTERACTION

ICT improves customer service as it provides faster respond rate to customers such as boosting service delivery/request. Also, the interaction becomes world-wide which organizations can easily interact with their customers and even business partners in any parts of the territories or country base through e-mail, social media, and video conferencing.

8.5.6 INCREASE PRODUCTION RATE OR QUALITY OF GOODS

ICT helps to increase the productivity of the business and be able to maintain a high quality of output by using the necessary software that can detect faulty goods or Total Quality Management (TQM) as an example or any other quality checking services.

8.6 SHIFTING PATIENT ENHANCEMENTS: AN IMPACT

Despite the huge amount of money spent on ICT, productivity has not matured as expected. ICT supports most re-engineered processes; however, ICT should not be neglected. In order for ICT to play a technology leadership role, ICT requires innovative and visionary thinking. In BPR, ICT serves as a tool to support and redesign business processes and assists the progress of cross-functional workflow. Technology implementation, however, is a complex specialty. Process orientation in a BPR environment demands building information systems along the cross–functional processes. Building systems to support the newly re-engineered business processes requires strategic planning of information systems across the organization. If BPR mandates the organization of business tasks from the cross-functional process perspective, a top–down approach for information systems development is called for. An efficient and relevant IT system is an essential

BPR enabler. Without such system, it is not possible to keep a check on all factors affecting the change. Mobile health as part of the process re-engineering, particularly in the healthcare organizations that empower patient by increasing its understanding of the diseases when facing with the symptoms.

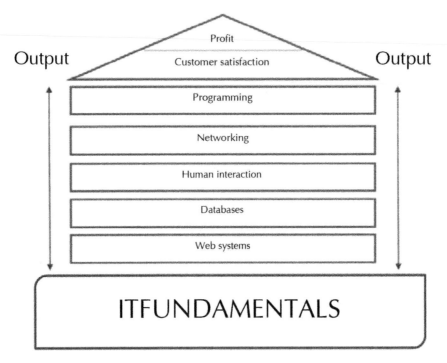

FIGURE 8.1   mHealth and ICT as enabler for re-engineering process.

Moreover, development and treatment of diseases take place in daily life outside of traditional clinical settings. To determine and adjust treatment for these diseases, clinicians depend heavily on patient reports of symptoms, side effects, and functional status. Here, mobile health makes it feasible for patients to collect and share relevant data at any time, not only at the time when they visit a clinic, allowing more rapid convergence to optimal treatment. For example, a patient with epilepsy can self-report on drugs and dosages taken and the number and severity of seizures and side effects. The app sends this data in real time to the clinician, who can look for patterns of response and guide the patient to titrate his medications over weeks instead of months. mHealth apps can contribute to a rapid, learning health system, but this may be difficult if each app is built as a closed application with its own proprietary data format, management, and analysis. Such a "stove-pipe" or "siloed" approach fundamentally limits the potential of mHealth by impeding data-sharing with other apps and with EHRs and PHRs. Inefficiencies and lack of innovation plague HIS that are closed and rigid. The mHealth architecture, design, and configuration are fine-tuned to the client-server system. Client-server applications are a rapidly growing trend in the interconnected environment, especially in an organization with national and global capacities. As such, mHealth is expected to be the center point for the patient disease monitoring.

An open architecture built around shared data standards and the global communication network already in place to support interoperable voice and data transfer can promote the scaling, coherence, and power of mHealth. Such architecture should complement broader ongoing developments for scalable and sustainable HIS, including various national and international initiatives. Open Architecture Benefits In an open architecture, components have well-defined, published interfaces that allow interconnection and use in ways other than as originally implemented or intended. They allow interested parties to expand the functionality of the system without modifying existing components.

Currently, patient is the recipient of the information as a medical decision. This study enhances patient by increasing its understanding of health status and improving patient's confidence when facing the symptoms of disease. Mobile Health is different from the existing methods for helping patient comply with health status in two ways. First, the software is multiuser and it is developed from the framework. The architecture, design, and configuration are fine-tuned to the client-server system. Client-server applications are a rapidly growing trend in the interconnected environment, especially in an organization with national and global capacities. As such, mH-SFM h is expected to be the center point for the disease investigation and real-time monitoring. Second, mobile health is a tool that could help patient to perform self-assessment, monitor, and maintain their disease. Mobile health can quickly provide compliance information with regards to swam fever symptom controls. mH-SFM also monitors current circumstances associated with swam fever issues, including the symptom, effects, and recovery scenario.

ICT and mobile health emerging technology seen as a driver of BPR to run reengineering projects as mobile health has a central role in the completion of the activities throughout the BPR life cycle as well as in the redesign of the processes and their supporting systems in the area of healthcare organizations. Mobile health should be a partner that can provide education, support, and leadership when appropriate. The ICT function has to be repositioned to facilitate BPR by investing in Mobile health tools and provide required end-user skills. Moreover, it contributes to BPR in four ways: (1) Managing BPR project, (2) providing technological vision, (3) developing process oriented, (4) process management. BPR projects are risky, thus, those involved in BPR need necessary ICT skills to manage them. Without powerful ICT tools and ICT expertise in system implementation, the engineering initiatives could not proceed beyond modest prototypes. IS can help BPR teams control some of the other challenges and risks in managing large–scale projects, including the complexity, resistance to change, lack of experience and understanding migration, hence it is important that the BPR project is managed properly.

8.7 CONCLUSIONS

The study is expected to contribute in developing mobile health monitoring tools comprehensively and their implication toward extending the role of patient and organization process re-engineering, through ICT and SMD as enabler. It is seen that the roles of SMD and ICT play in BPR are almost limitless, taking note of their constantly advancing nature. This has led to the dependence of BPR and its methodology of ICT and SMD in various ways. The four major roles of ICT include managing BPR projects, providing technological vision, developing process-oriented and process management. This study helps patients by increasing their understanding of health status and improving patients' confidence when facing the symptoms of disease. Mobile health is different from the traditional existing methods for helping patient comply with health status in two ways. First, the software model is multiuser and it is developed from the framework. The architecture, design, and configuration are fine-tuned to the client-server system. Client-server applications are a rapidly growing trend in the interconnected environment, especially in an organization with national and global capacities. As such, mobile health is expected to be the center point for the disease investigation and real-time monitoring. Second, mobile health is a tool that could help patient to perform self-assessment, monitor, and maintain their disease. Mobile health can quickly provide compliance information with regards to swam fever symptom controls. Mobile health also monitors current circumstances associated with diseases issues, including the symptom, effects, and recovery scenario.

8.8 RECOMMENDATIONS

The future direction of this study can also accommodate and customize mobile health monitoring tools to fit with special symptom of disease. Mobile health was designed as an adaptable framework that can be implemented and extended to other type of disease by customizing its controls. Mobile health could possibly be implemented to others by following mapping stages through the grouping of controls to respective domains in each of these diseases. The ideas for further enhance the functionality of Mobile health in future developments is rising. It is suggested the integration of more complete and robust support features, particularly the integration of disease monitoring and DSS, an expert system and a disease pattern recognition system, complemented with a knowledge inference and learning system to emulate the decision making ability of a human expert. This software ability could solve compliance barriers, create an early warning system for suspected symptom, and help enhance strategic planning of patients.

KEYWORDS

- **smart mobile devices**
- **mobile health**
- **information and communication technology**
- **business process re-engineering**

REFERENCES

1. Khalil, O. Implications for the Role of Information Systems in a Business Process Reengineering Environment. *Int. J. Tech. Res. Appl.* **1997,** *2* (3), 149–156. Retrieved from http://www.ijtra.com/view/information-handing-through-information-technology-introduction.pdf
2. Chan, S. L. Information technology in Business Processes. *Bus. Process Manag. J.* **2000,** *6* (3), 224–237. doi: 10.1108/14637150010325444
3. Attaran, M. Exploring the Relationship between Information Technology and Business Process Reengineering. *Inf. Manage.* **2004,** *41* (5), 529–684.
4. Radman, A. E. H. S. A. A model for improving the role of IT in BPR. *Bus. Process Manage. J.* **2008,** *14* (5), 629–653. doi: 10.1108/14637150810903039
5. Krishnankutty, S. S. T. The role of BPR in the Implementation of ERP Systems. *Bus. Process Manag. J.* **2009,** *15* (5), 653–668. doi: 10.1108/14637150910987892

6. Asgarkhani, M.; Patterson, B. *Information and Business Process Re-Engineering through Application of Information and Communication Technologies (ICTs)*. Paper Presented at the International Conference on Recent Trends in Computer and Information Engineering (ICRTCIE'2012), Pattaya. Retrieved from http://psrcentre.org/images/extraimages/412149.pdf

7. Najjar, L;, Huq, Z.; Aghazedah, S. M.; Hafeznezami, S. Impact of IT in Process Improvement. *J. Emerging Trends Comput. Inf. Sci.* **2012,** *3* (1). Retrieved from http://www.cisjournal.org/journalofcomputing/archive/vol3no1/vol3no1_5.pdf

8. Panda, M. IT Enabled Business Process Reengineering. *Int. J. Inf. Technol. Manag. Inf. Syst.(IJITMIS).* **2013,** *4* (3), 85–95. Retrieved from https://www.academia.edu/5371023/IT_ENABLED_BUSINESS_PROCESS_REENGINEERING

9. De la Vara, J. L.; Sánchez, J.; Pastor, O. Business Process Modelling and Purpose Analysis for Requirements Analysis of Information Systems. *Adv. Inf. Sys. En.* **2008,** *5074,* 213–227. doi: 10.1007/978-3-540-69534-9_17.

10. Ramirez, R.; Melville, N.; Lawler, E. Information Technology Infrastructure, Organizational Process Redesign, and Business Value: An Empirical Analysis. *J. Decis. Support Syst.* **2010,** *49* (4), 417–429.

PART III
Analytical, Computational, and Experimental Techniques

CHAPTER 9

PRINCIPAL COMPONENT, CLUSTER, AND META-ANALYSES OF SOYA BEAN, SPANISH LEGUMES, AND COMMERCIAL SOYA BEAN

FRANCISCO TORRENS[1*] and GLORIA CASTELLANO[2]

[1]*Institut Universitari de Ciència Molecular, Universitat de València, Edifici d'Instituts de Paterna, P.O. Box 22085, València E-46071, Spain*

[2]*Departamento de Ciencias Experimentales y Matemáticas, Facultad de Veterinaria y Ciencias Experimentales, Universidad Católica de Valencia San Vicente Mártir, Guillem de Castro-94, València E-46001, Spain*

*Corresponding author. E-mail: torrens@uv.es

CONTENTS

ABSTRACT

Different types of soya bean, Mediterranean legumes, and commercial *soya bean* are classified by principal component analyzes (PCAs) of proximate, amino acids (AAs) and isoflavones contents, and legumes cluster analyses (CAs), which agree. Samples group into two classes. Compositional PCA and legumes CA allow classification and concur. The first axis explains 65%, the first two, 82%, the first three, 93% of the variance, etc. Legumes according to proximates, etc. are different depending on *energy* and AAs (Lys, Met, Trp). Spanish and Asian legumes result in separate classes. Macronutrients appear unconnected. Soya bean is compared to virgin olive oil in terms of iso/flavones and γ/α-tocopherol. It presents adequate proximate, AAs and isoflavones contents, good antioxidant capacity and may be used as a functional food. It represents a legume useful as a natural source for nutraceutical formulations.

Hymowitz discussed the domestication of Chinese soya bean *Glycine max* (L.) Merrill.[1] Gómez Garay reviewed its virtues in human food and health.[2] He compared its composition with the legumes most commonly used in Mediterranean diet (chickpea *Cicer arietinum*, bean *Phaseolus vulgaris* L., lentil *Lens culinaris*) and those commercialized as *soya bean* (genus *Vigna*: Indian mung bean *V. radiata* L.). Proximate contents of legumes in Spanish diet (cf. Table 9.1) do not show a soya bean advantage as regards proteins amount, lower than some beans, although it is greater than lentils and chickpeas. However, soya bean adds more fats to diet

TABLE 9.1 The Results of the Proximate and Isoflavone Compositions of the Analyzed Dry Legumes.

Species	Energy[a]	Fat	Protein	Fiber	Ca	Vit-A	Daidzein	Genistein
1. Chickpea *Cicer arietinum*	200.0	3.0	12.5	9	160	4.5	101.500	141.500
2. Bean *Phaseolus vulgaris* L.	225.0	0.2	37.5	15	48	8.5	23.500	268.000
3. Lentil *Lens culinaris*	275.0	1.0	15.0	10	60	1.9	6.500	13.000
4. Soya bean *Glycine max* L.	362.5	17.7	37.5	15	226	2.0	33.250	55.450
5. Mung bean *Vigna radiata* L.	340.0	1.6	27.0	5	110	3.0	25.500	34.000

[a] Compositions: proximates (i_1, energy in kcal/100g; i_2, fat in %; i_3, protein; i_4, fiber; i_5, Ca in mg/100g; i_6, Vit-A), AAs (i_7, Phe; i_8, Tyr; i_9, Ile; i_{10}, Leu; i_{11}, Lys; i_{12}, Met; i_{13}, Cys; i_{14}, Thr; i_{15}, Trp; i_{16}, Val), and isoflavones (i_{17}, daidzein; i_{18}, genistein).

(17.7%) and greater calorie contribution (till 375 kcal/100 g) than habitual legumes. It presents more Ca than the rest of legumes. Notwithstanding, Ca absorption is hindered because they contain substances (e.g., oxalates and phytates) interfering with it: oxalates content in soya bean is 3–6-fold that of beans and lentils.

In the dendrogram of Mediterranean diet legumes and soya bean according to proximate contents (cf. Fig. 9.1), chickpea groups with soya bean; bean, lentil, and mung bean class together. Spanish legumes turn out to be in separate groupings, as well as Asian beans. The greatest similarity is detected between bean and lentil.

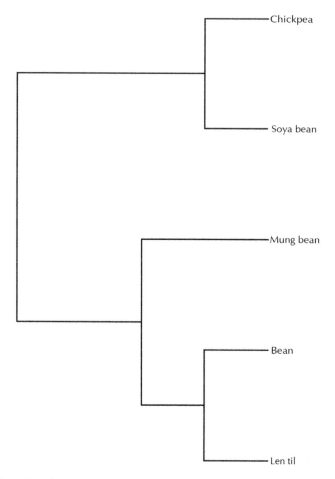

FIGURE 9.1   Dendrogram of Mediterranean diet legumes and soya bean according to proximate contents.

Composition in essential amino acids (AAs) of analyzed dry legumes is reported in Table 9.2.

TABLE 9.2 Results of Essential Amino Acids Compositions of the Analyzed Dry Legumes in mg/100 g.

Entry	Phe	Tyr	Ile	Leu	Lys	Met	Cys	Thr	Trp	Val
1	1151	589	891	1505	1376	209	238	756	0	913
2	1154	559	927	1685	1593	234	188	878	0	1016
3	1266	789	1045	1847	1739	194	221	960	0	1211
4	2055	1303	1889	3232	2653	525	552	1603	532	1995
5	1259	556	941	1607	2145	458	113	736	432	989

In dendrogram of Spanish legumes and soya bean with regard to proximates and AAs contents (cf. Fig. 9.2), Mediterranean chickpea, bean, and lentil class together; Asian soya and mung beans belong to the same grouping. Again, maximum resemblance results between bean and lentil.

Other components of soya bean are isoflavones [*daidzein* (cf. Fig. 9.3a), *genistein* (Fig. 9.3b), glycitein (Fig. 9.3c)], and lecithin (Fig. 9.3e).[3–6] The former is similar to 17β-estradiol [estrogen (ES) (Fig. 9.3d)]; 100 g of soya bean contains 130–500 mg of isoflavones.[7,8]

Earlier publications in *Nereis* classified yams[9] and lactic acid bacteria[10] by principal component (PCA), cluster (CA), and meta-analyses. The molecular classifications of 33 phenolics derived from the cinnamic and benzoic acids from *Posidonia oceanica*,[11] 74 flavonoids,[12] 66 stilbenoids,[13] 71 triterpenoids/steroids from *Ganoderma*,[14] 17 isoflavonoids from *Dalbergia parviflora*[15], and legumes[16] were reported. A tool for interrogation of macromolecular structure was published.[17] Mucoadhesive polymer hyaluronan favors transdermal penetration absorption of caffeine.[18,19] Endocrine disruptor diethylstilbestrol (DES, Fig. 9.3h), bisphenol-A (similar to DES), polycarbonate and epoxy–silica nanocomposites were reported.[20,21] The main aim of the present report is to develop code learning potentialities and, since legumes nutrients are more naturally described via varying size-structured representation, find approaches to structured information processing. In view of legumes nutritional benefits, the objective was to categorize them with PCA/CA to differentiate vegetables groups and identify characteristic compounds of various plants. The next section presents the method. Following that, two sections illustrate and discuss the results. Finally, the last section summarizes our conclusions.

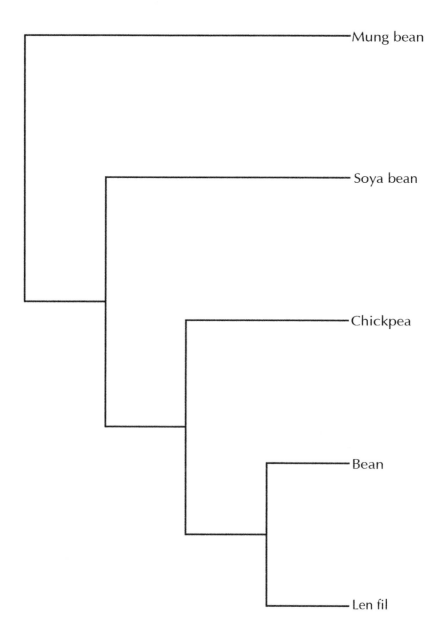

FIGURE 9.2 Dendrogram of Mediterranean diet legumes and soya bean according to proximates and AAs.

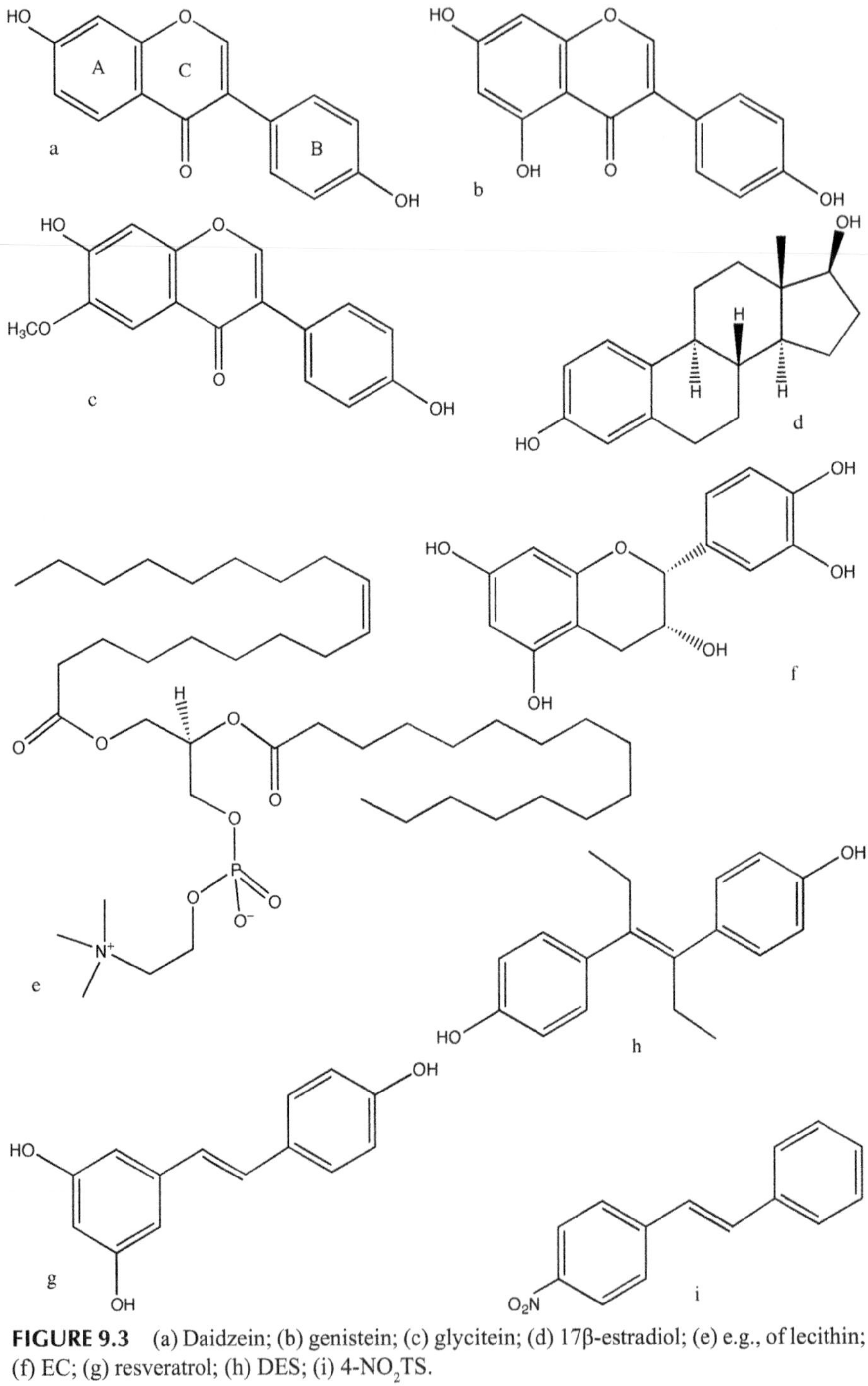

FIGURE 9.3 (a) Daidzein; (b) genistein; (c) glycitein; (d) 17β-estradiol; (e) e.g., of lecithin; (f) EC; (g) resveratrol; (h) DES; (i) 4-NO$_2$TS.

9.1   COMPUTATIONAL METHOD

Principal components analysis (PCA) is a dimension reduction technique.[22-27] From original variables X_j, PCA builds orthogonal variables $\tilde{P}_j$ linear combinations of mean-centered ones $\tilde{X}_j = X_j - \bar{X}_j$ corresponding to eigenvectors of sample covariance matrix $S = 1/(n-1) \sum_{i=1}^{n} (x_i - \bar{x})(x_i - \bar{x})'$. For every loading vector $\tilde{P}_j$, matching eigenvalue $\tilde{l}_j$ of S tells how much data variability is explained: $\tilde{l}_j = \mathrm{Var}(\tilde{P}_j)$. Loading vectors are sorted in decaying eigenvalues. First k PCs explain most variability. After selecting k, one projects p-dimensional data on to subspace spanned by k loading vectors and computes coordinates vs. $\tilde{P}_j$, yielding scores:

$$\tilde{t}_i = \tilde{P}'(x_i - \bar{x}) \tag{9.1}$$

for every $i = 1, \ldots, n$ having trivially zero mean. With respect to original coordinate system, projected data point is computed fitting:

$$\hat{x}_i = \bar{x} + \tilde{P}\tilde{t}_i \tag{9.2}$$

Loading matrix $\tilde{p}$ ($p \times k$) contains loadings column-wise and diagonal one $\tilde{L} = (\tilde{l}_j)_j$ ($k \times k$), eigenvalues. Loadings k explains variation:

$$\left(\sum_{j=1}^{k} \tilde{l}_j \right) / \left(\sum_{j=1}^{p} \tilde{l}_j \right) \geq 80\% \tag{9.3}$$

Cluster analysis (CA) encompasses different classification algorithms.[28,29] Starting point is $n \times p$ data matrix X containing p components measured in n samples. One assumes data were pre-processed to remove artifacts, and missing values, imputed. The CA organizes samples into small number of clusters such that samples within cluster are similar. Distances l_q between samples $x,x' \in \mathfrak{R}^p$ are:

$$\|x - x'\|_q = \left(\sum_{i=1}^{p} |x_i - x'_i|^q \right)^{1/q} \tag{9.4}$$

(e.g., Euclidean l_2, Manhattan l_1 distances). Comparing samples, *Pearson's correlation coefficient* (PCC) is advantageous:

$$r(x-x') = \frac{\sum_{i=1}^{p}(x_i - \bar{x})(x'_i - \bar{x}')}{\left[\sum_{i=1}^{p}(x_i - \bar{x})^2 \sum_{i=1}^{p}(x'_i - \bar{x}')^2\right]^{1/2}} \tag{9.5}$$

where $\bar{x} = \left(\sum_{i=1}^{p} x_i\right)\Big/ p$ is measure mean value for sample x.[30–36]

9.2 CALCULATION RESULTS

Energy, proximate, Ca, vitamin (Vit)-A, AAs, and isoflavones contents of five legumes were taken from Gómez Garay. Total isoflavones in soya bean are: 37% *daidzein*, 57% *genistein*, and 6% glycitein. The PCC matrix **R** was computed between legumes; the upper triangle turns out to be:

$$\mathbf{R} = \begin{pmatrix} 1.000 & 0.993 & 0.994 & 0.985 & 0.948 \\ & 1.000 & 0.991 & 0.980 & 0.955 \\ & & 1.000 & 0.989 & 0.953 \\ & & & 1.000 & 0.949 \\ & & & & 1.000 \end{pmatrix}$$

All correlations between pairs of legumes are high, for example, chickpea–lentil $R_{1,3} = 0.994$, etc. Chinese soya bean is close to Indian mung bean $R_{4,5} = 0.949$ and, especially, Mediterranean legumes $R_{1,4} \approx R_{2,4} \approx R_{3,4} \approx 0.985$. All are illustrated in the partial correlation diagram (PCD), which could contain high ($r \geq 0.75$), medium ($0.50 \leq r < 0.75$), low ($0.25 \leq r < 0.50$), and *zero* ($r < 0.25$) partial correlations. All 10 pairs of legumes present high partial correlations (cf. Fig. 9.4, red, grayscale). The corresponding interpretation is that all vegetables present similar composition. The results are in agreement with previous outcomes (Figs. 9.1 and 9.2).

　　The dendrogram of legumes according to nutrients content (cf. Fig. 9.5) shows different behavior depending on energy, and AAs *lysine* (Lys), *methionine* (Met), and *tryptophan* (Trp). Two classes are clearly recognized:

(1,2,3)　　　　　　　　　　　　　　　　　　　(4,5)

Asian beans taken in class 2 are separated from grouping 1: Mediterranean chickpea, bean, and lentil are relatively low in *energy*, etc., and group into class 1; Asian soya and mung beans are high in *energy*, Lys, Met, and Trp, and form cluster 2. Maximum similarity results between chickpea and lentil. Legumes in the same grouping appear highly correlated in PCD in qualitative agreement with previous results (Figs. 9.1, 9.2, and 9.4).

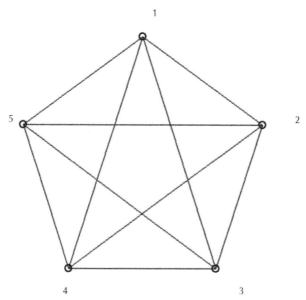

FIGURE 9.4 Partial correlation diagram showing all 10 high partial correlations (*red, grayscale*).

The radial tree (cf. Fig. 9.6) shows different behavior of legumes depending on *energy*, etc. Both classes above are recognized in qualitative agreement with previous results, PCD, and dendrogram (Figs. 9.1, 9.2, 9.4, and 9.5). Again, maximum resemblance results between chickpea and lentil.

Splits graph (cf. Fig. 9.7) for five legumes in Table 9.1 reveals conflicting relationships between the two classes above because of interdependences.[37] It indicates spurious relations between groupings resulting from base-composition effects. It illustrates different behavior of legumes depending on *energy*, etc. in qualitative agreement with previous results, PCD, and binary/radial trees (Figs. 9.1, 9.2, and 9.4–9.6). One more time, maximum similarity results between chickpea and lentil.

PCA allows *summarizing* information contained in **X**-matrix. It decomposes **X**-matrix as product of matrices **P** and **T**. *Loading matrix* **P** with information about variables contains a few vectors: PCs, which are obtained as linear combinations of original *X*-variables. In *score matrix* **T** with

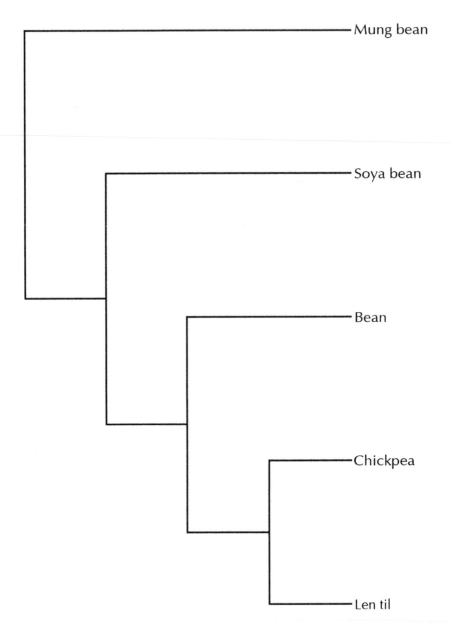

FIGURE 9.5 Dendrogram of legumes and soya bean according to proximates, AAs, and isoflavones.

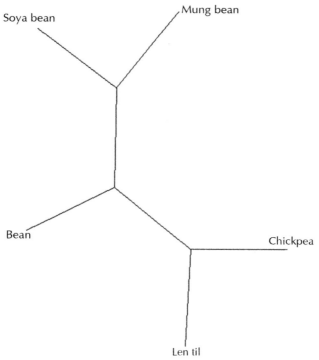

FIGURE 9.6 Radial tree of legumes and soya bean according to proximates, AAs, and isoflavones.

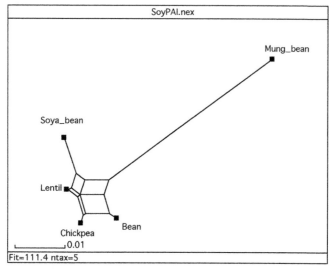

FIGURE 9.7 Splits graph of legumes and soya bean according to proximates, AAs, and isoflavones.

information about objects, every object is described by projections on to PCs instead of original variables: $X = TP' + E$, where ' denotes transpose matrix. Information not contained in matrices remains *unexplained X-variance in residual matrix* E. Every PC_i is a new coordinate expressed as linear combination of old x_j: $PC_i = \Sigma_j b_{ij} x_j$. New coordinates PC_i are *scores (factors)* while coefficients b_{ij} are *loadings*. Scores are ordered by information content vs. total variance among objects. *Score–score plots* show compounds positions in new coordinate system while *loading–loading plots* show location of features that represent compounds in new coordination. Properties of PCs follow: (1) they are extracted by decaying importance; (2) every PC is orthogonal to each other. A PCA was performed for legumes. Importance of PCA factors F_{1-18} for nutrients (cf. Table 9.3) shows that F_1 explains 65% variance (35% error), $F_{1/2}$, 82% variance (18% error), F_{1-3}, 93% variance (7% error), etc.

TABLE 9.3 Importance of PCA Factors for Proximates, AAS and Isoflavones of Legumes, and Soya Bean.

Factor	Eigenvalue	Percentage of Variance	Cumulative Percentage of Variance
F_1	11.68617550	64.92	64.92
F_2	3.01383699	16.75	81.67
F_3	2.00182671	11.12	92.79
F_4	1.29816080	7.21	100.00
F_5	0.00000000	0.00	100.00
F_6	0.00000000	0.00	100.00
F_7	0.00000000	0.00	100.00
F_8	0.00000000	0.00	100.00
F_9	0.00000000	0.00	100.00
F_{10}	0.00000000	0.00	100.00
F_{11}	0.00000000	0.00	100.00
F_{12}	0.00000000	0.00	100.00
F_{13}	0.00000000	0.00	100.00
F_{14}	0.00000000	0.00	100.00
F_{15}	0.00000000	0.00	100.00
F_{16}	0.00000000	0.00	100.00
F_{17}	0.00000000	0.00	100.00
F_{18}	0.00000000	0.00	100.00

The PCA factors loadings are shown in Table 9.4.

TABLE 9.4 PCA Loadings for Proximates, AAS and Isoflavones of Legumes, and Soya Bean.

Property	PCA Factor Loadings			
	F_1	F_2	F_3	F_4
i_1	0.23396520	-0.26675036	-0.26734095	0.04625375
i_2	0.28037263	0.07795120	0.16146261	0.09147414
i_3	0.15305747	0.27965193	-0.44985794	0.25651819
i_4	0.12030396	0.50680120	-0.03821892	-0.20366142
i_5	0.21891392	-0.03081272	0.37534453	0.34562517
i_6	-0.15839295	0.42200456	-0.19596139	0.26796364
i_7	0.29119008	0.04184882	0.04283471	-0.01105093
i_8	0.27804266	0.06245325	0.11488903	-0.21208285
i_9	0.28779107	0.08281935	0.05501340	-0.06435958
i_{10}	0.28551579	0.10594567	0.02678921	-0.09648938
i_{11}	0.27077168	-0.12813952	-0.20383631	0.09010807
i_{12}	0.23583016	-0.15919062	-0.19435808	0.39061009
i_{13}	0.26084122	0.17725581	0.22574684	-0.07940130
i_{14}	0.27606581	0.15175079	0.04083742	-0.16797265
i_{15}	0.23642922	-0.20408153	-0.17162727	0.35355035
i_{16}	0.28416819	0.08283180	0.02757198	-0.16212852
i_{17}	-0.06384871	0.06566634	0.57607922	0.46027792
i_{18}	-0.12404526	0.48570660	-0.08078324	0.27211079

The PCA F_1–F_4 profile is listed in Table 9.5. For F_1, variable i_7 shows the greatest weight in the profile; however, F_1 cannot be reduced to two variables $\{i_7,i_9\}$ without 83% error. For F_2, variable i_4 presents greatest weight; notwithstanding, F_2 cannot be reduced to two variables $\{i_4,i_{18}\}$ without 51% error. For F_3 and F_4, variable i_{17} assigns greatest weight; nevertheless, F_3 and F_4 cannot be reduced to two variables $\{i_3,i_{17}\}$ and $\{i_{12},i_{17}\}$, respectively, without 47 and 64% errors, etc.

The PCA scores plot F_2–F_1 for legumes (cf. Fig. 9.8) shows that both clusters above are clearly distinguished: class 1 with three Mediterranean legumes ($F_1 < F_2 \approx 0$, *left*) and grouping 2 with two Asian ones ($F_1 >> F_2$, *bottom right*). Spanish legumes are closer between themselves than Asian soya to mung bean. Once more, maximum resemblance results between chickpea and lentil.

TABLE 9.5　Profile of Principal Component Analysis Factors for Proximates, AAS and Isoflavones of Legumes, and Soya Bean.

	$\%\,i_1$	$\%\,i_2$	$\%\,i_3$	$\%\,i_4$	$\%\,i_5$	$\%\,i_6$	$\%\,i_7$	$\%\,i_8$	$\%\,i_9$	$\%\,i_{10}$	$\%\,i_{11}$	$\%\,i_{12}$	$\%\,i_{13}$	$\%\,i_{14}$	$\%\,i_{15}$	$\%\,i_{16}$	$\%\,i_{17}$	$\%\,i_{18}$
F_1	5.47	7.86	2.34	1.45	4.79	2.51	8.48	7.73	8.28	8.15	7.33	5.56	6.80	7.62	5.59	8.08	0.41	1.54
F_2	7.12	0.61	7.82	25.68	0.09	17.81	0.18	0.39	0.69	1.12	1.64	2.53	3.14	2.30	4.16	0.69	0.43	23.59
F_3	7.15	2.61	20.24	0.15	14.09	3.84	0.18	1.32	0.30	0.07	4.15	3.78	5.10	0.17	2.95	0.08	33.19	0.65
F_4	0.21	0.84	6.58	4.15	11.95	7.18	0.01	4.50	0.41	0.93	0.81	15.26	0.63	2.82	12.50	2.63	21.19	7.40

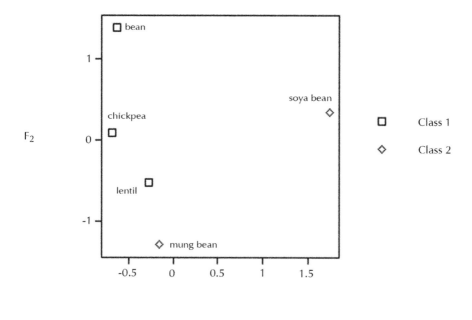

FIGURE 9.8 PCA scores plot of legumes and soya bean according to proximates, AAs, and isoflavones.

From PCA factors loading of legumes F_2–F_1 loadings plot (cf. Fig. 9.9) depicts 18 proximates, AAs and isoflavones contents. Proximate *fat*, and AAs Tyr, Ile, Leu, and Val collapse. Three clusters are clearly distinguished: class 1 with *energy*, and AAs Lys, Met, and Trp ($\{1,11,12,15\}$, $F_1 \gg F_2$, *bottom right*), grouping 2 with *fat*, Ca, and AAs *phenylalanine* (Phe), *tyrosine* (Tyr), *isoleucine* (Ile), *leucine* (Leu), *cysteine* (Cys), *threonine* (Thr), and *valine* (Val) ($\{2,5,7,8,9,10,13,14,16\}$, $F_1 > F_2 \approx 0$, *right*), and cluster 3 with *protein*, *fiber*, Vit-A, and isoflavones ($\{3,4,6,17,18\}$, $0 \approx F_1 < F_2$, *left*). Constituents in class 1 are closer between themselves than grouping 2 and cluster 3. Macronutrients *fat*, *protein*, and *fiber* result in separate classes 2 and 3: *protein* appears closer to *fiber* than to *fat*. Isoflavone *genistein* is closer to Vit-A than to *daidzein*. In addition, as a complement to scores diagram for loadings it is confirmed that Asian legumes in class 2, located in the bottom right, present a more pronounced contribution from nutrients in grouping 1 situated in the same side in Figure 9.8. Mediterranean legumes in cluster 1 in the left side show a contribution from nutrients in class 3 situated in the same side of Figure 9.8.

Instead of five legumes in the space R^{18} of 18 nutrients consider 18 nutrients in the space R^5 of five legumes. Matrix **R** upper triangle results:

$$R = \begin{pmatrix}
1.000 & 0.623 & 0.450 & -0.070 & 0.443 & -0.651 & 0.739 & 0.636 & 0.687 & 0.675 & 0.958 & 0.900 & 0.445 & 0.601 & 0.924 & 0.686 & -0.508 & -0.670 \\
 & 1.000 & 0.452 & 0.477 & 0.872 & -0.451 & 0.976 & 0.938 & 0.973 & 0.958 & 0.802 & 0.719 & 0.960 & 0.933 & 0.713 & 0.940 & 0.047 & -0.286 \\
 & & 1.000 & 0.609 & 0.143 & 0.338 & 0.514 & 0.376 & 0.514 & 0.544 & 0.590 & 0.593 & 0.386 & 0.529 & 0.523 & 0.499 & -0.424 & 0.351 \\
 & & & 1.000 & 0.141 & 0.366 & 0.473 & 0.534 & 0.544 & 0.587 & 0.177 & 0.000 & 0.641 & 0.661 & -0.060 & 0.567 & -0.155 & 0.502 \\
 & & & & 1.000 & -0.471 & 0.768 & 0.697 & 0.741 & 0.697 & 0.592 & 0.647 & 0.785 & 0.647 & 0.653 & 0.667 & 0.470 & -0.301 \\
 & & & & & 1.000 & -0.506 & -0.554 & -0.471 & -0.438 & -0.553 & -0.427 & -0.374 & -0.392 & -0.507 & -0.488 & 0.136 & 0.974 \\
 & & & & & & 1.000 & 0.967 & 0.995 & 0.989 & 0.886 & 0.760 & 0.930 & 0.964 & 0.759 & 0.982 & -0.166 & -0.372 \\
 & & & & & & & 1.000 & 0.981 & 0.980 & 0.784 & 0.584 & 0.955 & 0.981 & 0.593 & 0.990 & -0.189 & -0.405 \\
 & & & & & & & & 1.000 & 0.998 & 0.849 & 0.699 & 0.953 & 0.985 & 0.696 & 0.993 & -0.173 & -0.328 \\
 & & & & & & & & & 1.000 & 0.840 & 0.677 & 0.949 & 0.993 & 0.670 & 0.996 & -0.219 & -0.297 \\
 & & & & & & & & & & 1.000 & 0.933 & 0.656 & 0.779 & 0.938 & 0.837 & -0.409 & -0.515 \\
 & & & & & & & & & & & 1.000 & 0.506 & 0.587 & 0.996 & 0.650 & -0.198 & -0.405 \\
 & & & & & & & & & & & & 1.000 & 0.958 & 0.498 & 0.940 & 0.053 & -0.183 \\
 & & & & & & & & & & & & & 1.000 & 0.578 & 0.992 & -0.229 & -0.244 \\
 & & & & & & & & & & & & & & 1.000 & 0.650 & -0.203 & -0.489 \\
 & & & & & & & & & & & & & & & 1.000 & -0.261 & -0.352 \\
 & & & & & & & & & & & & & & & & 1.000 & 0.258 \\
 & & & & & & & & & & & & & & & & & 1.000
\end{pmatrix}$$

Correlation is greater between AAs (e.g., $R_{9,10} = 0.998$) than proximates (e.g., $R_{2,5} = 0.872$) and isoflavones *daidzein–genistein* $R_{17,18} = 0.258$. Some correlations between groups are high (e.g., *fat*–Phe $R_{2,7} = 0.976$, Vit-A–*genistein* $R_{6,18} = 0.974$). Macronutrient *protein* correlation with *fiber* $R_{3,4} = 0.609$ is greater than with *fat* $R_{2,3} = 0.452$. Correlations of Vit-A and isoflavones with most nutrients ($R_{i,6}$, $R_{i,17}$, and $R_{i,18}$) are negative, especially *energy* ($R_{1,6} \sim R_{1,17} \sim R_{1,18} \sim -0.6$). The dendrogram for 18 proximates, AAs and isoflavones of legumes (cf. Fig. 9.10) separates the three clusters above in agreement with PCA loadings plot (Fig. 9.9). Again, macronutrients appear in separate classes 2 and 3; *genistein* is closer to Vit-A than to *daidzein*.

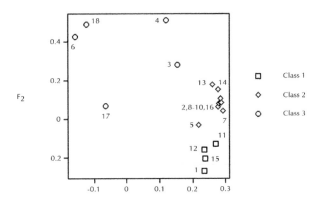

FIGURE 9.9 PCA loadings plot of legumes and soya bean according to proximates, AAs and isoflavones.

Radial tree for 18 proximates, AAs and isoflavones of legumes (cf. Fig. 9.11) separates three clusters above in agreement with PCA loadings plot and dendrogram (Figs. 9.9 and 9.10). Once more, macronutrients result in separate classes 2 and 3; *genistein* is closer to Vit-A than to *daidzein*.

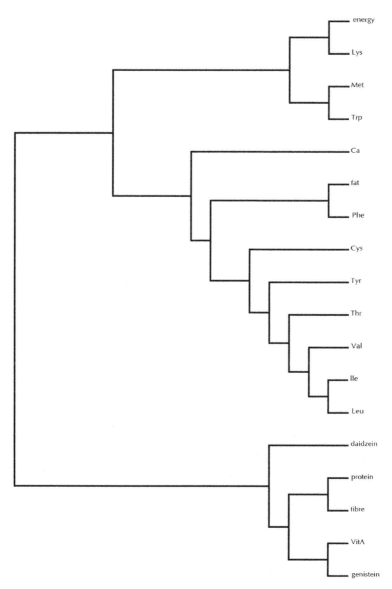

FIGURE 9.10   Dendrogram of proximates, AAs and isoflavones for Mediterranean legumes, and soya bean

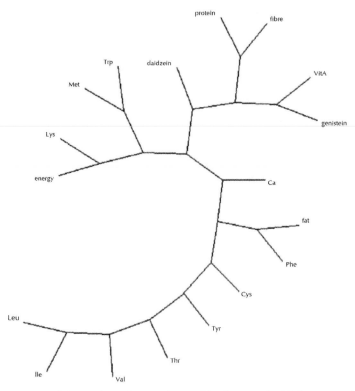

FIGURE 9.11   Radial tree of proximates, AAs and isoflavones for Mediterranean legumes, and soya bean.

Splits graph for 18 proximates, AAs and isoflavones of legumes (cf. Fig. 9.12) shows that nutrients 2, 7–11, 13, 14, and 16 collapse, as well as 12 and 15. It reveals conflicting relationships between classes. It separates the three clusters above in agreement with loadings plot and binary/radial trees (Figs. 9.9–9.11). Again, macronutrients split into classes 2 and 3; *genistein* is closer to Vit-A than to *daidzein*.

A PCA was performed for nutrients. Factor F_1 explains 98% variance (2% error), $F_{1/2}$, 99.3% variance (0.7% error), F_{1-3}, 99.8% variance (0.2% error), etc. Scores plot of PCA F_2–F_1 for nutrients distinguished the three clusters above: class 1 ($0 \approx F_1 < F_2$, cf. Fig. 9.13 *top*), grouping 2 ($F_1 >> F_2$, *bottom*), and cluster 3 ($F_1 < F_2 \approx 0$, *left*). Now constituents in class 3 result closer between themselves than grouping 1 and cluster 2. Again, macronutrients appear in separate classes 2 and 3. *Daidzein* and *genistein* are close to Vit-A. The results are in qualitative agreement with loadings plot, binary/radial trees and splits graph (Figs. 9.9–9.12).

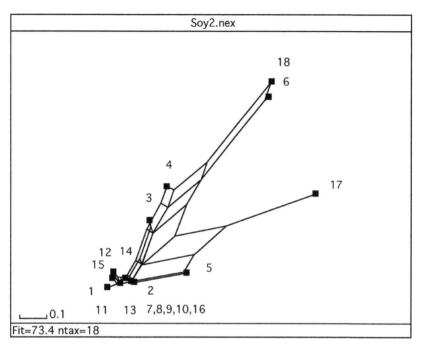

FIGURE 9.12   Splits graph of proximates, AAs and isoflavones for Mediterranean legumes, and soya bean.

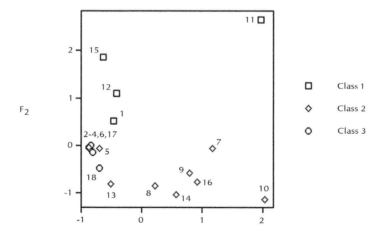

FIGURE 9.13   PCA scores plot of proximates, AAs and isoflavones for legumes, and soya bean.

Soya bean oil is in disadvantage vs. Mediterranean diet: although contents in oleic or linoleic acid of sunflower (*Helianthus annuus*) oil be similar to soya bean one, sunflower oil lacks linolenic acid, which is easily oxidized and the component that makes inadvisable to make fried foods with soya bean oil. Disadvantages vs. virgin olive *Olea europaea* L. (VOO) or high-oleic sunflower oil are greater. Soya bean oil is not a source of isoflavones because it does not contain them. The VOO classes with high-oleic sunflower oil, and sunflower, with soya bean oil (cf. Fig. 9.14).

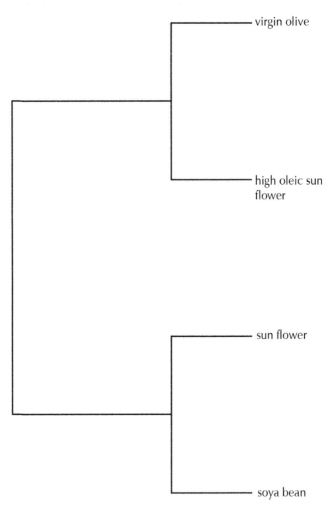

FIGURE 9.14　Dendrogram of Mediterranean diet and soya-bean oils according to unsaturated oily acids.

The VOO contains 36 phenolic compounds, which are responsible for its health benefits.[38–42] Its hydrophilic phenols are phenolic acids and alcohols, flavonoids, secoiridoids, and lignans.[43–46] Main VOO hydrophilic phenols are secoiridoids [oleuropein (cf. Fig. 9.15e), ligstroside, elenolic acid], flavones [luteolin (Fig. 9.15a), apigenin (Fig. 9.15b)], and phenolic alcohols/derivatives (hydroxytyrosol, its acetate). Comparison of VOO flavones with soya bean isoflavones (Fig. 9.3abc) shows that both hydrophilic polyphenols are planar, unsaturated in ring C and electron-rich. Structure of lipophilic methylated phenol Vit-E α-tocopherol (α-TCP, Fig. 9.15c) of VOO and sunflower oils[47] is close to γ-TCP (Fig. 9.15d) of soya bean and maize oils.[48] Phenolic antioxidants are used as preservatives for edible oils and fats to prevent rancidity.[49] Nutraceutical properties of VOO polyphenols (cultures, animals, and humans) were reviewed.[50] Functional properties of BEP flour preserved with neem/moringa seed oil mixtures were analyzed.[51]

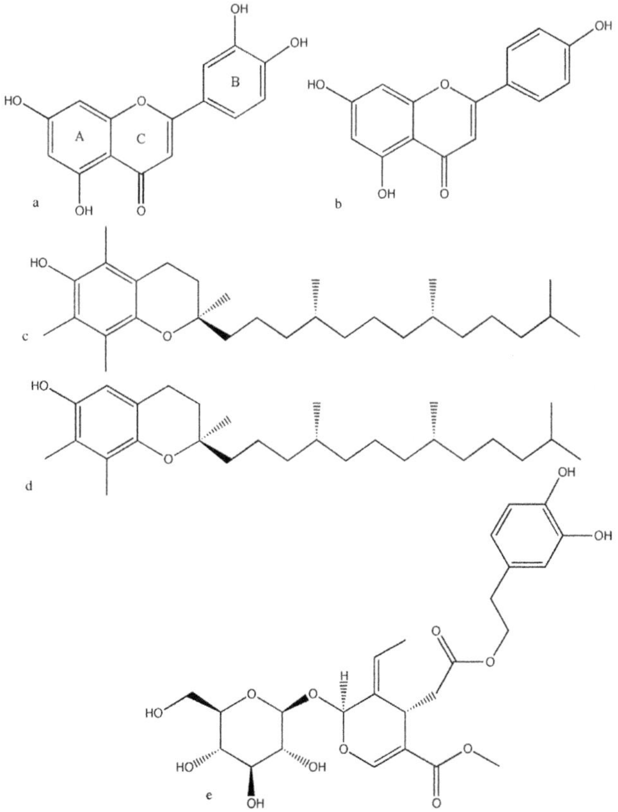

FIGURE 9.15 VOO flavones: (a) luteolin and (b) apigenin; (c) α- and (d) γ-tocopherol; (e) oleuropein.

9.3 DISCUSSION

Soya bean is planted all over the world and its cold resistance gained interest. Proteases inhibitors are plentiful in cereals and legumes, especially soya bean.[52] Because of protein nature, they are denatured by thermal treatments but residual activity remains. Their antinutritient activity is associated with growth delay and pancreatic hypertrophy. Soya bean trypsin inhibitors (TIs) produce inactivation and trypsin loss in the small intestine, causing cholecystokinine release and inducing pancreatic trypsin synthesis, and S-AAs-requirements rise. Low incidences of pancreatic cancer were observed in populations with high soya bean intake. For human consumption, soya bean must be cooked with *wet* heat to destroy TIs. Raw soya beans, for example, immature green form, are toxic to all monogastric animals. Amylase inhibitors (Bowman–Birk) derived from soya bean inhibit or prevent chemically induced cancers of liver, lung, colon, mouth, and esophagus.

Flavonoids form sulfated, small organic ligands with anticoagulant properties alike heparins. Isoflavonoids, for example, in soya bean, exhibit antioxidant activity. Besides functioning as antioxidant and anthelmintic, many isoflavones interact with animal and human ES receptors causing effects similar to primary female sexual hormone ES. Main flavonoid group with ES activities is isoflavones, for example, *genistein*, which follows from earlier recognition that an ewes dietary disease in Australia was caused by isoflavone constituents of pasture clover. Specific phytostrogens (PESs, similar to ESs) possess their beneficial properties but they are not feminizing; for example, *genistein*, one of the most abundant PESs in soya bean, reproduces ES antioxidant effects via a mechanism: ESs activate receptors that induce signal cascades, which bring on the expression of antioxidant genes associated with longevity. *Daidzein* can be converted to its end metabolite S-equol in some humans based on intestinal bacteria. S-equol presents health benefits. Isoflavones are polyphenolic and electron-rich, acting as substrate inhibitors for cytochrome P450 (CYP) enzymes and inducing detoxification enzymes, for example, CYP-dependent monooxygenases (MOs).[53] Some polyphenols penetrate the blood–brain barrier (BBB) into regions mediating cognition.[54] The ES, PESs, and soya bean products cause brain health benefits/risks and affect functions. Flavan-3-ol epicatechin (EC, Fig. 9.3f) is able to cross BBB better than more hydrophilic stilbenoid resveratrol (Fig. 9.3g). Some antioxidants [e.g., stilbenoids DES, resveratrol, 4-nitrostilbene (4-NO_2TS, Fig. 9.3i)] show ES activity. However, DES induces vaginal adenocarcinoma. Resveratrol and 4-NO_2TS exhibit minor ES activity but both are cytotoxic in Michigan Cancer Foundation type-7 (MCF-7) human breast cancer (BC) cell line at higher concentrations. Soya bean interacts with medications for depression [monoamine oxidase inhibitors (MAOIs)], antibiotics, ESs, tamoxifen, warfarin, and drugs changed by liver (CYP2C9 substrates).

Soya bean-lecithin role in cholesterol regulation is controversial, and results have more in common with presence in oil of unsaturated fatty acids (UFAs), which contain one or more double bonds in its chain while saturated ones do not. To prevent coronary diseases, to use an oil with a greater UFAs proportion is better; so one will choose VOO, maize or soya bean oil rather than butter. Fats from mammals are saturated while those from vegetables, fish and chicken are UFAs. Examples of foods that were used in fungal-mediated secondary metabolite modification are soya bean products, which present potential for antimicrobial and antioxidant activities.[55,56] Studies reported that new phytochemicals are synthesized during fungal fermentation of food materials, for example, soya bean.[57] It is necessary before claiming benefit from consuming antioxidant-rich foods to establish that antioxidant activity they exhibit in vitro occur in vivo: it sometimes does, for example, dark soya sauce.[58,59] A vegetable-fruit-soya diet protects vs. BC.[60] Differential influence of dietary soya intake on recurrent BC risk related to human epidermal growth factor receptor (EGFR) type-2 (HER2) status.[61] To assess soya bean role in human food and health is essential. Current market offer exceeds attributes that are conferred to soya bean, which should not substitute other legumes habitual in Mediterranean diet but diversify offer. Research, from basic studies in laboratory to agronomic and clinical ones, will be responsible for answers. From knowledge, producers, food and agriculture industry, health professionals, and consumers could make a dependable use of soya bean, benefiting from qualities.

9.4   FINAL REMARKS

From the present results and discussion the following final remarks can be drawn.

1.   Never blindly trust what you get (do your results have chemical sense?). Mathematical and statistical models are not the panacea (always ask yourself three questions: Why am I doing it? What the results might be? Will I be successful?). Use this knowledge in data analysis to guide your investigation or experimentation, not as an end in itself.
2.   The hypothesis that variance in classification learning can be expected to decay as training-set size rises was examined and confirmed.

3. Some criteria reduced the analysis to a manageable quantity from the enormous set of legumes compositions: they refer to proximates, AAs, and isoflavones. The meta-analysis was useful to rise the number of samples and variety of analyzed data. Different behavior of legumes depends on *energy*, and AAs *lysine*, *methionine*, and *tryptophan*. Mediterranean legumes and Asian beans are detected in separate classes. With regard to components, macronutrients result unconnected in two groupings and *protein* was closer to *fiber* than to *fat*. Soya bean presents adequate proximate, AAs and isoflavones contents, good antioxidant capacity, and may be used as a functional food. It represents a legume useful as a natural source for nutraceutical formulations.

4. PSAs of compositions and legumes cluster analyses allowed classifying them and agreed. Phytochemistry, cytochemistry, and understanding of computational methods are essential for tackling associated *data mining* tasks.

5. A need exists to increase the knowledge and investigation on the characteristics and properties that make soya bean a benefit from the viewpoint of nutrition and health. Soya bean presents characteristic profiles that could be welcome in a Mediterranean diet but more detailed research is necessary. In addition, further analyses are essential for a more complete evaluation of this cultivar.

ACKNOWLEDGMENTS

The authors thank support from the Spanish Ministerio de Economía y Competitividad (Project No. BFU2013-41648-P), EU ERDF and Universidad Católica de Valencia *San Vicente Mártir* (Project No. PRUCV/2015/617).

KEYWORDS

- soya bean
- Mediterranean diet
- *daidzein*
- *genistein*
- legumes

REFERENCES

1. Hymowitz, T. On the Domestication of Soya Bean. *Econ. Bot.* **1970,** *24,* 408–421.
2. Gómez Garay, A. La Soja y Sus Virtudes en Alimentación y Salud Humana. *Alim. Nutr. Salud.* **2006,** *13,* 91–96.
3. Yang, C. Y.; Hsu, C. H.; Tsai, M. L. Effect of Crosslinked Condition on Characteristics of Chitosan/Tripolyphosphate/Genipin Beads and Their Application in the Selective Adsorption of Phytic Acid from Soybean Whey. *Carbohydr. Polym.* **2011,** *86,* 659–665.
4. Chang, Y. L.; Liu, T. C.; Tsai, M. L. Selective Isolation of Trypsin Inhibitor and Lectin from Soybean Whey by Chitosan/Tripolyphosphate/Genipin Co-Crosslinked Beads. *Int. J. Mol. Sci.* **2014,** *15,* 9979–9990.
5. Maria John, K. M.; Jung, E. S.; Lee, S.; Kim, J. S.; Lee, C. H. Primary and Secondary Metabolites Variation of Soybean Contaminated with *Aspergillus sojae. Food Res. Int.* **2013,** *54,* 487–494.
6. Maria John, K. M.; Enkhtaivan, G.; Lee, J.; Thiruvengadam, M.; Keum, Y. S.; Kim, D. H. Spectroscopic Determination of Metabolic and Mineral Changes of Soya-Chunk Mediated by *Aspergillus sojae. Food Chem.* **2015,** *170,* 1–9.
7. Kaufman, P. B.; Duke, J. A.; Brielmann, H.; Boik, J.; Hoyt, J. E. A Comparative Survey of Leguminous Plants as Sources of the Isoflavones, Genistein and Daidzein: Implications for Human Nutrition and Health. *J. Altern. Complement. Med.* **1997,** *3,* 7–12.
8. Uifălean, A.; Schneider, S.; Gierok, P.; Ionescu, C.; Iuga C. A.; Lalk, M. The Impact of Soy Isoflavones on MCF-7 and MDA-MB-231 Breast Cancer Cells Using a Global Metabolomic Approach. *Int. J. Mol. Sci.* **2016,** *17,* 1443.
9. Torrens-Zaragozá, F. Molecular Categorization of Yams by Principal Component and Cluster Analyses. *Nereis.* **2013,** *5,* 41–51.
10. Torrens-Zaragozá, F. Classification of Lactic Acid Bacteria against Cytokine Immune Modulation. *Nereis.* **2014,** *6,* 27–37.
11. Castellano, G.; Tena, J.; Torrens, F. Classification of Polyphenolic Compounds by Chemical Structural Indicators and its Relation to Antioxidant Properties of *Posidonia Oceanica* (L.) Delile. *MATCH Commun. Math. Comput. Chem.* **2012,** *67,* 231–250.
12. Castellano, G.; González-Santander, J. L.; Lara, A.; Torrens, F. Classification of Flavonoid Compounds by Using Entropy of Information Theory. *Phytochemistry.* **2013,** *93,* 182–191.
13. Castellano, G.; Lara, A.; Torrens, F. Classification of Stilbenoid Compounds by Entropy of Artificial Intelligence. *Phytochemistry.* **2014,** *97,* 62–69.
14. Castellano, G.; Torrens, F. Information Entropy-Based Classification of Triterpenoids and Steroids from *Ganoderma. Phytochemistry.* **2015,** *116,* 305–313.
15. Castellano, G.; Torrens, F. Quantitative Structure–Antioxidant Activity Models of Isoflavonoids: A Theoretical Study. *Int. J. Mol. Sci.* **2015,** *16,* 12891–12906.
16. Torrens, F.; Castellano, G. From Asia to Mediterranean: Soya Bean, Spanish Legumes and Commercial *Soya Bean* Principal Component, Cluster and Meta-Analyses. *J. Nutr. Food Sci.* **2014,** *4* (5), 98–98.
17. Torrens, F.; Castellano, G. A Tool for Interrogation of Macromolecular Structure. *J. Mater. Sci. Eng. B.* **2014,** *4* (2), 55–63.
18. Torrens, F.; Castellano, G. Mucoadhesive Polymer Hyaluronan as Biodegradable Cationic/Zwitterionic-Drug Delivery Vehicle. *ADMET DMPK.* **2014,** *2,* 235–247.

19. Torrens, F.; Castellano, G. Computational Study of Nanosized Drug Delivery from *Cyclo*dextrins, Crown Ethers and Hyaluronan in Pharmaceutical Formulations. *Curr. Top. Med. Chem.* **2015**, *15*, 1901–1913.

20. Torrens Zaragozá, F. Polymer Bisphenol-A, the Incorporation of Silica Nanospheres into Epoxy–Amine Materials and Polymer Nanocomposites. *Nereis.* **2011**, *3*, 17–23.

21. Torrens, F.; Castellano, G. Bisphenol, Diethylstilbestrol, Ppolycarbonate and the Thermomechanical Properties of Epoxy–Silica Nanostructured Composites. *J. Res. Updates Polym. Sci.* **2013**, *2*, 183–193.

22. Hotelling, H. Analysis of a Complex of Statistical Variables into Principal Components. *J. Educ. Psychol.* **1933**, *24*, 417–441.

23. Kramer, R. *Chemometric Techniques for Quantitative Analysis*; Marcel Dekker: New York, 1998.

24. Patra, S. K.; Mandal, A. K.; Pal, M. K. J. State of Aggregation of Bilirubin in Aqueous Solution: Principal Component Analysis Approach. *Photochem. Photobiol. Sect. A.* **1999**, *122*, 23–31.

25. Jolliffe, I. T. *Principal Component Analysis*; Springer: New York, 2002.

26. Xu, J.; Hagler, A. Chemoinformatics and Drug Discovery. *Molecules.* **2002**, *7*, 566–600.

27. Shaw, P. J. A. *Multivariate Statistics for the Environmental Sciences*; Hodder-Arnold: New York, 2003.

28. IMSL. *Integrated Mathematical Statistical Library (IMSL)*; IMSL: Houston, TX, 1989.

29. Tryon, R. C. J. A Multivariate Analysis of the Risk of Coronary Heart Disease in Framingham. *Chronic Dis.* **1939**, *20*, 511–524.

30. Priness, I.; Maimon, O.; Ben-Gal, I. Evaluation of Gene-Expression Clustering via Mutual Information Distance Measure. *BMC Bioinformatics.* **2007**, *8*, 111–112.

31. Steuer, R.; Kurths, J.; Daub, C. O. Weise, J.; Selbig, J. The Mutual Information: Detecting and Evaluating Dependencies between Variables. *Bioinformatics.* **2002**, *18* (Suppl. 2), S231–S240.

32. D'Haeseleer, P.; Liang, S.; Somogyi, R. Genetic Network Inference: From Co-Expression Clustering to Reverse Engineering. *Bioinformatics.* **2000**, *16*, 707–726.

33. Perou, C. M.; Sørlie, T.; Van Eisen, M. B.; De Rijn, M.; Jeffrey, S. S.; Rees, C. A.; Pollack, J. R.; Ross, D. T.; Johnsen, H.; Akslen, L. A.; Fluge, O.; Pergamenschikov, A.; Williams, C.; Zhu, S. X. Lønning, P. E.; Børresen-Dale, A. L.; Brown, P. O.; Botstein, D. Molecular Portraits of Human Breast Tumours. *Nature (London).* **2000**, *406*, 747–752.

34. Jarvis, R. A.; Patrick, E. A. Clustering Using a Similarity Measure Based on Shared Nearest Neighbors. *IEEE Trans. Comput.* **1973**, *C22*, 1025–1034.

35. Page, R. D. M. *Program Tree View*; University of Glasgow: Glasgow, UK, 2000.

36. Eisen, M. B.; Spellman, P. T.; Brown, P. O.; Botstein, D. Cluster Analysis and Display of Genome-Wide Expression Patterns. *Proc. Natl. Acad. Sci. USA.* **1998**, *95*, 14863–14868.

37. Huson, D. H. Splits Tree: Analyzing and Visualizing Evolutionary Data. *Bioinformatics.* **1998**, *14*, 68–73.

38. Cicerale, S.; Conlan, X. A.; Sinclair, A. J.; Keast, R. S. J. Chemistry and Health of Olive Oil Phenolics. *Crit. Rev. Food Sci. Nutr.* **2009**, *49*, 218–236.

39. Martín-Peláez, S.; Covas, M. I.; Fitó, M.; Kußar, A.; Pravst, I. Health Effects of Olive Oil Polyphenols: Recent Advances and Possibilities for the Use of Health Claims. *Mol. Nutr. Food Res.* **2013**, *57*, 760–771.

40. Íaroli, M.; Gugi, M.; Tuberoso, C. I. G.; Jerkovi, I.; Íuste, M.; Marijanovi, Z.; Kuß, P. M. Volatile Profile, Phytochemicals and Antioxidant Activity of Virgin Olive Oils from

Croatian Autochthonous Varieties *Maßnjaça* and *Krvavica* in Comparison with Italian Variety *Leccino*. *Molecules*. **2014**, *19*, 881–895.

41. Parkinson, L.; Keast, R. Oleocanthal, a Phenolic Derived from Virgin Olive Oil: A Review of the Beneficial Effects on Inflammatory Disease. *Int. J. Mol. Sci.* **2014**, *15*, 12323–12334.

42. Kalogeropoulos, N.; Tsimidou, M. Z. Antioxidants in Greek Virgin Olive Oils. *Antioxidants*. **2014**, *3*, 387–413.

43. El Riachy, M.; Priego-Capote, F.; León, L.; Rallo, L.; Luque de Castro, M. D. Hydrophilic Antioxidants of Virgin Olive Oil. Part 1: Hydrophilic Phenols: A Key Factor for Virgin Olive Oil Quality. *Eur. J. Lipid Sci. Technol.* **2011**, *113*, 678–691.

44. El Riachy, M. Priego-Capote, F.; León, L. Rallo, L.; Luque de Castro, M. D. Hydrophilic Antioxidants of Virgin Olive Oil. Part 2: Biosynthesis and Biotransformation of Phenolic Compounds in Virgin Olive Oil as Affected by Agronomic and Processing Factors. *Eur. J. Lipid Sci. Technol.* **2011**, *113*, 692–707.

45. Ouni, Y.; Flamini, G.; Issaoui, M.; Nabil, B. Y.; Cioni, P. L.; Hammami, M.; Douja, D.; Zarrouk, M. Volatile Compounds and Compositional Quality of Virgin Olive Oil from Oueslati Variety: Influence of Geographical Origin. *Food Chem.* **2011**, *124*, 1770–1776.

46. Ouni, Y.; Taamalli, A.; Gómez-Caravaca, A. M.; Segura-Carretero, A.; Fernández-Gutiérrez, A.; Zarrouk, M. Characterisation and Quantification of Phenolic Compounds of Extra-Virgin Olive Oils According to their Geographical Origin by a Rapid and Resolutive LC–ESI–TOF MS Method. *Food Chem.* **2011**, *127*, 1263–1267.

47. Wagner, K. H.; Kamal-Eldin, A.; Elmadfa, I. Gamma-tocopherol—An Underestimated Vitamin? *Ann. Nutr. Metab.* **2004**, *48*, 169–188.

48. Jiang, Q.; Christen, S.; Shigenaga, M. K.; Ames, B. N. γ-Tocopherol, the Major Form of Vitamin E in the US Diet, Deserves More Attention. *Am. J. Clin. Nutr.* **2001**, *74*, 714–722.

49. Barbaro, B.; Toietta, G.; Maggio, R.; Arciello, M.; Tarocchi, M.; Galli, A.; Balsano, C. Effects of the Olive-Derived Polyphenol Oleuropein on Human Health. *Int. J. Mol. Sci.* **2014**, *15*, 18508–18524.

50. Rigacci, S.; Stefani, M. Nutraceutical Properties of Olive Oil Polyphenols. An Itinerary from Cultured Cells through Animal Models to Humans. *Int. J. Mol. Sci.* **2016**, *17*, 843.

51. Ilesanmi, J. O. Y.; Gungula, D. T. Functional Properties of Cowpea (*Vigna unguiculata*) Flour Preserved with Mixtures of Neem (*Azadichata indica* A. Juss) and Moringa (*Moringa oleifera*) Seed Oils. *Int. J. Sci.* **2016**, *5*, 142–151.

52. Sarría Chueca, A. Antinutrientes De Los Alimentos: Efectos Perjudiciales y Beneficiosos Para La Salud. *Alim. Nutr. Salud.* **2007**, *14*, 81–87.

53. Stahl, W.; Ale-Agha, N.; Polidori, M. C. Non-Antioxidant Properties of Carotenoids. *Biol. Chem.* **2002**, *383*, 553–558.

54. Jäger, A. K.; Saaby, L. Flavonoids and the CNS. *Molecules*. **2011**, *16*, 1471–1485.

55. Fan, J.; Zhang, Y.; Chang, X.; Saito, M.; Li, Z. Changes in the Radical Scavenging Activity of Bacterial-Type Douchi, a Traditional Fermented Soybean Product, During the Primary Fermentation Process. *Biosci. Biotechnol. Biochem.* **2009**, *73*, 2749–2753.

56. Kim, J.; Choi, J. N.; Kang, D.; Son, G. H.; Kim, Y. S.; Choi, H. K. Correlation Between Antioxidative Activities and Metabolite Changes During Cheonggukjang Fermentation. *Biosci. Biotechnol. Biochem.* **2011**, *75*, 732–739.

57. Shyur L. F.; Yang, N. S. Metabolomics for Phytomedicine Research and Drug Development. *Curr. Opin. Chem. Biol.* **2008**, *12*, 66–71.

58. Lee, C. Y. J.; Isaac, H. B.; Wang, H.; Huang, S. H.; Long, L. H.; Jenner, A. M.; Kelly, R. P.; Halliwell, B. Cautions in the Use of Biomarkers of Oxidative Damage; the Vascular and Antioxidant Effects of Dark Soy Sauce in Humans. *Biochem. Biophys. Res. Commun.* **2006,** *344,* 906–911.

59. Wang, H.; Jenner, A. M.; Lee, C. Y. J.; Shui, G.; Tang, S. Y.; Whiteman, M.; Wenk, M. R.; Halliwell, B. The Identification of Antioxidants in Dark Soy Sauce. *Free Radical Res.* **2007,** *41,* 479–488.

60. Butler, L. M.; Wu, A. H.; Wang, R.; Koh, W. P.; Yuan, J. M.; Yu, M. C. A Vegetable-Fruit-Soy Dietary Pattern Protects against Breast Cancer among Postmenopausal Singapore Chinese Women. *Am. J. Clin. Nutr.* **2010,** *91,* 1013–1019.

61. Woo, H. D.; Park, K. S.; Ro, J.; Kim, J. Differential Influence of Dietary Soy Intake on the Risk of Breast Cancer Recurrence Related to HER2 status. *Nutr. Cancer.* **2012,** *64,* 198–205.

CHAPTER 10

ELECTROKINETIC SOIL REMEDIATION: AN EFFICIENCY STUDY IN CADMIUM REMOVAL

CECILIA I. A. V. SANTOS[1,*], ANA C. F. RIBEIRO[1], DIANA C. SILVA[1], VICTOR M. M. LOBO[1], PEDRO S. P. SILVA[2], CARMEN TEIJEIRO[3], and MIGUEL A. ESTESO[3]

[1]*Department of Chemistry and Coimbra Chemistry Centre, University of Coimbra, Coimbra 3004535, Portugal*

[2]*Department of Physics, CFisUC, Universidade de Coimbra, Rua Larga, P-3004516 Coimbra, Portugal*

[3]*U.D. Quнmica Fнsica, Facultad de Farmacia, Universidad de Alcalб, 28871 Alcalб de Henares, Madrid, Spain*

Corresponding author. E-mail: Cecilia.iav.santos@mail.com

CONTENTS

ABSTRACT

The growth of industrial activity, along with the improper and uncontrolled disposal of waste generated by our society, is linked to the incidence of heavy metal (mainly Cr, Cu, Pb, and Ni) and/or organic pollutants (pesticides and other chlorine compounds) soil and groundwater contamination. Pollution of groundwater by non-biodegradable heavy metals is a matter of the greatest concern to public health. The governments of the industrialized countries spend substantial amounts of money trying to solve, or at least diminish, these pollution problems, the main difficulty laying both in its characterization (due to the complexity of the chemical composition of the soil), and the assessment of decontamination risks. The problem of decontamination of soils polluted with heavy metals is of great actuality in our country, particularly due to inactive or abandoned mining areas, like the ones found in Iberian Pyrite Belt in Alentejo (Portugal). They remain a risk mainly due to the unstable nature of the minerals exploited (essentially sulfides), that can cause very acidic waters and release potentially toxic metals such as lead, arsenic, mercury, cadmium, antimony, selenium, etc.

In the past two decades, research has shown that the electrokinetic soil remediation is a viable, simple, and inexpensive technique that can be applied successfully to the removal of various heavy metals, in certain types of soils. Moreover, being a technology that encompasses natural physicochemical processes, it respects aspects such as sustainability and environmental ethics.

10.1   INTRODUCTION

One of the most common forms of contamination comes from industrial pollutants, due to inefficient waste handling techniques that corporations often give to these residues or hazardous waste leakage. Depending on the industry's activity in question, it is possible to find significant quantities of heavy metals and/or organic contaminants in industrial waste that make a huge impact on the quality of groundwater, soil, and related ecosystems.[1]

During the past decades, several novel and original solutions for efficient contaminant removal from soils have been investigated. Soil decontamination methods can be done by in situ or ex situ, the first being preferred because of their lower cost and less exposure to the environment and to those who must carry out decontamination tasks. Among the options available (bioremediation, pump, etc.), some have been proven to be ineffective in remediating heavy metal-contaminated fine-grained soils due to the low hydraulic conductivity of the soil and strong interactions between heavy metal contaminants and soil particle surfaces.[2] Electrokinetic decontamination, first observed in 1809 by Reuss[3,4] and also called electroremediation has become a big focus of attention as promising method for removing heavy metal contaminants from fine-grained soils and sludges.[5-7]

Fundamentally, electroremediation is a method that involves the application of direct current (dc) through electrodes placed in the soil, creating an electric field for the mobilization and removal of contaminants. The driving mechanisms contributing to the transport of contaminants in the liquid phase of a contaminated soil when a dc electrical potential is imposed across the soil are electro-osmosis and electromigration, electrolysis and geochemical reactions. When the soil is subjected to a dc electric field, migration of pore fluid is generated by electro-osmosis and movement of ions relative to the pore fluid is caused by ionic migration. The combined effects of these two contaminant removal mechanisms control the movement of ionic and non-ionic contaminants in the liquid phase. However diffusion can be significant when electro-osmosis is suppressed.[7]

Extraction and removal of heavy metals are generally achieved by electroplating, precipitation, or ion exchange. Currently, several studies at laboratorial scale demonstrate the applicability of electroremediation also for the removal of organic compounds or as technical aid in optimization of biodegradation of these compounds.[8,9]

This technology presents as main advantages that can be implemented in situ with minimal disruption, to be effective in fine-grained soils and heterogeneous media (where other techniques are unsuccessful), and high performance both in transport and contaminant removal. However, electrokinetic remediation processes are highly dependent on acidic conditions during the application, which favors the release of the heavy metal contaminants into the solution phase. That implies that the effect of electrolysis of the sample is a very important process, to be monitored in conjunction with the transport of ions, as it can interfere appreciably on the efficiency of removal of contaminants. Such electrolysis causes the appearance of H^+ and gaseous O_2 at the anode and gaseous H_2 and OH^- in the cathode. H^+ cations migrate from the anode toward the cathode through the soil, while the same occurs but in opposite direction, with the OH^- anions generated at the cathode. These migrations cause pH changes in different areas of the soil, which in turn affect the main physical and chemical processes occurring within this soil, such as adsorption–desorption, precipitation–dissolution and redox processes involved the partition and speciation of the contaminants.[10] Intensity of the migration of these ions depends mainly on the regulatory capacity of soils.

The main goal of electrokinetic remediation is to influence the migration of subsurface contaminants in an imposed electric field via electro-osmosis, electromigration, and electrophoresis. Electromigration is the transport of ions or polar molecules, in the interstitial fluid (between the soil pores) under the influence of an electric field toward the electrode of opposite charge. Thus, the negative ions are transported to the positively charged electrode (anode) and positive ions to the negatively charged electrode (cathode) and are separated according to their charge. As a result of this movement the electro-osmotic flow, generated by electro-osmosis phenomenon, that is, the movement of water, increases the migration of certain ions delaying the other (oppositely charged). It is also important to consider the movement of H^+ and OH^- ions. Such movement is particularly important in low permeability soils because it contributes largely to the decontamination process, for example, reinforcing the movement of H^+ ions to the cathode. Finally, electrophoresis is the transport of charged particles or colloids under the influence of an electric field; contaminants bound to mobile particulate matter can be transported in this manner. In a compact system, this process has less influence as the solid phase has restricted movement.

Of the above mechanisms, electromigration is the dominant transport mechanism for the ionic species.[6,11] Some characteristics of soils, like high surface area and possible presence of surface functional groups, assist a high adsorption of heavy metals by the formers. Different studies show that soil properties, especially the ratio of oxides and hydroxides of Fe, Al, and Mn, and the content of clay and organic matter, control the adsorption of heavy metals.[12] The study of the influence of electro-osmosis mechanism during decontamination treatment is very relevant, given the close relationship between the electro-osmotic flow and soil characteristics. For example, it is known that an increase in the amount of water in the soil causes an increase in electro-osmotic flow, and if the electrolyte concentration increases, this increase results in a reduction of electro-osmotic flow. These effects are most noticeable in the presence of clays with high capacity for anionic retention. In this regard, in recent years it has been proposed a modified theory on electro-osmotic velocity. In this theory, it is believed that the "true" electro-osmotic flow is proportional to the current carried by the charged solid surfaces in soil.[13,14]

The study of decontamination of soils and sludges at laboratorial scale, can handle various parameters that are pertinent for understanding the transport mechanisms of each pollutant. Thus, with an electroremediation cell apparatus, it is possible to determine the amounts of contaminant present in different areas of soil, the local variation of pH and redox potential, the variation of the current intensity, the electro-osmotic flow, etc.[15]

This work describes the use of electroremediation as in situ methodology to implement the treatment of highly contaminated soil and waters by heavy metals, specifically, by cadmium. This is a very important metal at industrial level, widely used in the production of batteries, paints, and plastics and a common contaminant in abandoned industrial sites. Accumulated cadmium in soils is harmful to human health by contaminating groundwater and agriculture activity. Earlier, electroremediation studies were conducted toward the viability of the electrokinetic method to remove different contaminants.[16,17] Because its effectiveness is proven, we are actually concerned in understanding the physicochemical foundations of this technique so the current work is directed to the increasing the efficiency of the removal and optimization the various experimental parameters (pH, intensity of the applied electric field, electrolysis time, soil characteristics, and use of adsorbent materials) that are the basis for modeling of this electrochemical process.

10.2   EXPERIMENTAL

The use of in situ electroremediation technique can be made by applying a low dc (in the order of mA/cm^2 of the cross-sectional area) or a low potential gradient (in the order of a few V/cm of the distance between the electrodes) to electrodes that are inserted into the contaminated soil region. When this system is triggered, contaminants migrate through the porous matrix, accumulating around the oppositely charged electrodes and are subsequently extracted at the end of the remediation. Under laboratory-scale conditions experiments are usually performed with an apparatus that includes an electroremediation cell.

10.2.1   EXPERIMENTAL APPARATUS

For this work a cylindrical "Teflon" electroremediation cell was designed and built in a way to allow the determination of the redox potential inside soil samples of 350 g and possibility of extracting soil samples—at varying distances from the electrodes—for subsequent analysis.

This cell had a length of 12 cm with 6 cm internal diameter and with two removable chambers of 2.8 cm in length, for receiving the electrolyte. The cell was connected to a constant dc power supply Freak EP-613 which allows applying potential differences between 0 V and 30 V. Both of the electrode compartments were filled with the deionized water. Schematic of the cell and the respective mounting scheme are presented in Figures 10.1 and 10.2.

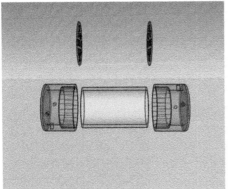 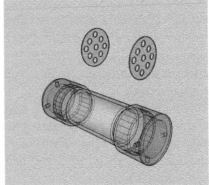

FIGURE 10.1   Electroremediation cell.

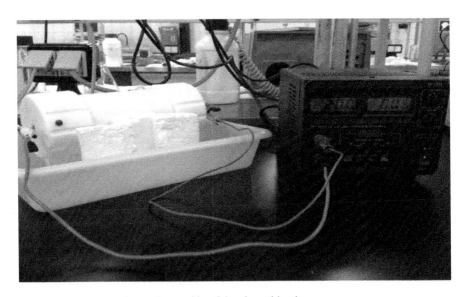

FIGURE 10.2   Experimental assembly of the electrokinetic test apparatus.

Initially in the electrochemical cell graphite working electrodes were used. However, by applying potential above 10 V these electrodes were altered significantly so they were replaced with stainless steel electrodes. The choice of current density and work potential was done following the guidelines established on literature.[18,19] The potential applied varied between 5 V and 20 V.

10.2.2   SOIL PREPARATION

Transport rate of a certain pollutant, and the effectiveness of its elimination, depends mainly on the composition and characteristics of the soil (presence of certain ion-adsorbent substances, water content, percentage of clay, plasticity index, grain size, etc.) and its porosity.[20]

We have worked with several samples of synthetic soil (kaolinite), with constant grain size. These samples were contaminated in the laboratory with known aqueous solutions of Cd (II) concentration, prepared in acetate buffer. Previous tests without contamination (humidified ground only with the acetate buffer) were done to evaluate whether there was any contamination caused by release of the components of the electrodes (stainless steel electrodes, i.e., an alloy of metals) when applying a voltage.

10.2.3   MEASUREMENT AND ANALYSIS

10.2.3.1   ELECTROKINETIC PROCEDURE

The electroremediation cell was filled with about 350 g of soil contaminated with aqueous $CdSO_4$, moisturized to about 56%. The anodic and cathodic compartments (side chambers) were filled with buffer solution (electrolyte). Electrodes were introduced into the side chambers and connected to the power supply by clamps. A constant potential gradient was applied to the cell and different test voltage between 0.42 V/cm and 1.7 V/cm were completed. Experiments were conducted for 48 h.

10.2.3.2   POST-ELECTROREMEDIATION TREATMENT

Once the electrochemical treatment was completed, soil was recovered and fractionated into three sections of 4 cm long each; one next to the cathode (fraction 1), a central fraction (fraction 2), and another near the anode (fraction 3). Electrolytes in each of the chambers and the respective electrodes are also recovered.

For quantification of metals, soil samples were extracted at different distances from the electrodes, that is, from each of the previous sections, a central 1 g sample was recovered and dissolved in distilled-water at a ratio of 1:10 (fragment mass, water volume), stirred for 30 min and then allowed to stand for 24 h. The supernatant was then analyzed by differential pulse voltammetry (with hanging mercury drop electrode (HMDE)). Also, the cell potential and soil pH and electrolyte pH were monitored.

Stainless steel electrodes used in the tests within the electrochemical cell, and deposits that have arisen from the electroremediation treatment were examined using X-ray fluorescence (XRF) analysis.

10.3   RESULTS AND DISCUSSION

As mentioned above, the electroremediation consists in applying a continuous potential difference between two electrodes placed in a contaminated soil sample. As a result of the applied potential, the ions of the contaminant(s) move to the corresponding electrode (electromigration) and can be eliminated using appropriate methods.

10.3.1   POST-REMEDIATION SOIL ANALYSIS

Process performance, that is the assessment of heavy metal removal, is described in terms of displacement (mobility) of the metal ion to the corresponding electrode (generally the cathode) and can be determined from the metal distribution curve (calibration curve) obtained by measuring the change in concentration of such metal as a function of electrolysis time and distance to the electrode.

Experimental techniques such as atomic absorption spectrometry or inductive coupled plasma are commonly used to determine the concentration of metals in soil. However, in this work alternative methods were pursued to achieve the same results. The effectiveness of the use of differential pulse voltammetry was tested due to potential advantages such as speed, simplicity, selectivity, and inexpensiveness for qualitative and quantitative determinations of heavy metals.[21,22] Also this technique is characterized by its low detection limits, and working even in the presence high salt concentrations, allowing metal speciation and differentiation between free and complexed metal ions.

Voltammetric determination of cadmium was performed using Metromh 663 VA Stand equipment from Autolab with three-electrode system consisting of a HMDE working electrode, an auxiliary platinum (Pt) electrode, and an Ag/AgCl/KCl (3 mol/L) reference electrode. Working parameters were previously optimized for these determinations and their values are presented in Table 10.1.

TABLE 10.1 Operating Parameters for the Determination of Cadmium by DPA (Differential Pulse Voltammetry).

Parameter	Description
Working electrolyte	HDME
Number of replications	2
Drop size	2
Stirrer speed	1000 rpm
Mode	Diferential pulse
Initial purge time	600 s
Deposition potential	−1.15 V
Deposition time	90 s
Equilibration time	10 s
Pulse amplitude	0.05 V
Start potential	−1.3 V
End potential	0.05 V
Voltage step	0.006 V
Peak potential (Cd)	−0.403 V

At the start, the calibration curve for aqueous solutions of Cd (II) was built. MiliQ water (11 mL) and 1 mL of acetate buffer were placed in the polarographic vessel and the measure was performed under the parameters described in Table 10.1. At the end this voltammogram was recorded as reference (blank). All the following analyses were done with automatic blank subtraction feature of instrument. The subsequent steps consisted in placing in the polarographic vessel 10 mL milliQ water, 1 mL of acetate buffer solution, and 1 mL of Cd (II) of known concentration. The respective sample voltammograms were collected (example voltammogram shown in Fig. 10.3). All measurements were performed at least two times. From the intensity ratio vs. concentration of the samples, the calibration curve was obtained.

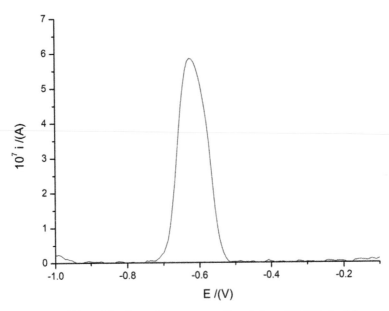

FIGURE 10.3   Differential pulse voltammogram of a solution of CdSO$_4$ 1 mM.

The determination of the concentration of Cd (II) in the soil samples, after the electroremediation treatment, was done also by differential pulse voltammetry. Each sample (1 mL of the supernatant) was placed in the polarographic vessel together with 10 mL of milliQ water and 1 mL of acetate buffer. The final concentration of Cd (II) remaining in the soil sample was obtained from the intensity measured at the maximum of the peak voltammogram of the sample. The results obtained in each of the tests are presented in Table 10.2.

We can see from the results of Table 10.2 that almost all the cadmium present in the soil can be removed with efficiencies very close to 100%. Indeed, in the end of the treatment, large deposits of cadmium were observed precipitated both over the electrode and in the cathodic chamber, possibly due to the high pH existing therein.

To better understand the outcome of applying a potential gradient over the cadmium distribution in the soil, Table 10.2 values have been converted into V/cm—regarding distance to anode—and are plotted against the ratio between final and initial concentration of cadmium (Fig. 10.3).

The effect of the voltage gradient between electrodes is important as it appears that the migration of cadmium does not occur evenly (Fig. 10.4). For the first test, after 48 h remediation treatment there was still

TABLE 10.2 Results of the Electrorremediation Treatment on Cadmium (II) Contamined Soil Samples.

Test	Potential Gradient/(V/cm)	Soil Fraction	Residual Average Concentration/(g/L)	Standard Deviation/10^{-5}	Elimination (%)
1	0.42	1	0.000615	6.51	98.7
		2	0.000204	4.06	99.6
		3	0.000000	5.95	100.0
2	0.83	1	0.000352	7.84	99.3
		2	0.000685	1.95	98.6
		3	0.000000	1.43	100.0
3	1.25	1	0.000270	2.52	99.4
		2	0.000518	2.52	98.9
		3	0.000000	0.56	100.0
4	1.67	1	0.000267	6.65	99.5
		2	0.000000	5.95	100.0
		3	0.000000	0.13	100.0

Note: The soil fraction 1 is the nearest fraction of the anode, the fraction 3 is the closest to the cathode, and fraction 2 is an intermediate distance.

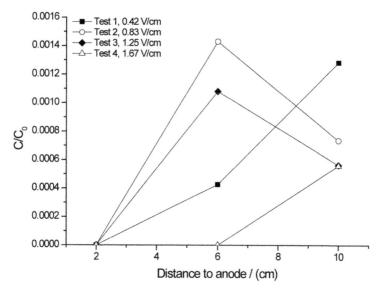

FIGURE 10.4 Effect of the voltage gradient between electrodes on concentration ratio (C/C_0) cadmium distribution in the soil, where each curve corresponds to a test shown in Table 10.2, and the curves 1, 2, 3, and 4 corresponding to the voltages applied 0.42, 0.83, 1.25, and 1.67 V/cm, respectively.

adsorbed cadmium throughout the soil (kaolinite). In following tests the increase in the voltage gradient applied between electrodes produced an increase in cadmium migration toward the cathode, even though this was partially precipitated in the central region, possibly due to creation of a pH front resulting from the electrolysis of water. In test 4, where the voltage gradient applied between electrodes was greater, practically all cadmium was removed. Furthermore, the results obtained in this study are in very good agreement with those that can be found in the literature for similar tests.[23] In fact, despite having divided the soil in just three fractions, the behavior is similar and the results are consistent, with cadmium elimination attained in similar extent.

Changes in pH, in both the soil and in the electrolytic solutions present in the cell chambers, were monitored for all measurements. The results of each test are presented in Table 10.3.

TABLE 10.3   Monitoring of Soil pH and Anodic/Cathodic Chambers Electrolytic Solution pH.

Test [a]	Soil Fraction	Soil pH	Chamber pH	
			Cathode	Anode
0	1	4.9–5.5	9.4–9.5	2.3–2.5
	2	4.8–4.9		
	3	3.5–4.0		
1	1	4.9	5.6	4.4
	2	5.6		
	3	4.1		
2	1	5.1	6.1	2.0
	2	5.2		
	3	4.4		
3	1	5.4	9.7	1.7
	2	5.2		
	3	2.9		
4	1	5.6	12.3	1.6
	2	3.6		
	3	2.8		

Note: (a) The test 0 is the reference (blank). The tests 1, 2, 3, and 4 correspond to applied voltages of 5, 10, 15, and 20 V, respectively. The soil fraction 1 is the nearest fraction of the anode, the fraction 3 is the furthest, and fraction 2 is an intermediate distance.

Acidification generally causes an increase of the solubility of the mineral species present in the soil, leading to an increase in the ionic strength and electrical conductivity of such soil. If the acidification of the soil (as a consequence of the electrolytic production of H^+ in the remediation process) is not controlled, a part of the electrical potential applied to the electroremediation cell will be used in "moving" such protons (from the anode the cathode), with the consequent loss of performance of the remediation process. Moreover, and similarly in the cathode, the OH^- ions generated by the electrolytic process must be controlled to avoid yield losses due remediation process "movement" of such hydroxyl ions toward the anode the cell. This is the reason why it is so important to monitor pH inside the electroremediation cell.

10.3.2 POST-REMEDIATION ELECTRODES AND DEPOSITS ANALYSIS

XRF analysis was used post-remediation to examine the stainless steel electrodes used in the tests together with the precipitate deposits that have arisen from the application of different values of voltage between the electrodes. These studies were performed to identify the constituent elements of both electrodes and deposits and, in the case of the latter, demonstrate that the residues were actually cadmium. In the experiments, as described in Table 10.2, the potential gradients used between the electrodes were of 10, 15, and 20 V for the reference test, and 5, 10, 15, and 20 V for artificial soils tests, contaminated with 10 g/L of cadmium sulfate.

The studies were performed in the Trace Analysis and Imaging Laboratory (TAIL)-UC analytical platform using an XRF spectrometer equipped with a SSD Vortex detector and a source XR Mo (50 kV, 50 W), Hitachi SEA6000VX.

10.3.2.1 X-RAY FLUORESCENCE ANALYSIS OF THE ELECTRODES (POST-TREATMENT OF A NON CONTAMINATED SOIL)

As can be seen from Figure 10.5 the electrode is spotless after the applied voltage of 10 V for 48 h. It is observable that it has in its composition the metals Cr, Mn, Fe, and Ni, being the most prevalent Fe and Mn.

From the analysis of Figure 10.6 is observed that higher voltages may compromise the structure of the electrode. At 15 V some changes are already clear in the electrode surface.

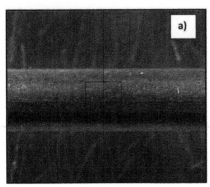
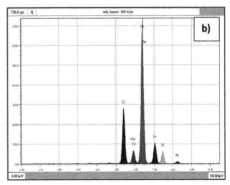

FIGURE 10.5 Anode, after applying a 10 V voltage gradient. (a) Representative anode illustration after applying a 10 V voltage in uncontaminated soil. (b) XFR spectrum of the anode after applying a 10 V voltage in uncontaminated soil.

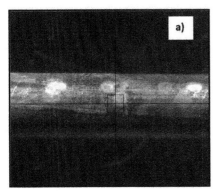
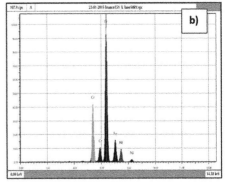

FIGURE 10.6 Cathode, after applying a 15 V voltage gradient. (a) Representative cathode illustration after applying a 15 V voltage in uncontaminated soil. (b) XFR spectrum of the cathode after applying a 15 V voltage in uncontaminated soil.

10.3.2.2 X-RAY FLUORESCENCE ANALYSIS OF THE ELECTRODES (POST-TREATMENT OF A CONTAMINATED SOIL)

As shown in Figure 10.7 and priors, the composition of the electrodes does not vary significantly, and as it can be seen, the electrodes do not have in its constitution the soil contaminant metal, cadmium, so there is no risk that they can contaminate the samples.

On the other side it is clear that high voltages can make great damage to the electrodes but other conditions can be in the origin of this damage, as it is the case of the sudden changes in pH. In the case of this electrode the surrounding pH reached 1.6, a very acidic environment.

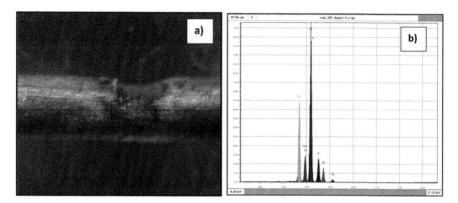

FIGURE 10.7 Anode after applying a 20 V voltage (contaminated soil). (a) A representative illustration of the anode after applying a 20 V voltage in contaminated soil. (b) XFR spectrum of the anode after applying a 20 V voltage in contaminated soil.

10.3.2.3 X-RAY FLUORESCENCE ANALYSIS OF THE PRECIPITATE DEPOSITS (POST-TREATMENT OF A CONTAMINATED SOIL)

Figure 10.8 shows the residue deposited in the cathodic chamber where it is expected that cadmium may precipitate. Indeed the residue has in its constitution mostly cadmium, demonstrating both feasibility and effectiveness of this electrokinetic method for the removal of heavy metals from soil.

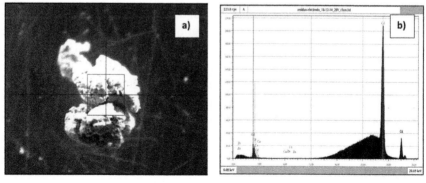

FIGURE 10.8 Analysis of the residue deposited in the cathodic chamber after applying a 20 V voltage (contaminated soil). (a) A representative illustration of the residue deposited in the cathode chamber after applying a 20 V voltage in contaminated soil. (b) XFR spectrum of the residue deposited on the cathode chamber after applying a 20 V voltage in contaminated soil.

In Figure 10.9, it is perceived that the residue that was directly deposited on the cathode has as major constituent cadmium. Other components may be identified, probably as a result of taking this residue directly from the electrode and fragments of the latter may have been also removed. Equally this can also be a consequence of the high degradation of the electrode due to high pH.

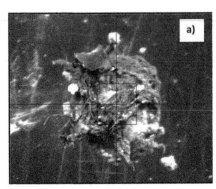 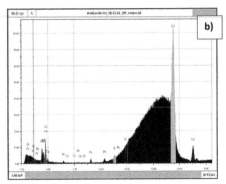

FIGURE 10.9 Analysis of the residue deposited on the cathode after applying a 20 V voltage (contaminated soil). (a) A representative illustration of the residue deposited on the cathode after applying a 20 V voltage in contaminated soil. (b) XFR spectrum of the residue deposited on the cathode after applying a 20 V voltage in contaminated soil.

10.4 CONCLUSIONS

Electroremediation is a decontamination technique which is based on well-established electrochemical principles, and has a very high potential for application of soils. Among the most common applications of this technique are the removal of metals such as cadmium, lead, and zinc, with clearance rates that may exceed 90%. The research here described was done with the major goal of testing the removal efficiency of cadmium from a synthetic contaminated by an electroremediation process. Differential pulse voltammetry was used as an alternative method for the quantification of metals in the soil. The use of voltammetry as detection method proved appropriate, allowing identify residual concentration of cadmium in the soil in the order of 10^{-5} g Cd/kg of soil.

After the analysis of soil samples, subject to 48 h electroremediation inside the electrochemical cell, it is noticeable that the migration of cadmium is not uniform. In fact, increasing voltage gradient between the electrodes increased migration of cadmium toward the cathode, even though this was partially precipitated in the central region, possibly due to the pH front formed by electrolysis of water. When comparing these results with the other literature trials the results are in very good agreement. XRF analysis allowed confirming the composition of the precipitated deposit to be predominantly cadmium, and also reassuring there was no contamination by the electrodes used in the electroremediation process.

It was concluded that the removal of cadmium in soil can be achieved with efficiencies close to 100%.

ACKNOWLEDGMENTS

The authors are grateful for funding from the Coimbra Chemistry Centre, which is supported by the *Fundação para a Ciência e a Tecnologia* (FCT), Portuguese Agency for Scientific Research, through the programs UID/QUI/UI0313/2013 and COMPETE. C.I.A.V.S. also thanks the FCT for support through Grant SFRH/BPD/92851/2013 and to Universidad de Alcalá is grateful for financial support through "Ayudas Postdoctorales-2014."

KEYWORDS

- **organic pollutants**
- **electrokinetic soil remediation**
- **heavy metals**
- **electroremediation**
- **electro-osmosis**

REFERENCES

1. Cherry, K. F. *Plating Waste Treatment;* Ann Arbor Sciences Publications Inc.: Ann Arbor, MI, 1982.
2. Yeung, A. T. In *Remediation Technologies for Contaminated Sites,* Proceedings of the International Symposium on Geoenvironmental Engineering (ISGE 2009), Hangzhou, 2009; 328–369.
3. Yeung, A. T. Electrokinetic Flow Processes in Porous Media and Their Applications. In *Advances in Porous Media;* Corapcioglu, M. Y., Ed.; Elsevier: Amsterdam, 1994; Vol. 2, pp 9–395.

4. Yeung, A. T. Contaminantxtractability by Electrokinetics. *Environ. Eng. Sci.* **2006,** *23* (1), 202–224.

5. Hamed, J.; Acar, Y. B.; Gale, R. J. *Pb(I) Removal from Kaolinite by Electrokinetics, J. Geotech. Eng. Div. ASCE.* **1991,** *117* (2), 241–271.

6. Pamukcu, S.; Wittle, J. K. Electrokinetic Removal of Selected Heavy Metals from Soil. *Environ. Prog.* **1992,** *11* (3), 241–249.

7. Probstein, R. F.; Hicks, R. E. Removal of Contaminants from Soils by Electric Fields. *Science.* **1993,** *260,* 498–503.

8. Lageman, R.; Pool W.; Seffinga, G. A. Electro-Reclamation: State-of-the-Art and Future Developments. In *Contaminated Soil 90;* Arendt, F., Hinsenveld, M., van den Brink, W. J., Eds., Kluwer Academic: Dordrecht, The Netherlands, 1990; pp 1071–1078.

9. Shapiro, A. P.; Probstein, R. F. Removal of Contaminants from Saturated Clay by Electroosmosis. *Environ. Sci. Technol.* **1993,** *27* (2), 283–291.

10. Reddy, K. R.; Parupudi, U. Removal of Chromium, Nickel and Cadmium from Clays by in-situ Electrokinetic Remediation. *J. Soil. Contam.* **1997,** *6* (4), 391–407.

11. Acar, Y. B.; Alshwabkeh, A. N. Principles of Electrokinetic Remediation. *Environ. Sci. Technol.* **1993,** *27* (13), 2638–2647.

12. Gong, C.; Donahoe, R. J. An Experimental Study of Heavy Metal Attenuation and Mobility in Sandy Loam Soils. *Appl. Geochem.* 1997, *12,* 243–254.

13. Gray, D. H.; Mitchell, J. K. Fundamental Aspects of Electro-Osmosis in Soils. *JSMFD ASCE.* **1967,** *93* (SM6), 209–236.

14. Wise, D. L.; Trantolo, D. J. *Remediation of Hazardous Waste Contaminated Soils;* CRC Press: Boca Raton, FL, 1994.

15. Tobón, D. A. Q.; Vázquez, M. V. Estudio Del Movimiento De Iones NO3 y Pb3 en Andisoles Sometidos a Campo Eléctrico. *Afinidad.* **2001,** *58* (496), 437–441.

16. Pazos, M.; Rosales, E.; Alcántara, T.; Gómez, J.; Sanromán, M. A. Decontamination of Soils Containing PAHs by Electroremediation: A Review. *J. Hazard. Mater.* **2010,** *177,* 1–11.

17. Virkutyte, J.; Sillanpää, M.; Latostenmaa, P. Electrokinetic Soil Remediation – Critical Overview. *Sci. Total Environ.* **2002,** *289,* 97–121.

18. Vengris, T.; Binkiene, R.; Sveikauskaite, A. Electrokinetic Remediation of Lead-, Zinc- and Cadmium-Contaminated Soil. *J. Chem. Technol. Biotechnol.* **2001,** *76,* 1165–1170.

19. Raj, J.; Mohineesh, A. R.; Dogra, T. D. In *Direct Determination of Zinc, Cadmium, Lead, Copper Metal in Tap Water,* E3S Web of Conferences, 2013.

20. Page, M. M; Page, C. L. Electroremediation of Contaminated Soils. *J. Environ. Eng.* **2002,** *128* (3), 208–219.

21. Selehattin, Y.; Sultan, Y.; Gulsen, S.; Murat, S. Direct Determination of Zn Heavy Metal in Tap Water of Canakkale (TURKEY) by Anodic Stripping Voltammetry Technique. *Int. J. Electrochem. Sci.* **2009,** *4,* 288–294.

22. Ostapczuk, P.; Valenta, P.; Rützel H.; Nürnberg, H. W. Application of Differential Pulse Anodic Stripping Voltammetry to the Determination of Heavy Metals in Environmental Samples. *Sci. Total Environ.* **1987,** *60,* 1–16.

23. Wenxin, S.; Zhiwei, Z.; Liye, Z.; Tunmin, Y. Cadmium Removal and Electroosmotic Flow during Electrokinetic Remediation of Contaminated Soil. *Adv. Mat. Res.* **2010,** *113–114,* 2237–2240.

CHAPTER 11

NANO-SIZED NICKEL BORATE PREPARATION AND CHARACTERIZATION

MERVE TÜRK, BARIŞ GÜMÜŞ, FATMA USTUN, and
DEVRIM BALKÖSE*

Department of Chemical Engineering, İzmir Institute of Technology, Gulbahce, Urla İzmir, Turkey

Corresponding author. E-mail: devrimbalkose@gmail.com

CONTENTS

ABSTRACT

Nano-sized nickel borate hydrate were precipitated from equimolar mixtures of dilute nickel nitrate and borax solutions at 25°C. Produced nickel borate samples were characterized by TGA, DSC, FTIR spectroscopy, X-ray diffraction, SEM, Elemental Analysis (EDX), Titration (nickel determination by EDTA, B_2O_3 determination by NaOH), Particle Size Distribution, and Dehydration. The particles with 55 nm, 80 nm and 70 nm sizes were obtained for the cases without template, with span 60 and PEG 4000 in the reaction mixture respectively. The empirical formula of the vacuum dried precipitates were $NiO.1.3B_2O_3.5.6\ H_2O$, $NiO.1.2B_2O_3.5.6\ H_2O$ and $NiO.1.0B_2O_3.5.4\ H_2O$ for the cases without template, with span 60 and PEG 4000. The density of the nickel borate hydrates was around 2 g/ml and they had a color described by 157, 199 and 158 in RGB color scale. The nickel borate hydrates were amorphous in structure and no sharp peaks related to a crystal structure was present in their x-ray diffraction diagram. The effect of presence of span 60 and PEG 4000 were not significant on the particle size and chemical composition of the nanoparticles.

11.1 INTRODUCTION

Metal borates are used in several applications such as fuel cells, nonlinear optical and laser devices. They are also known for their interesting magnetic, catalytic and phosphorescence properties.[1] The borate glasses are useful for design and applications for gamma and neutron shielding.[2] Nickel borate thin films were used as oxygen evolution catalyst.[3]

Hydrated borates of heavy metals can be prepared by mixing aqueous solutions or suspensions of the metal oxides, sulfates, or halides and boric acid or alkali metal borates such as borax. The precipitates formed from basic solutions are often sparingly-soluble amorphous solids having variable compositions. Crystalline products are generally obtained from slightly acidic solutions.[4]

Among all the borates, nickel borate is of special interest because of its fuel cell efficiency under hard conditions. Nickel is a metallic element that is naturally present in the earth's crust. Due to unique physical and chemical properties, metallic nickel and its compounds are widely used in modern industry[5]. Nickel borate has been identified as a good catalyst and fuel cell material. It was reported that thin nickel borate catalyst films that were electrodeposited from dilute Ni^{2+} solutions in borate electrolyte at pH 9.2 exhibit electro catalytic water oxidation properties. Oxygen electrode efficiency increased ten times when nickel borate was used as the electrode. As an alternative to Co-OEC catalyst used in solar energy storage applications and water purification, nickel borate films can be prepared with precise thickness control. The Ni based oxide catalyst exhibits long term stability in water with no observed corrosion and they may permit energy storage in solar energy applications to be performed with devices that are inexpensive and highly manufacturable.[6] Water oxidation catalysis proceeds at the domain edge of NiO_6 octahedra as indicated by the XAFS studies by Yoshida et al.[7] Nickel borate has antimicrobial properties against gram positive bakteria and has high elecrocapacitance.[8]

Nickel borate has been prepared by solid state route at high temperature (~1200°C). Knyrim et al.[9] synthesized transition metal borates, β -MnB_4O_7, β -NiB_4O_7, and β -CuB_4O_7 at high pressure and temperature. Starting reagents were molar mixtures of boron oxide and the corresponding metal oxide (MnO_2, NiO, CuO), which were transformed at 7.5 GPa to the borates β-MB_4O_7 (M = Mn, Ni, and Cu) at temperatures of 1000, 1150, and 550°C, respectively.

The shape and size of nickel borate particles can be controlled by adding sodium chloride and a surfactant. For examle, rod-like micron sized $Ni_3(BO_3)_2$ particles were synthesized via a facile flux NaCl and surfactant poly(oxyethylene)9 nonylphenolether (NP-9) assisted thermal conversion route at 750–800°C for 2.0 h. The introduction of NaCl and NP-9 was favorable for the formation of the $Ni_3(BO_3)_2$ particles with rod-like morphology and smaller size, respectively.[10] When calcined at 750°C in the presence of 8.0 mL of NP-9 and with the molar ratio of Ni:B:Na=3:8:3,high aspect ratio rod-like micron $Ni_3(BO_3)_2$ particles were obtained.

Nickel orthoborate $Ni_3(BO_3)_2$ nanoribbons were synthesized via a facile solid state reaction from H_3BO_3 and $Ni(NO_3)_2$.[7] Nickel borate ($Ni_3(BO_3)_2$ nanorods were synthesized by the sol-gel method using nickel nitrate and boric acid as reactants, citric acid as the foaming agent.[10] The length and diameter of the nickel borate($Ni_3(BO_3)_2$) nanorods were controlled by adjusting the molar ratio of $Ni(NO_3)_2/H_3BO_3$.The diameters of the nanorods were in the range of 200–300 nm and the lengths were in the range of 2–3 μm with $nNi(NO_3)_2$:nH_3BO_3 = 1:3. The reaction of citric acid and nickel ions forms coordination compound in mesh structure, thus prompting even dispersion of nickel between the grid and providing favorable response space for the formation of nickel borate nanorods.[11]

The growth of nickel borate, $Ni_3(BO_3)_2$ nanowhiskers was successfully achieved by simply heating a mixture of nickel and boron oxide (B_2O_3)

powders at 950°C in air.[12] The synthesized nanowhiskers, with diameters of 150–500 nm and lengths of 10–30 mm, possessed high aspect ratio, and were found to grow along the [012] crystallographic direction. The nanowhiskers partially decomposed after being treated at 1200°C. $Ni_3(BO_3)_2$ nanowhiskers were successfully used for bonding nickel parts in air.[12]

Different from borate materials traditionally prepared under high temperature/pressure solid-state conditions, hydrothermal approach has been recently proved to be very effective in the construction of metal borate compounds.[13] Two amorphous hydrated nickel borates were synthesized by Hong-Yan et al.[14] with nickel chloride hexahydrate and borax decahydrate as reactants at mole ratios of 1:2 and 1:8, respectively. The pH of the aqueous solution of the mixture at 50°C was regulated with H_3BO_3 to 8.8 and 7.9. The chemical compositions of these nickel borates were determined to be $NiO.0.8B_2O_3.4.5H_2O$ and $NiO.B_2O_3 \cdot 3H_2O$ through thermogravimetry-derivative thermogravimetry (TG-DTG) and chemical analysis, in which the main anions were determined to be $B_3O_3(OH)_5^{2-}$ and $B_2O(OH)_6^{2-}$, respectively by Raman spectroscopy. The the local structure around the Ni atom has a similar structure to that of the $Ni_3B_2O_6$ crystal as shown by EXAFS study. The neighboring coordination atoms of Ni in these amorphous nickel borates are O, B, and Ni. The first shells of Ni^{2+} in $NiO·0.8B_2O_3·4.5H_2O$ and $NiO·B_2O_3·3H_2O$ are octahedral with six oxygen atoms.[14]

To obtain nanocrystalline nickel borate particles at low temperatures micro emulsions was used. In this method, reverse micelles are surfactant aggregates where reactants are confined. The size, shape and homogeneity of the nanostructure obtain depends on the dynamics of the micellar exchange through their interfaces. Firstly a nickel-boron precursor was synthesized using reverse micellar route at room temperature. Then, decomposition of this precursor in air at 800°C led to parrot green nickel borate product with ~30 nm particle size. Sodium borohydrate was used as reducing agent and to reduce metal ions like Ni^{2+} by H_2 production.[15]

In this study, it was aimed to produce nanosized nickel borate by the reaction of nickel nitrate and sodium borate solutions at ambient conditions. Effects of surface active agents Span 60 or PEG 4000 in size of the particles were also investigated. Nickel borate samples were characterized by SEM, x-ray diffraction, thermal gravimetry, differential scanning calorimetry and infrared spectroscopy. The particle size distributions of hyrosols of nickel borate were determined by Zeta-Sizer. The chemical analysis of samples was made EDX and by titration experiments.

11.2   MATERIALS AND THE METHOD

1.2.1   MATERIALS

In this study, nickel (II) nitrate hexa hydrate (Panreac), sodium tetraborate (Sigma Aldrich), span 60 (Sigma), PEG 4000 (Merck), EDTA (Merck, 99 %), NaOH (Riedel, 99 %), murexide (Merck), phenolpthalein (Merck), methyl orange (Merck), D-mannitol (Sigma), pH buffer NH_4Cl+NH_3 (NH_4Cl (Merck, 99.8%), NH_4OH (Riedel, 26 %)), HCl (6 M) (Riedel, 37 %), boric acid (Sigma) were used. Distilled water was used for all solutions and treatment processes.

11.2.2   METHODS

Overall, 50 mL of 0.1 M sodium borate solution was added instantly to 50 mL of 0.1 M nickel nitrate solution and mixed at 600 rpm for 2 h at ambient temperature of 23°C. Experiments were repeated by adding 1 mL of 0.002 M span 60 and 1 mL of 0.4 g PEG 4000 in 100 mL to see the templating effect of surfactants on the products. While the formed hydrosols have been mixed, their temperature and pH values were recorded.

The precipitates were separated by centrifugation of the hydrosols at 4000 rpm for 5 min, washed with ethanol and centrifuged at 4000 rpm for 5 min. The gelatinous samples were dried under vacuum for 18 h at 25°C to obtain nanoparticles.

Produced nickel borate samples were characterized by TGA, DSC, FTIR spectroscopy, X-ray diffraction, SEM, Elemental Analysis (EDX), Titration (nickel determination by EDTA, B_2O_3 determination by NaOH), Particle Size Distribution, and Dehydration.

Thermal gravimetric analysis of the samples was made on SHIMATZU TGA-51 under nitrogen 40mL/min flow and heating rate of 10°C /min up to 600°C. Differential scanning calorimetry of the samples was performed on SHIMADZU DSC-50 under nitrogen media with a flow rate of 50 mL/min and heating rate of 10°C/min up to 400°C. IR spectra of the raw materials and the samples were obtained by KBr disc method by using SHIMADZU FTIR-8400S. DRIFT FTIR spectroscopy measurements were carried out in a praying mantis diffuse reflection attachment (Harrick Scientific Products Inc.) equipped with a high-temperature, low-pressure reaction chamber fitted with CaF_2 windows (HVC-DRP, Harrick Scientific Products Inc.) using FTS 3000 MX spectrophotometer (Digilab Excalibur Series). Then

0.1 g sample was placed in the sample holder and after taking its spectrum at room temperature and 101.3 kPa pressure, the sample chamber was evacuated down to 0.1 Pa pressure and heated up to 500°C at 2°C/min rate under vacuum. The DRIFT spectrum of the sample was obtained at different temperatures during its dynamic heating. X-ray diffraction diagrams of the samples were obtained by using SHIMADZU XRD-6000 diffractometer employing Ni-filtered Cu K-α radiation. The morphology of the samples was examined using QUANTA 250F scanning electron microscope (SEM). EDX analysis for elemental composition was made with the same instrument.

The nickel content of the samples was determined by analytical titration of the dissolved samples in HCl. Murexide indicator was used to determine the end point. The B_2O_3 content of the samples were determined by titrating of the solutions with 0.1M NaOH after complexing nickel ions with equivalent amount of EDTA. The solutions were neutralized first with NaOH and the end point was monitored by methyl orange indicator. Titration with NaOH was continued after adding phenolphthalein indicator and mannitol. The amount of NaOH consumed between the turning points of methyl orange and phenolphthalein was equivalent to B_2O_3 contents of the samples.

The particle size distribution of the particles formed in water was carried out by Malvern Zetasizer 3000 HSA. The density of the particles was determined by immersion of 0.5 g of particles in 25 mL water and measuring the replaced volume using 25 mL glass pycnometer. The color of the hydrosols was measured in RGB scale using ImageJ program using their photographs.

11.3 RESULTS AND DISCUSSION

11.3.1 RAW MATERIAL CHARACTERIZATION

FTIR spectrum of the raw materials used to produce Nickel Borate is represented in Figures 11.1 and 11.2. Template materials PEG 4000 and Span 60 were also analyzed within FTIR spectrum and they are presented in Figures 11.3 and 11. 4.

In Figure 11.1, FTIR spectrum of Nickel Nitrate, there was a broad peak between 3550–3200 cm^{-1} that represents the hydrogen bonded O—H groups. The peak at 1620 cm^{-1} represented the H—O—H bending vibrations and the peak at 1400cm^{-1} belongs to asymmetric stretching of O—NO_2.[16]

FTIR spectrum of other raw material for nickel borate production, sodium tetraborate (borax) is shown in Figure 11.2. The broad peak between 3200 and 3700 cm^{-1} represented hydrogen bonded O—H groups. H—O—H bending vibration is observed as a small shoulder at 1640 cm⁻¹. The bands at 1380 and 1000 cm⁻¹ might be due the asymmetric and symmetric stretching vibrations of B(3)—O respectively; the bands around 1100 and 830 cm⁻¹ were the asymmetric and symmetric stretching of B(4)—O, respectively. B(3)—O had small peaks also at 980 and 720 cm⁻¹ while B(4)—O had small peaks at 1100and 830 cm⁻¹.[17]

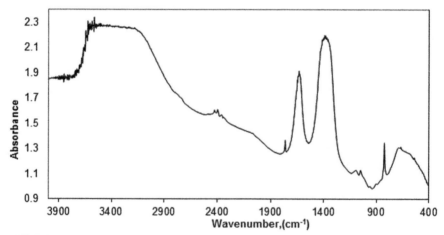

FIGURE 11.1   FTIR spectrum of nickel nitrate hexahydrate.

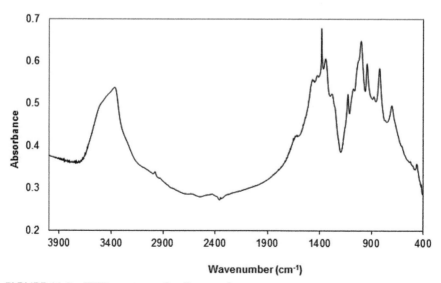

FIGURE 11.2   FTIR spectrum of sodium tetraborate.

Figure 11.3 shows the FTIR spectrum of PEG4000. The broad peak between 3600 and 3200 cm^{-1} wavenumber represent the hydrogen bonded O—H groups. At 2890 cm^{-1} there was a strong peak that belongs to C—H bonds. C—O—C group was represented with a sharp peak at 1100 cm^{-1}, two small peaks at 840 and 960 cm^{-1} and a small peak of asymmetric C—O—C group. The small peak at 1470 might be due to bending vibration of CH$_2$ groups.

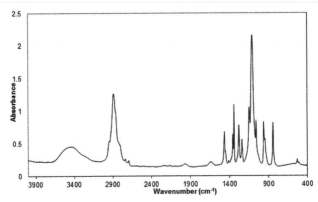

FIGURE 11.3   FTIR spectrum of PEG 4000.

Hydrogen bonded O—H group vibrations gave a broad peak again at 3430 cm^{-1} in the FTIR spectrum of Span 60 (Fig. 11.4). The peak at 1730 cm^{-1} is attributed to C=O of esters, whereas the band at 1467 cm^{-1} belogs to CH$_2$ bending vibrations. CH$_2$ asymmetric and symmetric vibrations are observed at 2930 and 2860 cm^{-1}.

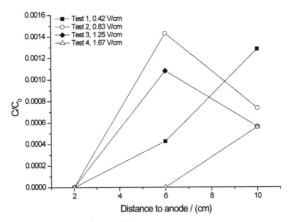

FIGURE 11.4   FTIR spectrum of span 60.

11.3.2 REACTION OF AQUEOUS BORAX AND NICKEL NITRATE SOLUTIONS

The nickel borate precipitation reaction is expected to occur as given in eq 11.1

$$(y/2)B_4O_7{}^{2-} (aq) + xNi^{2+}(aq) + zH_2O \rightarrow xNiO.yB_2O_3.zH_2O \text{ (s)} \qquad (11.1)$$

There are other simultaneous reactions in the reaction medium, such as

$$2Ni^{2+}(aq) + 3OH^-(aq) + NO_3{}^-(aq) \rightarrow Ni_2(OH)_3(NO)_3(s) \qquad (11.2)$$

$$Ni^{2+}(aq) + 2OH^-(aq) \rightarrow Ni(OH)_2(s) \qquad (11.3)$$

The pH changes during reaction of aqueous borax and nickel nitrate solutions. The measured initial temperature and pH values of the solutions are shown in Table 11.1. As seen in the Table while the borax solution was at pH 9.3, the nickel nitrate solution is at pH 7.09. The change of pH and temperature of the reaction medium after mixing the reactant and template solutions are shown in Figure 11.5.

TABLE 11.1 pH and Temperature Values of Reactant Solutions.

Solutions	pH	T(°C)
0.1 M of $Na_2B_4O_7$	9.31	23.1
0.1 M of $Ni(NO_3)_2.6H_2O$	7.09	23.2
0.002 M of Span 60	8.1	23.4
0.001 M PEG 4000	8.42	23.3

The pH was 7.90, 7.96, and 7.99 immediately after mixing the reactants and it was lowered to 7.78, 7.80, and 7.86 for mixtures without template, with span 60 and with PEG 4000 respectively during the course of the reaction indicating the release of protons during the course of the reaction. The lowering of the pH values could be attributed to the precipitation of $Ni_2(OH)_3(NO)_3$[15] or $Ni(OH)_2$ by reaction of Ni^{2+} ions with OH ions initially present in borax solution. However, the change in Ni^{2+} concentration due to these reactions could be up to only 0.01% due to low concentrations of OH⁻ groups in initial solutions.

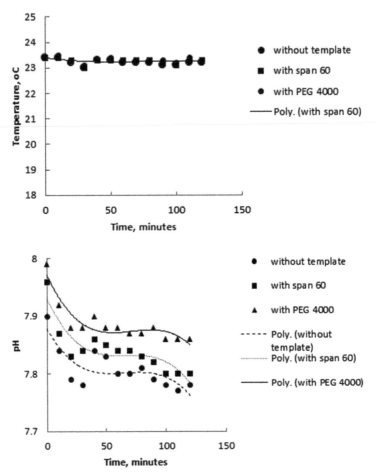

FIGURE 11.5   (a) The pH and (b) temperature change during reaction.

11.3.3 *DENSITY AND COLOR OF THE PRECIPITATES*

The precipitates obtained had nearly 2 g/mL density; however, the density was highest 2.14 g/mL for the sample without template as seen in Table 11.2.

Hydrosols of nickel borate were formed by mixing nickel nitrate and borax solutions. The color of the hydrosols was green-blue in color as seen in Figure 11.6. The RGB values measured by using the ImageJ program on the photograph of the hydrosols are shown in Table 11.2 for quantitative description of their color.

TABLE 11.2 The Density of the Samples Obtained and the Color of the Initial Hydrosols Formed.

Sample	Density, g/mL	Color of Hydrosols		
		R	G	B
Without template	2.14	157	199	158
With span 60	2.06	159	204	160
With PEG 4000	1.94	159	204	160

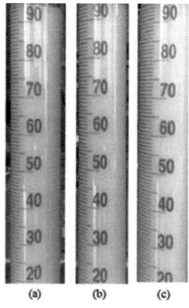

(a)   (b)   (c)

FIGURE 11.6 The hydrosols of nickel borates placed in graduate cylinders (a) without template, (b) with span 60, and (c) with PEG 4000.

11.3.4 FUNCTIONAL GROUPS BY FTIR SPECTROSCOPY

The FTIR spectra of the samples without any template and with templates span 60 and PEG 4000 are very similar to each other as seen in Figure 11.8. There is a broad peak at 3416 cm^{-1} due to hydrogen bonded O—H group vibrations. Asymmetric B(3)—O vibrations were observed at 1383 and at 1350 cm^{-1}. H—O—H bending vibration was observed at 1624 cm^{-1}. At 977 cm^{-1} a peak for symmetric B(3)—O, vibration was present. Out of plane bending vibration of B(3)—O gave a small peak at 684 cm^{-1}.[17-18]

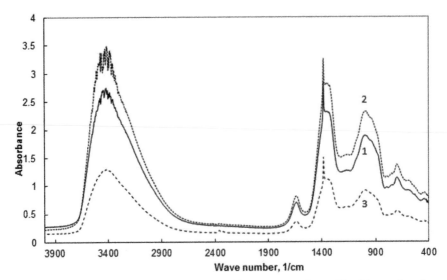

FIGURE 11.7   FTIR spectra for sample (1) without any templates, (2) with template Span 60, and (3) with template PEG 4000.

11.3.5   DRIFT FTIR SPECTRA CHANGE ON HEATING THE PRODUCTS

DRIFT FTIR analyses were carried out to observe changes occurring on dynamic heating of the samples. Between this interval, O—H, H—O—H, B(3)—O and B—O—H groups can be detected. The DRIFT spectra of each sample was recorded at the samples were illustrated at ambient conditions (25 °C, 101.3 kPa) and at 50, 125, 205, 300, 400, and 500°C under vacuum (0.1 Pa) and are shown in Figures 11.8–11.10. The DRIFT spectra give information about the surface of the particles thus they are different than the FTIR absorption spectrum shown in Figure 11.7. The main peaks in DRIFT spectra were at 3350 cm^{-1} belonging to hydrogen bonded OH groups, at 1600 cm^{-1} belonging to H—O—H bending vibration and 1350 and 1310 cm^{-1} belonging to B(3)—O group. B—O—H group is represented by the peak at 1200 cm^{-1}.

As the temperature increases, intensities of the peaks at 3450 and 1650 cm^{-1} decreases indicating that water was eliminated from the sample. The peak at 1200 cm^{-1} belonging to B—OH group also decreases in intensity with increasing temperature.

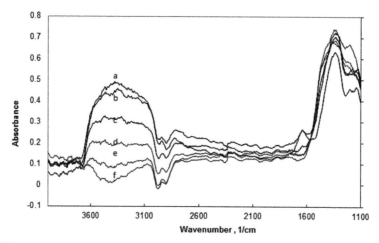

FIGURE 11.8 DRIFT spectra for nickel borate without template (a) before vacuum at 25°C, (b) under vacuum at 50°C, (c) at 125°C, (d) at 205°C, (e) at 300°C, and (f) at 500°C.

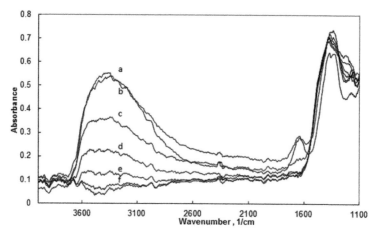

FIGURE 11.9 DRIFT spectra for nickel borate with span 60 (a) before vacuum at 25°C, (b) under vacuum at 50°C, (c) at 125°C, (d) at 205°C, (e) at 300°C, (f) at 400°C, and (g) at 500°C.

B(3)—O group is represented by the broad peak around 1350 cm^{-1} wavenumber at each DRIFT spectra figure. Increase in the temperature did not affect intensity of this molecule significantly. However, at 500°C, B(3)—O had weaker intensity because some decompositions occur at that temperature. There is a very little change between 50 and 400°C. There is a very weak peak at 2900 cm^{-1}, which may be attributed to CH$_2$ stretching vibrations. The source of the CH$_2$ groups in the samples could be the strongly adsorbed ethanol which was used in washing the precipitates.

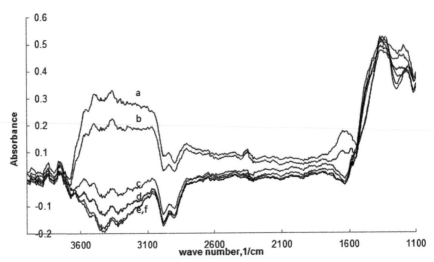

FIGURE 11.10   DRIFT spectra for nickel borate with PEG 4000 (a) before vacuum at 25°C, (b) under vacuum at 125°C, (c) at 205°C, (d) at 300°C, (e) at 400°C, and (f) at 500°C.

The absorbance values of O—H and H—O—H vibrations were normalized by dividing the initial absorbance values at 25°C and 1 atm to find relative intensity for each temperature.

The relative intensity versus temperature graphs are shown in Figures 11.11–11.13 for each sample.

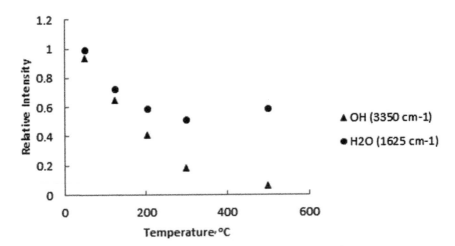

FIGURE 11.11   Change in relative band intensity with temperature for nickel borate without any templates.

For all samples, the relative intensities of the 3350 and 1625 cm^{-1} band decreased with increasing temperature (Figures 11.11–11.13) indicating hydrogen bonded OH groups which may be present as H_2O or B—OH and H_2O molecules were removed by heating the hydrated nickel borates by heating. In situ DRIFT study showed that by heating up to 500°C, all OH groups were removed and anhydrous nickel borate was obtained.

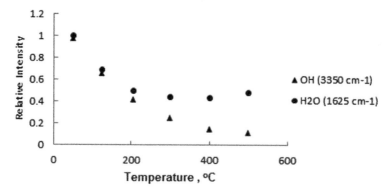

FIGURE 11.12   Change in relative band intensity with temperature for nickel borate with span 60.

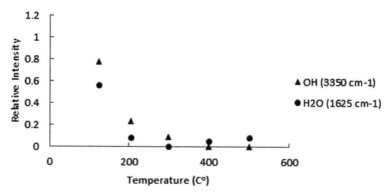

FIGURE 11.13   Change in relative band intensity with temperature for nickel borate with PEG 4000.

11.3.6 TG ANALYSIS AND WATER CONTENT OF THE PRECIPITATED PRODUCTS

TG curves of produced nickel borate samples are represented in Figure 11.14. As seen in Figure 11.14 all the samples have very similar thermal behavior.

Mass loss of the samples was due to removal of water as the FTIR DRIFT studies has shown. It generally started at 44°C and continued until 600°C. The sample without any templates had 65.7% remaining mass, sample with Span 60 had 65.85% remaining mass and third sample with PEG 4000 had 66.67% remaining mass at 600°C at which H_2O removal from the structure was almost completed. On heating the samples, free water, loosely bound water and tightly bond water were removed as water vapor thus the mass decreases with increasing temperature. The temperature ranges are 25–85, 85–285, and 285–1000°C for external, loosely bound and tightly bound water respectively.[19] All the samples contained very similar amount of water, around 34% of which 4% free water, 24% loosely bound water and 6% tightly bound water as reported in Table 11.3. The total water content of hydrated nickel borate prepared at 50°C was slightly lower than that of the present study in Hong-Yan et al.'s study.[14]

TABLE 11.3 Remaining Mass Percentage at Different Temperatures and Percentage of the Types of Water in the Samples.

Sample	Remaining Mass, %			Water, %			
	85°C	285°C	600°C	Free Water 25–85°C	Loosely Bound Water 85–285°C,	Tightly Bound Water 285–600°C	Total Water
Without template	96	71.7	65.7	4	24.3	6	34.3
With span 60	95.9	71.6	65.9	4.1	24.2	5.8	34.1
With PEG 400	96.3	72.2	66.7	3.7	24.6	5.5	33.8

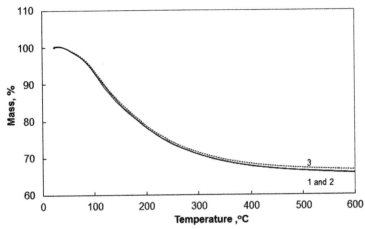

FIGURE 11.14 TG curves for samples (1) without template, (2) with template span 60, and (3) with template PEG 4000.

11.3.7 CHEMICAL ANALYSIS BY ENERGY DISPERSIVE X-RAY SPECTROSCOPY (EDX) AND ANALYTICAL TITRATION

The EDX analysis of the samples indicated that mainly B, Ni, and O elements were present in the samples. A representative EDX spectrum of sample without any template is shown in Figure 11.15. EDX analysis gives the composition on H element free basis since H could not be determined by EDX analysis.

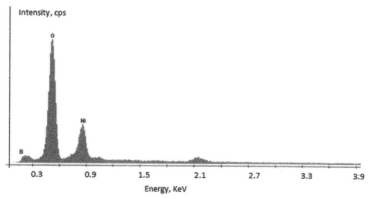

FIGURE 11.15 EDX spectrum of sample without template.

Elemental compositions of the samples determined by analytical titration and EDX analysis are shown in Table 11.4. The B content determined by the two methods was close to each other, but the Ni content found by EDX was higher than analytical titration. The end point of nickel titration with EDTA using Murexide was not very sharp and the EDX analysis gives information only about the surface composition of the particles.

TABLE 11.4 Elemental Composition of Produced Nickel Borate Samples by EDX Analysis and Titration.

Sample	% B		% Ni		% O
	EDX	Titration	EDX	Titration	EDX
Without template	18.08	15.0	35.33	25.42	44.37
With span 60	12.24	12.9	40.33	24.00	44.70
With 4000	11.04	12.34	39.9 3	26.12	46.28

Even if the analysis of the products has some errors, still the empirical formula of the products could be approximated. Considering the oxide content determined by analytical titration and water content by TG analysis the empirical formula of the nickel borates were determined and were shown in Table 11.5. Empirical formulas were found as $NiO.1.3B_2O_3.5.6\ H_2O$ for sample without template, $NiO.1.2B_2O_3.5.6\ H_2O$ for sample with span 60 and $NiO.B_2O_3.5.4\ H_2O$ for sample with PEG 4000. Nearly all the Ni ions in the initial mixture were present in the products as shown in Table 11.5. By changing the reactants ratio, pH and temperature different nickel borate products could be obtained. For example, Menaka et al[13] obtained $NiO.0.8B_2O_3.4.5H_2O$ and $NiO.B_2O_3\ 3H_2O$ from the mixture with Ni/Borax ratio 1:2 and 1:8, at 50°C with pH regulated with boric acid to 8.8 and 7.

TABLE 11.5 Composition of Nickel Borate Formed According to Analytical Titration and TG Analysis and the Yield of the Reaction with Respect to Ni Element Present in the Products.

Sample	%NiO	%B$_2$O$_3$	%H$_2$O	Empirical Formula	Mass of the Product, g	Yield, % (Based on Ni)
Without template	25.42	30.03	34.31	$NiO.1.3B_2O_3.5.6\ H_2O$	0.8192	89
With span 60	24	25.9	34.15	$NiO.1.2B_2O_3.5.6\ H_2O$	0.7760	89
With PEG 4000	26.12	24.69	33.33	$NiO.1.0B_2O_3.5.4\ H_2O$	0.8277	100

11.3.9 MORPHOLOGIES OF THE POWDERS

The SEM images of the produced nickel borate samples are seen in Figure 11.16. The SEM images showed that nanoparticles were obtained. Average particle size was determined by choosing ten random particles for each sample and having their average size.

(a)   (b)   (c)

FIGURE 11.16 SEM micrographs of the powders obtained (a) without any template, (b) with template span 60 and (c) with template with PEG 4000.

According to the SEM image figures it is seen that sample obtained without surface active agent had smaller particle size (55 nm) than the samples with surface active agents Span 60 and PEG 4000 having 55 nm and 70 nm sizes, respectively. However, the nanoparticles were agglomerated to give larger particles.

1.3.10 PARTICLE SIZE DISTRIBUTION OF THE INITIALLY FORMED HYDRATED NICKEL BORATES

Size distribution of prepared nickel borate samples in sol state in water are shown in Figure 11.17. Since the samples were analyzed in hydrosols, particles had larger diameters due to their hydrated state. The sample without any template had average size of 153 nm, The sample templated with span 60 has 261 nm average size. The sample templated with PEG 4000 had a bidisperse distribution with 508 and 2400 nm average sizes. It is shown that contrary to the expectations, the templates used did not lower the size of the particles, on the contrary they were agglomerated to higher sizes in the presence of the templates.

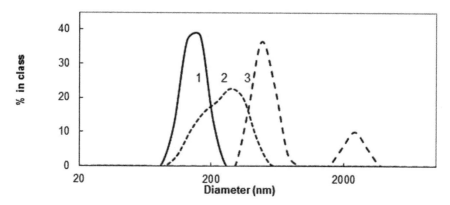

FIGURE 11.17 Size distributions from Zeta-Sizer for sample (1) without any templates, (2) with template span 60, and (3) with template PEG 4000.

While SEM shows the size of the particles in dry state, the zeta sizer indicated the size of the initially formed and hydrated particles.

11.3.11 X-RAY DIFFRACTION ANALYSIS (XRD)

The x-ray diffraction diagrams of the samples are shown in Figure 11.18. As seen in the Figure, there is not any sharp diffraction peak. Only broad peaks are observed at 2θ values of 10° and 25° indicating that all the samples were amorphous. Similar amorphous diffraction diagram was observed in hydrated nickel borates prepared from nickel chloride and borax decahydrate by Hong-Yan et al.[14]

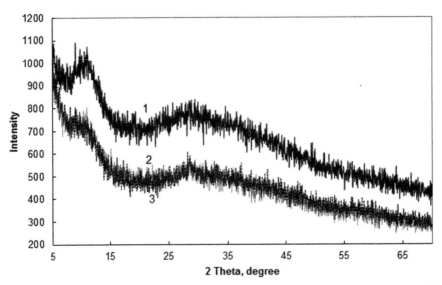

FIGURE 11.18 X-ray diffraction diagrams of sample (1) without any templates, (2) with template span 60, and (3) with template PEG 4000.

11.3.12 DIFFERENTIAL SCANNING CALORIMETRY (DSC)

DSC curves for the obtained nickel borates are shown in Figure 11.19. There was an endothermic transition due to release of water from the structure of nickel borates. This transition was not completed up to 400°C as the DSC curve did not return to the base line. Further endothermic processes would occur, if the samples were heated up to higher temperatures. The endothermic peak maxima were observed at 126.4, 125.2 and 120.1°C for sample without template, with span 60 and with PEG 4000, respectively. The enthalpy change of this transition was found from the area of the endothermic peak. The peak maxima, enthalpy change of transition per gram of sample, enthalpy change for removal of per gram water and mass loss values found from TG analysis are shown in Table 11.6. The enthalpy change for removal of water from the samples was calculated using eq 11.4.

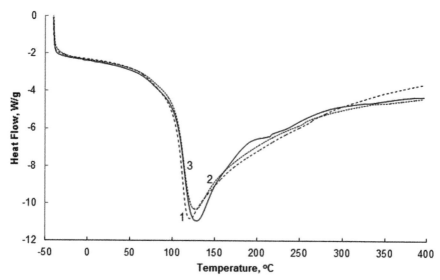

FIGURE 11.19 DSC curves for samples (1) without template, (2) with template span 60, and (3) with template PEG 4000.

TABLE 11.6 DSC and TGA Data of Produced Nickel Borates.

Samples	DSC			TGA
	peak Maximum, °C	Enthaly Change of Transition, J/g Sample	Enthalpy of Removal of Water, J/g Water	Mass Loss, % at 400°C
Without template	126.4	608.6	1769.2	32.0
With span 60	125.2	683.6	1998.8	32.3
With PEG 4000	120.1	768.4	2307.5	31.7

$$\Delta H \text{ for removal of water} = (\Delta H \text{ for the transition/water} \qquad (11.4)$$
$$\text{loss } \% \text{ of sample}) \times 100$$

The enthalpy values for the removal of water were at the same order with the heat of evaporation of liquid water, 2257 J/g, indicating the high contribution of the enthalpy change of the evaporation of water to the enthalpy of the removal of water from the sample.

11.4 CONCLUSIONS

Nickel borate hydrates with oxide formula of $NiO.1.3B_2O_3.5.6\ H_2O$, $NiO.1.2B_2O_3.5.6\ H_2O$ and $NiO.1.0B_2O_3.5.4\ H_2O$ were prepared from equimolar mixtures of 0.1 molar nickel nitrate and 0.1 molar borax solution at 25 °C without template, with span 60 and with PEG 4000 templates, respectively. The reactants were mixed instantaneously, to prevent the growth of the particles formed and dried under vacuum at 25°C. Thus nanoparticles with 55, 80, and 70 nm sizes were obtained for the cases without template, with span 60 and PEG 4000. The density of the nickel borate hydrates was around 2 g/mL and they had a color described by 157, 199 and 158 in RGB color scale. They had H_2O molecules, B—OH and B(3)—O groups in their structure as shown by FTIR spectroscopy. On heating the samples up to 500°C free H_2O (around 4%), lightly bond H_2O (around 24 %), and tightly bond H_2O (around 6%) were removed as shown by TG analysis and FTIR DRIFT study. Anhydrous nickel borate was formed on heating the samples up to 500°C. Water was removed from the system with an enthalpy close to the enthalpy of heat of evaporation of the samples as shown by DSC studies. The nickel borate hydrates were amorphous in structure and no sharp peaks related to a crystal structure was present in their x-ray diffraction diagram. The effect of presence of span 60 and PEG 4000 were not significant on the particle size and chemical composition of the nanoparticles.

KEYWORDS

- nanoparticles
- nickel borate
- metal borates
- hydrated borates
- oxygen evolution catalysts

REFERENCES

1. Tian, Y.; He, Y.; Yu, L.; Deng, Y.; Zheng, Y.; Sun, F.; Liu, Z.; Wang Z. In Situ and One-step Synthesis of Hydrophobic Zinc Borate Nanoplatelets, *Colloids and Surfaces A: Physicochem. Eng. Aspects.* **2008**, *312*, 99–103.

2. Singh, V. P.; Badiger, N. M. Shielding Efficiency of Lead Borate and Nickel Borate Glasses for Gamma Rays and Neutrons. *Glass Phys. Chem.* **2015,** *41* (3), 276–283.

3. Bediako, D. K.; Surendranath, Y.; Junko Yano, J.; Vittal, K.; Yachandra,V.K.; Daniel, G.; Nocera, D.G. Mechanistic Studies of the Oxygen Evolution Reaction Mediated by a Nickel-Borate Thin Film Electrocatalyst. *J. Am. Chem. Soc.* **2013,** *135* (9); 3662–3674.

4. Bulutektin, M. B. *"Economics of Boron: State of Turkey on World Born Market"*, II. National Finance Congress, İzmir-Turkey, 2008.

5. Denkhaus, E.; Salnikow, K. Nickel Essentiality, Toxicity, and Carcinogenicity. *Crit. Rev. Oncol. Hematol.* **2002,** *42,* 35–56.

6. Dinca, M.; Surendranath ,Y.; Nocera, D. G. Nickel-borate Oxygen-evolving Catalyst that Functions under Benign Conditions. *Proc. Natl. Acad. Sci. USA.* **2010,** *107* (23), 10337–10341.

7. Yoshida, M.; Mitsutomi, Y.; Mineo, T.; Nagasaka,M.; Yuzawa H;Kosugi, N.; Kondo, H. Direct Observation of Active Nickel Oxide Cluster in Nickel Borate Electrocatalyst for Water Oxidation by In Situ O K-Edge X-ray Absorption Spectroscopy. *J. Phys. Chem. C.* **2015,** *119* (33), 19279–19286.

8. Pang, H.; Lu, Q. Y.; Changyun Chen, C.; Xiaoran Liu, X.; Feng Gao, F. Facile Synthesis of $Ni_3(BO_3)_2$ Nanoribbons and Their Antimicrobial, Electrochemical and Electrical Properties. *J. Mater. Chem.* **2011,** *21* (36), 13889–13894.

9. Knyrim, J. S.; Friedrichs, J.; Neumair S.; Roeßner, F.; Floredo, Y.; Jakob S.; Johrendt, D.; Glaum, R.; Huppertz, H. High-pressure Syntheses and Characterization of the Transition Metal Borates b-MB4O7 (M = Mn^{2+}, Ni^{2+}, Cu^{2+}). *Solid State Sci.* **2008,** *10,* 168–176.

10. Liu, X. F.; Zhu W. C.; Cui, X.; Liu T.; Zhang, Q. Facile Thermal Conversion Route Synthesis, Characterization, and Optical Properties of Rod-like Micron Nickel Borate. *Powder Technol.* **2012,** *222,* 160–166.

11. Chen, A. M. ; Hu, F. C. ; Gu, P. ; Ni, Z. M. Sol-gel Synthesis, Characterization of Nickel Borate Nanorods. *Chin. J. Inorganic Chem.* **2011,** *27,* 30–34.

12. Cao, J.; Wang, Y.; Songab, X.; Fenga, J. One-dimensional Nickel Borate Nanowhiskers: Characterization, Properties, and a Novel Application in Materials Bonding. *RSC Adv.* **2014,** *4,* 19221–19225.

13. Menaka, Sharma, S.; Ramanjachary, K. V.; Loftland, S. E.; Ganguli, A. K. Controlling the Size and Morphology of Anisotropic Nanostructures of Nickel Borate Using Microemulsions and Their Magnetic Properties. *J. Colloid Interface Sci.* **2011,** *360,* 393–397.

14. Hong-Yan, L. ; Chun-Hui, F.; Yan, F.; Yong-Quan, Z.; Fa-Yan, Z.; Hai-Wen, G.; Zi-Xiang, Y.; Yu-Ling, T. EXAFS Study of the Structure of Amorphous Nickel Borate Acta Phys. *Chim. Sin.* **2014,** *30* (11), 1979–1986.

15. Menaka, Sharma, S.; Ramanjachary, K. V.; Loftland ,S. E.; Ganguli, A. K. A New Low Temperature Methodology to Obtain Pure Nanocrystalline Nickel Borate. *J. Organometallic Chem.* **2010,** *695,* 1002–1005.

16. Biswick, T.; Jones, W.; Pacu, A.; Ewa Serwicka, E. Synthesis, Characterisation and Anion Exchange Properties of Copper, Magnesium, Zinc and Nickel Hydroxy Nitrates. *J. Solid State Chem.* **2006,** *179* (1), 49–55.

17. Jun, L.; Shupping, X.; Shiyang, G. FT-IR and Raman Spectroscopic Study of Hydrated Borates, *Spectrochim. Acta Mol.Biomol. Spectrosc.* **1995,** *51,* 519–532.

18. Zhihong, L.; Bo, G.; Mancheng, H.; Shuni, L.; Shuping, X. FT-IR and Raman Spectroscopic Analysis of Hydrated Besium Borates and Their Saturated Aqueous Solution. *Specrochim. Acta.* **2003,** *59,* 2741–2745.

19. Knowlton, G. D.; White, T. R. Thermal study of types of water associated with clinoptilolite. *Clays Clay Miner.* **1981,** *29* (5), 404–411.

CHAPTER 12

THE NEW EQUIPMENT FOR CLEARING OF TECHNOLOGICAL GASES

R. R. USMANOVA[1*] and G. E. ZAIKOV[2]

[1]Department of Resistance of Materials, Ufa State Technical University of Aviation, Ufa 450000, Bashkortostan, Russia

[2]N. M. Emanuel Institute of Biochemical Physics, Russian Academy of Sciences, Moscow 119991, Russia

*Corresponding author. E-mail: Usmanovarr@mail.ru; chembio@chph. ras.ru

CONTENTS

ABSTRACT

In this chapter new equipment for prevention of ecological pollution presented. New perspectives which correspond to industrial ecology requirements are discussed.

12.1   INTRODUCTION

The guideline of the European parliament 96/61/EC on prevention of ecological pollution establishes the list of requirements for the industrial enterprises. The basic direction of their development is creation of drainless technological systems on the basis of the existing enterprises and perspective methods of clearing of gas. The new equipment which corresponds to industrial ecology requirements is developed. One of the major advantages of configuration of systems of a gas cleaning is possibility of the closed cycle of an irrigation thanks to system internal circulation of a liquid in the apparatus. Technical and economic indicators of new devices considerably exceed known analogs.

12.2   THE INTERNATIONAL CONTROL AND THE GOVERNMENT QUALITY OF ENVIRONMENT

The international cooperation in the field of wildlife management is carried out normally under the scheme: carrying out of the international meetings—the conclusion of contracts—creation of the international organizations—working out and coordination of programs of environmental safety.

Intensive development of economic activities of people, failures, and accidents on industrial and defensive objects became destructive environmental impact and have led the nature to a condition of the crisis threatening by ecocatastrophe.

Therefore before humankind, there was a harmonious exploitation problem in a combination to effective decrease in deleterious effect of industrial production on environing natural habitat.

Prompt formation of the increasing quantity of a waste is a subject of anxiety European Parliament. Guideline European Parliament 96/61/EC on the integrated pollution prevention and the control over them have been accepted on September 24, 1996.[1]

The guideline establishes the list of ecological requirements for the industrial enterprises, which enterprises should carry out to obtain the permit to the activity.

Member countries of the European parliament undertake to take necessary measures to guarantee that during work of the enterprise of a condition.

Undertake all necessary preventive actions on environmental pollution prevention, in particular, by application of the best existing technologies; do not make considerable environmental pollution; prevent formation of a scrap in conformity; with the guideline on a scrap.

Technologies without waste and landlocked cycles—one of the most radical measures of protection of environment from pollution. Four basic directions of their development (according to the declaration on technology without scrap and reclamation of scrap materials)[2] are more low formulated:

1. Creation of drainless technological systems of different function on the basis of existing and perspective methods of clearing and a reuse of the cleared runoffs.
2. Working out and introduction of systems of processing industrial and a human refuse which are considered thus as secondary material resources.
3. Working out of technological processes of reception of traditional kinds of production by essentially new methods at which the greatest possible carrying over of substance and energy on finished goods is reached.
4. Working out and creation of industrial complexes with landlocked structure of material streams and production residue in them.

The analysis of the data about a condition of an environing natural habitat of the Russian Federation shows that the total quantity of atmospheric emissions from industrial sources in 2014 has made about 32 million tons of harmful substances. The composition of gas emissions is given in Table 12.1.

TABLE 12.1 Results of Posttest Examination.

Compound	Concentration at the Inlet, g/m³	Concentration after Clearing, g/m³	Quantity of Emissions, Mill Tons
Dust	0.02	0.00355	14.8
NO_2	0.10	0.024	7.6
SO_2	0.03	0.0005	9.2
CO	0.01	0.0019	3.5

In Figure 12.1 the scheme of rationing of admixtures in air taking into account their carrying over and dispersion to atmosphere[4-6] is introduced. Performance of condition $C_B \leq$ maximum concentration limit, where C_v is the concentration of harmful substance will be a condition of environmental safety for the person, mg/m^3. In a ground layer of atmosphere of human settlements maximum concentration limit which values are resulted in "the Sanitary code of designing of industrial enterprises SN 3917-05" are established.[3]

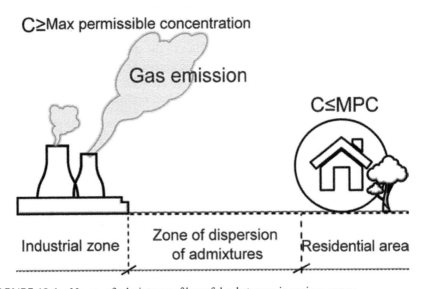

FIGURE 12.1 Norms of admixtures of harmful substances in various zones.

The total of the weighed particles arriving in atmosphere as a result of diverse human activity (according to experts European Parliament), becomes commensurable with quantity of pollution of a natural origin.

The dust in the gases departing from raw and cement dryers, mills, roasting furnaces, in air from transport devices, is a consequence of imperfection of the equipment and technological processes.

12.3 CLEARING AND PROCESSING OF TECHNOLOGICAL GASES

Environment protection against pollution includes special methods and cleaning equipment of gas, waste regaining and secondary use of warmth

and the maximum decrease in pollution. For this purpose develop technological processes and the equipment, meeting the requirements of industrial
ecology.

To clearing of gaseous emissions for the purpose of their detoxification
or extraction from them expensive and scarce ingredients apply the various
clearing equipment and corresponding processing methods.

Now methods of clearing of dusty gases classify inheriting groups;

1. "Dry" mechanical dust extractors.
2. Filters.
3. Electric separators.
4. "Wet" dust removal devices.

Mechanisms of dust collection. The basic operations in dust collection
by any device are separation of the gas-borne particles from the gas stream
by deposition on a collecting surface; retention of the deposit on the surface;
and removal of the deposit from the surface for recovery or disposal. The
separation step requires application of a force that produces a differential
motion of a particle relative to the gas and a gas retention time sufficient for
the particle to migrate to the collecting surface. The principal mechanisms of
aerosol deposition that are applied in dust collectors are gravitational deposition, flow-line interception, inertial deposition, diffusional deposition, and
electrostatic deposition.

Thermal deposition is only a minor factor in practical dust-collection
equipment because the thermo force is small. Two other deposition mechanisms, in addition to the six listed, may be in operation under particular
circumstances. Some dust particles may be collected on filters by sieving
when the pore diameter is less than the particle diameter. Except in small
membrane filters, the sieving mechanism is probably limited to surface-type
filters, in which a layer of collected oust is itself the principal filter medium.

The other mechanism appears in scrubbers. When water vapor diffuses
from a gas stream to a cold surface and condenses, there is a net hydrodynamic flow of the gas directed toward the surface. This flow, termed the
Stefan flow, carries aerosol particles to the condensing surface, and can
substantially improve the performance of a scrubber. However, there is a
corresponding Stefan flow directed away from a surface at which water is
evaporating, and this will tend to repel aerosol particles from the surface.

In addition to the deposition mechanisms themselves, methods for preliminary conditioning of aerosols may be used to increase the effectiveness of
the deposition mechanisms subsequently applied. One such conditioning

method consists of imposing on the gas high-intensity acoustic vibrations to cause collisions and flocculation of the aerosol particles, producing large particles that can be separated by simple inertial devices such as cyclones. This process, termed "sonic agglomeration," has attained only limited commercial acceptance.

Another conditioning method, adaptable to scrubber systems, consists of inducing condensation of water vapor on the aerosol particles as nuclei, increasing the size of the particles and making them more susceptible to collection by inertial deposition.

Most forms of dust-collection equipment use more than one of the collection mechanisms, and in some instances the controlling mechanism may change when the collector is operated over a wide range of conditions. Consequently, collectors are most conveniently classified by type rather than according to the underlying mechanisms that may be operating.[4]

Dust-collector design. In dust-collection equipment, most or all of the collection mechanisms may be operating simultaneously, their relative importance being determined by the particle and gas characteristics, the geometry of the equipment, and the fluid-flow pattern. Although the general case is exceedingly complex, it is usually possible in specific instances to determine which mechanism or mechanisms may be controlling. Nevertheless, the difficult of theoretical treatment of dust-collection phenomena has made necessary simplifying assumptions, with the introduction of corresponding uncertainties. Theoretical studies have been hampered by a lack of adequate experimental techniques for verification of predictions. Although theoretical treatment of collector performance, been greatly expanded in the period since 1980, few of the resulting performance models have received adequate experimental confirmation because of experimental limitations.

The best-established models of collector performance are those for fibrous filters and fixed-bed granular filters, in which the structures and fluid-flow patterns are reasonably well defined. These devices are also adapted to small-scale testing under controlled laboratory conditions. Realistic modeling of full-scale electrostatic precipitators and scrubbers is incomparably more difficult. Confirmation of the models has been further limited by a lack of monodisperse aerosols that can be generated on a scale suitable for testing equipment of substantial sizes. When a polydisperse test dust is used, the particle-size distributions of the dust both entering and leaving a collector must be determined with extreme precision to avoid serious errors in the determination of the collection efficiency for a given particle size.

The design of industrial-scale collectors still rests essentially on empirical methods, although it is increasingly guided by concepts derived from theory. Existing theoretical models frequently embody constants that must be evaluated by experiment and that may actually compensate for deficiencies in the models.

Mechanical centrifugal separators. A number of collectors in which the centrifugal field is supplied by a rotating member are commercially available. In the typical unit shown in Figure 12.2, the fan and dust collector are combined as a single unit. The blades are especially shaped to direct the separated dust into an annular slot leading to the collection hopper while the cleaned gas continues to the scroll.

As a result of scrubber laboratory researches operating modes at which its efficiency many times over exceeds efficiency of analogs apparatuses are installed.

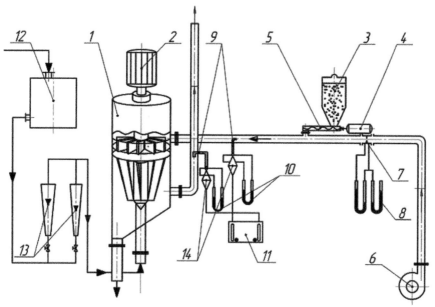

1—scrubber; 2—the actuator; 3—bunker dust; 4—the electric motor; 5—screw feeder; 6—fan; 7—the diaphragm; 8,10—differential pressure gauges; 9—the samplers; 11—the aspirator; 12—pressure tank; 13—flow meters; and 14—the samplers

FIGURE 12.2 Scheme of the experimental equipment.

Although no comparative data are available, the collection efficiency of units of this type is probably comparable with that of the single-unit high-pressure-drop-cyclone installation. The clearances are smaller and the centrifugal fields higher than in a cyclone, but these advantages are probably compensated for by the shorter gas path and the greater degree of turbulence with its inherent re-entrainment tendency. The chief advantage of these units lies in their compactness, which may be a prime consideration for large installations or plants requiring a large number of individual collectors. Caution should be exercised when attempting to apply this type of unit to a dust that shows a marked tendency to build up on solid surfaces, because of the high maintenance costs that may be encountered from plugging and rotor unbalancing.

They have been used to collect chemical incinerator fume and mist as well as sulfuric and phosphoric acid mists. The collection efficiency of a scrubber is highly dependent on the throat velocity or pressure drop, the liquid-to-gas ratio, and the chemical nature of the particulate. Throat velocities may range from 60 to 150 mps (200–500 ft/s). Liquid injection rates are typically 0.59–1.5 m³/1000 m³ of gas. A liquid rate of 1.1 m³/1000 m³ of gas is usually close to optimum, but liquid rates as high as 3.2 m³ have been used. Efficiency improves with increased liquid rate but only at the expense of higher-pressure drop and energy consumption. Pressure-drop predictions for a given efficiency are hazardous without determining the nature of the particulate and the liquid-to-gas ratio. In general, particles coarser than one micron can be collected efficiently with pressure drops of 20–65 cm of water. For appreciable collection of particles, pressure drops from 65 to 120 cm of water are usually required. When particles are appreciably finer than 0.7 microns, pressure drops of 180–280 cm of water have been used.

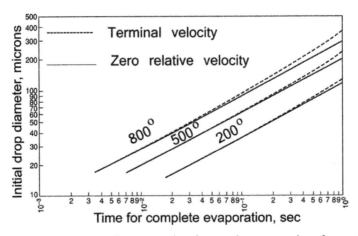

FIGURE 12.3　Effect of drop diameter on time for complete evaporation of water drops.

Figure 12.3 is easy to use to estimate the necessary spray dispersion. For typical conditions of 300 °F gas inlet temperature and a 20 °F approach to adiabatic saturation at the outlet, a 60 microns droplet will require about 0.9 s to evaporate, while a 110 microns droplet will require about 3 s. If the time available for drying is no more than 3 s, it therefore follows that the largest droplets in a humidifier spray should be no larger than 100 microns. If the material being dried contains solids, droplet top size will need to be smaller due to slower evaporation rates. Droplets this small not only require considerable expensive power to generate, but since they have inherent penetration distances, they require expensive dispersion arrangements to get good mixing into large gas flow's without allowing damp, stick particles to reach walls.

12.4   SELF-INDUCED SPRAY SCRUBBERS

Self-induced spray scrubbers form a category of gas-atomized spray scrubbers in which a tube or a duct of some other shape forms the gas–liquid-contacting zone. The gas stream flowing at high velocity through the contactor atomizes the liquid in essentially the same manner in a rotoklon. However, the liquid is fed into the contactor and later recirculated from the entrainment separator section by gravity instead of being circulated by a pump as in rotoklon.

Rotoklon represents the basin with water on which surface on a connecting pipe dusty gas arrives. Over water surface gas is developed, and a dust containing in gas by inertia penetrate into a liquid. Turn of shovels impeller is made manually, rather each other on a threaded connection. The slope of shovels was installed in the range from 25–45° to an axis. In rotoklon three pairs blades sinusoidal a profile, the adjustments of their position executed with possibility are installed. Depending on a dustiness of a gas stream the bottom blades by means of flywheels are installed on an angle matching to an operating mode of the device.

The scheme is well illustrated in Figure 12.4.

Although self-induced spray scrubbers can be built as high-energy units and sometimes are, most such devices are designed for only low-energy service. The principal advantage of self-induced spray scrubbers is the elimination of a pump for recirculation of the scrubbing liquid. However, the designs for high-energy service are somewhat more complex and less flexible than those for venturi scrubbers.

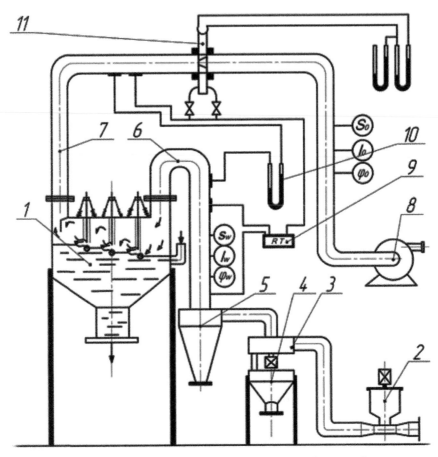

1—rotoklon; 2—batcher; 3—qualifier; 4—collector of a coarse dust;
5—cyclone; 6,7—gas pipeline; 8—fan; 9—potentiometer;
10—differential pressure gauge; and 11—diaphragm

FIGURE 12.4   Scheme of the experimental equipment.

One of the problems in predicting efficiency and required pressure drop of a rotoclon is the chemical nature or wettability of the particulate, which on 0.6 microns size particles can make up to a threefold difference in required pressure drop for its efficient collection. Nonatomizing froth scrubbing is described in tliose patents as occurring within defining boundaries on a new dimensionless velocity versus dimensionless liquid/gas ratio two-phase flow regime map, shown here as Figure 12.5.

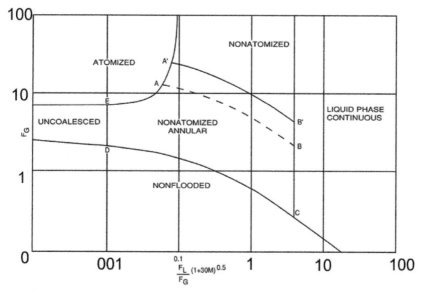

FIGURE 12.5 Spray scrubbing flow regime map.

As a result of researches dependence of water resistance dynamic scrubber on speed turbulent flow has been installed. Apparently from Figure 12.6 with growth of speed turbulent flow to 27 mps separation efficiency raises to 78%. The subsequent increase in speed over the range from 30 to 40 mps is accompanied by reduction of efficiency of separation to 63%.

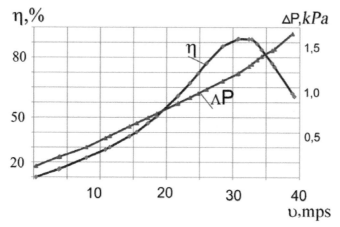

FIGURE 12.6 Dependence of efficiency of a gas cleaning η and pressure losses (ΔP) on entrance speed of a gas stream.

Calvert reports pressure drop through tube banks to be largely unaffected by liquid loading and indicates that s correlations for gas flow normal to tube banks or data for gas flow through heat-exchanger bundles can be used. However, the following equation is suggested:

$$\Delta P = 8.5 \cdot 10^{-3}\, \eta \rho U^2 \tag{12.1}$$

where ΔP is cm of water; η is the number of tubes; ρ is the gas density, g/cm³; and U is the actual gas velocity between tubes, cm/s. Calvert did find an increase in pressure drop of about 80–85% above that predicted by equation in vertical up flow of gas through tube banks due to liquid holdup at gas velocities above 40 mps.

Simultaneously with increase in speed of a gas stream the water resistance of the apparatus from 578 to 1425 Pascal raises also. Optimum result it is possible to see of 25 mps, in this case efficiency of clearing at speed makes 78%, at head losses no more than 1200 Pascal. By fractional efficiency η sizes of dispersion particles which are most effectively trapped by the apparatus have been defined.

General efficiency of a gas cleaning count by formula:

$$\eta = 1 - \exp\left[-\frac{\upsilon n W \theta}{57.3 U b \tan \theta} \right] \tag{12.2}$$

where η is the fractional primary collection efficiency; υ is the drop terminal centrifugal velocity in the normal direction, mps; U is the superficial gas velocity, mps; n is the number of rows of baffles or bends; θ is the angle of inclination of the baffle to the flow path; W is the width of the baffle, m; and b is the spacing between baffles in the same row, m.

For drop terminal centrifugal velocity of:

$$\upsilon = d^2 \rho a / 18 \mu$$

where d and ρ are the drop particle diameter, cm, and particle density, g/cm³, respectively; μ is the gas viscosity, p; and a is the acceleration due to centrifugal force. It is defined by the equation:

$$a = 2U^2 \sin \theta / W \cos^3 \theta$$

Efficiency of trapping of corpuscles in rotoklon depends as on characteristics of trapped corpuscles (a is size and density), and from operating conditions among which the most important is speed of a gas stream at passage of shovels impeller, thus, the formula (12.2) opens physical sense of the processes proceeding in contact channels rotoklon.

One of the major advantages of configuration of systems of a gas cleaning "Rotoklon" is possibility of the closed cycle of an irrigation thanks to system internal circulation of a liquid in the apparatus.

12.5   FEASIBILITY REPORT OF A CHOICE OF SYSTEM OF CLEARING OF GAS

Enormous scales of industrial activity of the person promoted a sharp decline of a state of environment that can cause far-reaching negative aftereffects for humankind. Last years the legislative deeds sharply raising the demands to protection of a free air[3] are published. The industrial factories are aimed to that the equipment for gas clearing was highly effective, reliable and inexpensive, that is would have high technical and economic indexes.

Designing of apparatuses for gas clearing in Russia in total loses world level. Work on elimination of this problem is made by slow rates. Protection of an aerosphere against pollution represents the important problem and its scientific implementation is still far from perfect.

The main task is decrease in volumes of the flying emissions during the basic process. Quite often economic benefit gained in sphere of the basic manufacture, is completely recoated by expenses for clearing of great volumes of gas emissions. The role of treatment facilities in system of provisions on aerosphere protection consists in liquidation of those emissions which cannot be prevented preventive measures.

The following was the primary goals which were put by working out of new apparatuses for gas clearing:

* to develop criteria the feasibility report of efficiency of clearing systems;
* create apparatuses for clearing of gas emissions of the basic industrial systems of a fine dust with a wide range of change of technological parameters.

The technical and economic estimation of installations for gas clearing is based on the comparative data. The installation of a gas cleaning is compared on technical and economic parameters with the best functioning analogs installation. The analog is led to the conditions comparable to conditions of sized up installation (power, separation efficiency, and conditions of production). Comparison is made on capital investments, productivity, and operational expenses.

The method of calculation of relative efficiency the installations has been developed, allowing to find the most rational constructive solutions for system of trapping of gas emissions. Generally, the damage from gas emissions can be defined as $Y = B \cdot M$.

If to express ecological efficiency of the developed design from the point of view of the least damage put to an aerosphere the ecological damage in this case will register in the form of $Y_m \to min$. Value Y_m decreases with magnitude growth:

$$E = \frac{\displaystyle\sum_{i=1}^{N} \eta_i \cdot C_{oi}}{B \cdot \displaystyle\sum_{i=1}^{N} A_i \cdot (1 - \eta_i) \cdot C_i} \qquad (12.3)$$

It is possible to consider magnitude E as criterion the technician of an economic estimation of installations for gas clearing. The criterion of ecological efficiency $\acute{\varepsilon}$ will register as a relationship of values E computed for new installation E_1 and base analog of E_o:

$$\acute{\varepsilon} = E_1 / E_o$$

Thus, for two installations of a gas cleaning of the concrete manufacture, discriminated in the separation extent, $\eta_1 \neq \eta_o$, the ecological system effectiveness is defined the technician in ecological parameter $\acute{\varepsilon} \to max$.

The prevented ecological damage Y_p is computed as a difference: $Y_p = Y_0 - Y_1$, between economic losses of competing systems.

The gained results have illustrative character and show possibility of application of criterion of feasibility report $\acute{\varepsilon}$ for the comparative analysis of apparatuses for gas clearing. Comparison of the developed gas apparatuses with other apparatuses applied now, on technical-and-economic indexes shows the big advantages of the first. New apparatuses are simple, low-cost, and effective gas apparatuses. At the best separation efficiency of gas from a dust their sizes, weight and cost it is less, than at the majority of other devices for gas clearing.

12.6 CONCLUSIONS

1. The solution of an actual problem on perfection of complex system of clearing of gas emissions and working out of measures on

decrease in a dustiness of air medium of the industrial factories for the purpose of betterment of hygienic and sanitary conditions of work and decrease in negative affecting of dust emissions given.

2. Designs on modernization of system of an aspiration of smoke gases of a flare with use of the new scrubber which novelty is confirmed with the patent for the invention are devised. Efficiency of clearing of gas emissions is raised. Power inputs of spent processes of clearing of gas emissions at the expense of modernization of a flowchart of installation of clearing of gas emissions are lowered.

3. Ecological systems and the result of the implementation of the recommendations is to a high degree of purification of exhaust gases and improve the ecological situation in the area of production. The economic effect of the introduction is of up to 3 million rubles/year.

KEYWORDS

- **gas emission**
- **industrial zone**
- **gas cleaning**
- **scrubber**
- **technical and economic efficiency**

REFERENCES

1. Directive 2000/76/EC of the European Parliament and of the Council of 24 December 1996 on the incineration of waste.
2. Decision 1600/2002/EC of the European Parliament and of the Council of 22 July 2002 laying down the Sixth Community Environment Action Programme.
3. GOST 17.2.3.02-05. Wildlife Management. Atmosphere, 2005.
4. Mel Pell; James, B. Dunson; Ted, M. Knowlton. *Gas-Solid Operations and Equipment. Handbook*; McGraw-Hill: NY, 2008.
5. Usmanova, R. R. "Dynamic Gas Washer." R. F. Patent 2339435, November 20, 2008.
6. Usmanova, R. R.; Zhernakov, V. S.; Panov A. K. "Rotoklon a Controlled Sinusoidal Blades." R. F. Patent 2317845, February 27, 2008.

CHAPTER 13

OPERATIONAL AND ENGINEERING ASPECTS OF PACKED BED BIOREACTORS FOR SOLID-STATE FERMENTATION

RAMÓN LARIOS-CRUZ[1], ARELY PRADO-BARRAGÁN[2], HÉCTOR A. RUIZ[1], ROSA M. RODRÍGUEZ-JASSO[1], JULIO C. MONTAÑEZ1, and CRISTÓBAL N. AGUILAR[1*]

[1]*Group of Bioprocesses an Bioproducts, Food Research Department, School of Chemistry, Universidad Autónoma de Coahuila, Saltillo, Coahuila 25280, México*

[2]*Department of Biotechnology, Universidad Autónoma Metropolitana Unidad Iztapalapa, Delegación Iztapalapa, Distrito Federal 09340, México*

Corresponding author. E-mail: cristobal.aguilar@uadec.edu.mx

CONTENTS

ABSTRACT

Solid-state fermentation (SSF) is a low cost and highly productive bioprocess for production of a wide range of bioproducts; however, several factors may limit its use at pilot and industrial level. Moreover, it is of a great importance the bioreactor geometry according to the microorganisms characteristics used for the fermentation. Filamentous fungi are used for several applications and their requirements in oxygen and unmixed beds make it suitable for production of high value metabolites in SSF bioreactors. Packed bed bioreactors are one of the most used bioreactors in SSF at Lab-scale because of it is possible to study variables of the process as temperature, airflow, particle size of the substrate, and humidity; but it is required a constant and very close monitoring to avoid negative inherent effects as temperature increment and humidity loss. For these reasons, there is relevant to evaluate other characteristics of the bioreactors in order to propose improvements to maintain the study variables during the fermentation process. Then, it is important to add heat exchangers with the bioreactor as solids behavior (degradation and compaction) in order to enhance the heat dissipation from the solids, and other techniques to maintain humidity during the fermentation as in drum bioreactors (adding water by spraying over the bed with agitation of the medium). Because of air force results insufficient to remove heat from the matrix, a combination of convection and conduction for cooling system result in a better way to maintain temperature over the fermentation time. In the present work, operational and engineering aspects of bioreactors are showed and their effect over SSF as humidity, temperature, and solid support. Also, it is mentioned the relationship of these aspects over SSF and the variables of the process.

13.1   INTRODUCTION

Solid-state fermentation (SSF) can be defined as a bioprocess for aerobic or micro-aerobic microbial culture using a solid porous matrix and also can be considered as a complex heterogeneous three-phase (gas–liquid–solid) system, with spaces between solid substrate particles where can be found a gas phase and certain water content to the development of biochemical processes. Solid matrix (i.e., agro industrial biomass) can act as support and/ or substrate. When the solid matrix is used as solid support, the material absorbs the nutrients from external medium (carbon and nitrogen sources, mineral salts and for aerobic processes the oxygen dissolved in water) that allow the appropriate growth of the organisms[1,2].

The SSF has gained importance over the past two decades because of high productivity yields and numeral applications including utilization and treatment of agro industrial by-products. Also these by-products are used to produce several metabolites. Agro industrial by-products could be used as feed for ruminant after SSF was applied for increment in protein content, total lipid, fatty acids, and digestibility for mentioned a few.[3] Uses of secondary metabolites from fermentation processes in food, pest control, and antibiotic field, among others, have become reaching a particular interest to develop more efficient procedures to improve quality and quantity of the products. Secondary metabolites produced in SSF are including pigments,[4] L-Lactic acid,[5] lovastatin,[6] lipopeptides,[7] poly(γ-glutamic acid,[8] and citric acid,[9] among others. Enzymes are more studied for many applications including value added to industrial by-products.[10] SSF Enzymes production are including pectinase,[11,12] cellulase,[13] inulinase,[14] lipase,[15] and chitinase,[16] among others (Table 13.1). The lack of knowledge on larger scale SSF bioreactors design limits the potential of industrial production; even if high productivity yields were obtained at Lab-scales. However, it is mentioned a successful case of industrial scale by Suryanarayan[17] with the application of a tray bioreactor in the production of secondary metabolites at Biocon India with a PlaFractor as lovastatin, mycophenolic acid, *Trichoderma viride* spores, and fungal proteases.

The majority of the SSF processes and experiments are developed at Lab-scale based on different types of bioreactors design, most of them on petri dishes, Erlenmeyer flasks.[18,19] Those works on Lab-scale have the characteristic of maintain stable the solids during the fermentation process, in which microorganisms sensitive to agitation forces do not suffer damages that could reduce their growth or even change its physiology or metabolic pathway.[20,21] Conversely, when scale up strategies is followed by researches, the disadvantages in the SSF processes appear as increment of set temperature and changes in the water content over the bed. Pilot-scale bioreactors are developed to produce metabolites as enzymes under SSF.[10] One of the most important aspects to control on the solid fermentation is temperature; it can reach levels so high that the microorganisms decrease their growth or died as fermentation time proceed or affect the performance.[22] The application of mathematical modeling, like energy balance and evaporative heat removal in tray bioreactors, is an important tool to predict the effect expected over the variables as temperature by the proposed experiments at Lab-scale; a few studies performed over packed bed and Zymotis have been focused in develop energy balances to describe the temperature over the bed.[23] Models development is an effective tool for large-scale bioreactors to evaluate the

TABLE 13.1 Enzyme Production in SSF with Different Microorganisms.

Microorganism	Enzyme	Substrate	Fermentation Conditions	Bioreactor	Reference
Fungi					
Aspergillus niger Aa-20	Pectinase	Lemon peel pomace	30°C; 70% humidity; 4 days	Column-tray	[12]
Aspergillus niger IMI 91881, *A. sojae* ATCC 20235, *A. sojae* IMI 191303, *A. sojae* CBS 100928	Pectinase (PMG[a] and PG[b]), Endo-pectinolytic	Wheat bran and orange Peel (70:30)	Room temperature; wet-ted 120% with 0.2 N HCl; 6 days	Erlenmeyer Flask	[11]
Aspergillus niger NS-2	Cellulase	Agro[c]- and kitchen residues	30°C; moisture ratio 1:1.5; 4 days	Erlenmeyer Flask	[13]
Eupenicillium javanicum	Endoglucanase, β-glucosidase, xylanase and PG[b]	Wheat bran, corn meal, rice straw, peanut shells, lotus seed shell	25, 28, 30, 32, 34°C; 40, 80, 120, 160, 200% (w/w) humidity; 2–6 days	Erlenmeyer Flask	[50]
Aspergillus awamori IOC-3914	Exoamylase, endoamy-lase, isoamylase, protease, cellulase, and xylanase	Babassu cake	22°C; 61.9% humidity; 8.3–10.2 l min^{-1} airflow; 7 days	Packed-bed bioreactor	[43]
Aspergillus awamori ATTC14331, *Aspergillus oryzae*	Glucoamylase	Bakery wastes (cake and bread)	30°C; 60–65% humidity; 11 days	Petri dish	[51]
Aspergillus sojae ATTC 20235	Polygalacturonase	Wheat bran	37°C; 62% humidity; 4 days	Tray type bioreactor	[52]
Yeast					
Kluyveromyces marxianus NRRL Y-7571	Inulinase	Sugarcane bagasse supple-mented with cane molasses, corn sleep liquor and soybean bran	27, 30, and 33°C of air temperature; 2.0, 2.4, and 3.0 m^3 h^{-1} airflow; 65% humidity; 1 day	Packed-bed	[14]

TABLE 13.1 (Continued)

Microorganism	Enzyme	Substrate	Fermentation Conditions	Bioreactor	Reference
Candida rugosa NCIM 3462	Lipase	Groundnut oil cake, sesame oil cake, coconut oil cake	30°C; moisture ratio 1:3.5 g mL^{-1}; 3 days	Erlenmeyer Flask	[15]
Bacteria					
Acidothermus cellulolyticus	Xylanase, Cellulase	Switchgrass	55°C; 80% humidity; 5–7 mL min^{-1} airflow; 6 days	250 mL Na-lgene cups	[53]
Oerskovia xanthineolytica NCIM 2839	Chitinase	Colloidal chitin on wheat bran	45°C; 60% humidity; 5 days	Erlenmeyer Flask	[16]

[a]Polymethylgalacturonase.

[b]Polygalacturonase.

[c]Corn cobs, carrot peelings, composite, grass, leaves, orange peelings, pineapple peelings, potato peelings, rice husk, sugarcane baggage, saw dust, wheat bran, wheat straw.

design and the best operation approach during the experiments.[24,25] Then, the main variables like temperature, humidity, and biomass estimation can be monitored in the SSF by multi-sensor outputs via online in order to increase the bioreactor productivity.[22,26]

In general, there is a production of metabolites as the main objective of the fermentation processes; however, other studies are focused in the variables that improve the fermentation process. Airflow, temperature, pH, agitation, particle size, and carbon source are some of the study variables that affects or improve the fermentation process, these variables can be classified as operational factors. On the other hand, the transfer phenomena represent the effect of the variable over heat and mass transfer. For example, a common problem in packed bed reactors is the small particle size of the solids that caused the formation of a compact bed by producing troubles in mass transfer overall the substrate (oxygen and water from the air) and heat transfer into the air, also in the whole system. The aim of this chapter is to describe the main operational and transfer phenomena involved in packed bed SSF bioreactors.

13.2　SSF BIOREACTORS: GENERAL ASPECTS AND PACKED BED REACTORS

SSF bioreactors are used according to the quantity of substrate/biomass and the type of system employed on laboratory-scale, pilot-scale, demo, and industrial-scale. On laboratory-scale the dry solid medium used begin from a few grams reaching a few kilograms, and the pilot and industrial-scale from kilograms reaching several tons of dry medium.[18] Wherever the scale, bioreactors could be on aerated or mixed system to improve quality of the process.[27]

The first step on the studies in SSF over a specific metabolite or growth of the microorganism is on Lab-scale. It is commonly developed by using Erlenmeyer flask or Petri dishes as bioreactors, because of the facilities on handling and control the fermentation.[12,18] Temperature is the main variable to control in this type of bioreactors; this can be achieved in an incubator, but an incubator do not form a part of the bioreactor in order to control temperature; there is not air force or substrate mixing as a characteristic on the operation mode of bioreactors,[18,27] but air force could be adding externally into the incubator. However, some important parameters as incubation temperature, medium pH, moisture content, substrate particle size, and porosity, can be analyzed at this scale (Erlenmeyer flask).[28,29] Moreover, on laboratory-scale, studies on SSF bioreactors are also made in systems as packed bed columns,[20,30] rotating drums,[20,21] or column-tray[12] where the operation conditions could be taking place to improve the productivity and purposes of further scaling-up.

On a pilot-scale, the studies are focused on going to industrial-scale; but it is hard to reproduce the events when volume of bioreactors increases. It is in this case where real troubles shows up because of the cooling system or operation system of the bioreactor do not offer enough capacity of controlling temperature. For example, metabolic heat generates a rise of temperature and it can reached values upper 20°C of the set temperature,[14,31,32] and the properties of the substrate as bulk density and bed porous could impact on the airflow direction and its diffusion over the substrate.[1] Also water of the medium can change the direction of the airflow or even modified the mycelium direction of the fungus.[33] The changes in the growth or even limitation of the growth induced by water in the medium can be attributed to the filling of the porous of the substrate as one of the principal effects of water distribution in the bed.[34,35]

According to the nature of the microorganism on the production of a specific metabolite, for example, filamentous fungi, it is critical to maintain the bed without movement because of possible ruptures of the mycelium. High productivity is the basis of the bioreactor choose[20] and the preference to use aerobic microorganisms on SSF[2] is taking into account to find the best bioreactor design to use.[18,19,31]

Packed bed and tray bioreactors are more common employed due to the static bed over fermentation time. Packed bed bioreactors are characterized by unmixed (or unfrequently mixed) the substrates by mechanical rotation or impellents, with forcefully air enter through the bottom of the column bed.[1,36] The design is regularly conformed by a column connected in the base with a humidifier for the inlet air.[18,37] Packed bed bioreactors at Lab-scale was described by Raimbault et al.[37] as a new culture method for studying the growth of *Aspergillus niger* over cassava meal.

Roussos et al.[38] designed a type of bioreactor called Zymotis with similar strategies of packed bed, and that consist in a column with panels inside the bed to act as heat exchanger. Packed bed bioreactor is one of the most common used in Lab-scale,[18] but one relevant disadvantage of this bioreactor is the high temperatures reached at the top of the column,[36] as long as the high increased. The same problem occurs for the majority of the bioreactors design; in general, Petri dishes and Erlenmeyer flasks are the bioreactors with advantages in controlling the process with external equipment as the use of incubators to control fermentation temperature trough fermentation time. However, there are variables that cannot be controlled using external and robust equipment, as the airflow. Packed bed bioreactors, tray bioreactors, and drum bioreactors, among others, include accessories for monitoring and controlling variables like temperature, airflow, pH, humidity, and CO_2 and O_2 concentration. For these reasons, to develop studies in those bioreactors for the production of a wide variety of metabolites can generate useful data to make fermentation in higher scales.

High temperatures reached in column bioreactors by metabolic heat are difficult to remove just with forced aerated because the unidirectional way of air through the bed,[24] and the geometric modifications and are principally based in the length size limited, because by making the column wider the heat elimination, is not significantly improved. Even if it is accompanied by a jacked, radial heat transfer would not be enough to dissipate metabolic heat it the radium is greater than the optimum. In this way, an improved in the design of the packed bed bioreactor was developed by Roussos et al.[38] using Zymotis bioreactor in a rectangular form with internal plates used as cooling system. In Zymotis bioreactor, the metabolic heat generated by microorganisms is removed efficiently, compared with the packed bed bioreactor (column). Mitchell et al.[39] modeled the Zymotis bioreactor in order to improve the heat transfer and decrease inhibition of the microorganism growth during fermentation processes. They determined that a distance of 5 cm between plates was suitable to gain heat transference inside the bioreactor.

13.3 OPERATIONAL FACTORS IN PACKED BED BIOREACTORS

The most relevant operational factors including in packed bed bioreactors are: airflow, particle size, and humidity. As it was mentioned above, operational factors affect other factors and facilitate the operation of the bioreactor. For example, airflow includes improvement in metabolites production. Rodríguez-Fernández et al.[21] studied the influence of airflow over phytase production with *A. niger* F3 when it was measured in VKgM (L air per kg wet mass per minute). Phytase production was recommended for 1 VKgM, the behavior showed that with less airflow the heat transfer is poor and with greater airflow phytase production was not significant and in the last conditions, it represents an increment in cost production. Phytase production resulted independent of the scale when the fermentation was from 2 to 20 kg of solids.[21] Moreover, VKgM value is a unit independent of mass; so, it is a useful tool that can be conserved in scale up.

The particle size and porosity of the substrate can improve metabolite production and the bioreactor performance. The solid content (substrate) during SSF is solubilized and degraded through the fermentation time; small particles present major degradation during fermentation process. Meanwhile, big particles are affected at the final fermentation time, this is due to enzymes release; enzymes help in components degradation and reduction of particles size. For example, in the case of wheat bran use as substrate, the type of sugars contained in this material will depend of the particle size; smaller particles contain the major quantity of sugars and carbohydrates hard to degrade (starch, hemicellulose, and cellulose). In studies of degradation of wheat bran substrate, the percentage of solids lost is up to 60% for smaller particle size (600–850 μm) and for grater particle up to 30% (1200–2057 μm).[40] Changes in porosity caused more influenced in the water content and absorption index. In case of the humidity was controlled in the solid phase, the intra-particle porosity of the medium was not in change. Besides, the growth of the microorganisms did not affect the porosity of the solids when other variables are controlled. Changes in porosity due by water are achieved by filling the porous with water.[35]

Water plays an important role in SSF due to its influence over microbial growth and the production of several metabolites. Water is important in the growth of the hyphae of fungi and its orientation, also impact in the sporulation and germination of spores.[41] Packed bed bioreactors present difficulties in water homogenization during the fermentation process, due to its lack of agitation system. Water in the bioreactor can have changes in both ways, increment by absorption from saturated air and decrement by evaporation, caused by dry air.[42] Water activity is a parameter affected by the height of the bed. According to the operation of the fermenter, at the top of the bioreactor, water can be greater than in the base. That implies humidification of air is not sufficient to control the humidity of the solids.[43] However, the water saturated airflow is the method used to controlled humidity over the bed in this kind of bioreactor. On the other hand, in bioreactors with agitation system water can be added during the fermentation process. Thus, water controlling could be well along the process by spraying it over the bed.[42,44]

13.4   TRANSFER PHENOMENA IN SOLID-STATE BIOREACTORS

Transfer phenomena comprise heat and mass transfer. The most important compounds in mass transfer are water, oxygen, carbon dioxide, and volatile compounds (as essential oils in citrus by-products). The overall mass transfer coefficient ($K_L a$) is a parameter to express the oxygen diffusion. Thibault et al.[45] considered a model for consume and diffusion of oxygen. However, the model considers that there was no microorganism penetration into the solid support. In fact, microorganism growth is at the biofilm surface over the solid support. The oxygen consumption is carried out in the liquid surface of the system employed. Also, different measure of the biofilm affects the oxygen diffusion.[45] Patel et al.[46] showed two methods for determination of $K_L a$, one dynamic method, and one stationary method. They also mentioned that $K_L a$ is a factor that changes along fermentation time because of the growth state of the microorganism and the air supply to the medium. $K_L a$ increments with the entrance increase of airflow to the bioreactor. This value is also affected with the agitation system of the bioreactor. $K_L a$ is grater at high agitation velocities and at minor biomass concentration. Also, some variations in the $K_L a$ values are associated to the chance of viscosity (at higher values), growth, and morphology of the microorganism. $K_L a$ goes from 2.2 min^{-1} (6.18 g/L of *Trichoderma reesei*) to 5.5 min^{-1} (2.29 g/L of *T. reesei*) at 500 rpm in a stirred tank bioreactor.[46]

Airflow and water are correlated in case of taken them as heat transfer applications. Airflow is used in SSF as heat exchanger for removing metabolic heat. It also removes and supplies water that can act as heat exchanger too. Evaporative cooling has been proved to be effective in the temperature control over SSF. Barstow et al.[42] showed the control of temperature in a rotatory drum bioreactor. In SSF without evaporative cooling, temperature goes from 36 to 42–45°C. Temperature was set to 36.9°C with evaporative cooling. Increment in airflow leads to evaporate more water to reduce medium temperature; loss of water compensates by water spray. Agitation of the medium helps to distribute water over the bed. Dry air feed to the medium increases water evaporation in order to control medium temperature.[42] Increment in airflow combined with dry air feed to the bioreactor control the temperature of the bioreactor with evaporative cooling. Then, water is controlled by spray and distributed with agitation of the medium. In case of not controlling water inlet an increment of temperature and decrease in humidity is observed during the fermentation process.[44] Even if evaporative cooling is not applicable in packed bed bioreactors, airflow acts as controlling water content and temperature over the bed. However, the airflow presents a positive effect in the heat removal. It helps keeping temperature in case of increment of heat by metabolism. In that case, the airflow has to be considered in scale up for heat removal.[21]

Organic matrix employed in SSF is little discussed in these kinds of fermentations. Poor heat conductivity of the solids support makes difficult the heat exchange between any cooling system used and the solid support.[17] Heat removal could be done by conduction or convection, in which conduction is more effective for cooling the support. Evaporative cooling is a conduction heat exchange form and it is represented in Figure 13.1, a common process in SSF. The operational gain of evaporative cooling over flow rate or jacket exchanger is due to the dispersion of water uniformly along the bed. Once the water vapor is produced by high temperature in the bed, its removal is important. The air force takes out the water vapor from the solid support. The main system for heat removal is not evaporative cooling in static bioreactors; even if it is a behavior that could be expected in some parts of the bed. For this type of bioreactors design, packed bed, Zymotis, and tray bioreactors, air is the cooling system employed. Heat transfer in a particle of solid support in a side of the wall of the bioreactor or a heat exchanger is showed in Figure 13.2. Air force feed to the bioreactor provides oxygen for the growth of aerobic microorganisms. It carries out the carbon dioxide and removes heat of the bed. The velocity of air feed is a critical step in the operation of the bioreactor. If the velocity of the air is too fast, the dry of the bed is expected. If the velocity of the air is too low and the growth of the microorganism is enough to increment the temperature of the bed, the air, even if it was saturated, is going to warm up. Consequence unsaturation of air and then the carry out of the water of the bed is expected too; finally, a dry of the bed is also occurring.

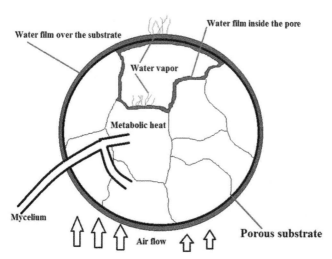

FIGURE 13.1 Evaporative cooling showed inside and outside of a porous substrate used as support in SSF. As metabolic heat is generated by microorganism growth, the evaporation of water removes heat from all particles of the substrate along the bioreactor.

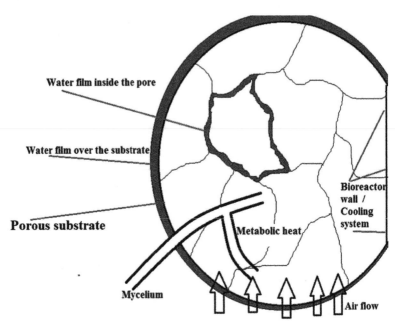

FIGURE 13.2 Convective heat transfer showed in a porous substrate used as support in SSF. Most of the heat is removed by the air feed to the bioreactor. The support close to the wall of the bioreactor or to the cooling plate the heat is also removed by convection by the water cooling at internal plates or at the jacked.

The inclusion of variables into heat transfer could improve the performance of packed bed bioreactors. Equations are useful due to description of the events. Simulation of operational conditions and productivity can be emulated for the purpose of depict the events for optimization or prediction of undesirable online ladings.[1]

Fanaei et al.[25] describe two models for temperature modeling in packed-bed bioreactors for SSF. One lumped model and one distributed model. The lumped model includes temperature determination considering one height of the bed of the bioreactor. The distributed model allows calculating temperature gradients at different heights of the bed. It is also showed a model for biomass estimation. This model includes a physiological factor with dependence of coefficients of synthesis and denaturalization in Arrhenius equation. In metabolic heat generation, convective and evaporative heat removal was considered into the lumped dynamic model. In contrast, axial conduction term is also considered at the distributed model.[25]

Sahir et al.[47] showed the energy balance for a packed bed bioreactor for SSF considering nth tanks. The bioreactor receives air from the bottom of (n−1)th tank and provide it to the (n+1)th tank. In fact, N-tanks design divides the column bioreactor in sections; thus, it makes possible to figure out temperature gradients by dividing the bed in parts. The profile of temperature demonstrates that as the height of bed increased, the maximum temperature in the bed along the fermentation time increased as well. The energy balance model became useful to different bioreactor design; this is for prediction of higher temperatures at different bed locations.[47,48] Considering the whole bioreactor to depict temperature during the fermentation process is limiting at higher scales in which the evaluation of the bed at different sections is recommended. The profile only shows the performance on the extremes of the bioreactor.[49]

13.5 FINAL COMMENTS

Operational and engineering factors are useful for improving the performance of a bioreactor in SSF. Packed bed bioreactors present some advantages in the use of different microorganisms due to its stable bed during the process. However, high temperatures in larger scale are one of the most important drawbacks. That is why airflow, humidity, and solid supports properties are considering very important in the success of any improvement in SSF. In particular for packed bed bioreactors, thus could enhance the temperature control during the fermentation, and could maintain humidity along the column bed. The inclusion of new materials with grater heat transfer coefficients, inclusion of heat exchangers, and several entrance of new humidify air along the column, are some features that can improve the design of packed bed bioreactors in order to apply it into industrial scale. The use of the majority of microorganisms into this bioreactor makes it suitable for several applications, and then a bioreactor design with several industrial applications.

ACKNOWLEDGMENTS

First author wants to acknowledge to the Consejo Nacional de Ciencia y Tecnología (CONACyT, Mexico) for the financial support in the form of scholarship with fellow number 279696.

KEYWORDS

- solid-state fermentation
- solid support
- packed bed bioreactor
- operational conditions
- engineering aspects

REFERENCES

1. Mitchell, D. A.; Krieger, N.; Berovic, M. *Solid-State Fermentation Bioreactors: Fundamentals of Design and Operation.* 1st ed.; Springer-Verlag: Germany, 2006.
2. Raimbault, M. General and Microbiological Aspects of Solid Substrate Fermentation. *Electron. J. Biotechnol.* **1998,** *1*, 26–27.
3. Graminha, E. B. N.; Goncalves, A. Z. L.; Pirota, R. D. P. B.; Balsalobre, M. A. A.; Da Silva, R.; Gomes, E. Enzyme Production by Solid-State Fermentation: Application to Animal Nutrition. *Anim. Feed. Sci. Technol.* **2008,** *144* (1–2), 1–22.
4. Nimnoi, P.; Lumyong, S. Improving Solid-State Fermentation of Monascus Purpureus on Agricultural Products for Pigment Production. *Food. Bioprocess.Technol.* **2011,** *4* (8), 1384–1390.
5. Phrueksawan, P.; Kulpreecha, S.; Sooksai, S.; Thongchul, N. Direct Fermentation of L(+)-Lactic Acid from Cassava Pulp by Solid State Culture of *Rhizopus oryzae. Bioprocess. Biosyst. Eng.* **2012,** *35* (8), 1429–1436.
6. Barrios-González, J.; Baños, J. G.; Covarrubias, A. A.; Garay-Arroyo, A. Lovastatin Biosynthetic Genes of *Aspergillus terreus* are Expressed Differentially in Solid-State and in Liquid Submerged Fermentation. *Appl. Microbiol. Biotechnol.* **2008,** *79* (2), 179–186.
7. Zhu, Z.; Zhang, G.; Luo, Y.; Ran, W.; Shen, Q. Production of Lipopeptides by *Bacillus Amyloliquefaciens* XZ-173 in Solid State Fermentation using Soybean Flour and Rice Straw as the Substrate. *Bioresour. Technol.* **2012,** *112* (0), 254–260.
8. Tang, B.; Xu, H.; Xu, Z.; Xu, C.; Xu, Z.; Lei, P.; Qiu, Y.; Liang, J.; Feng, X. Conversion of Agroindustrial Residues for High Poly(γ-Glutamic acid) Production by *Bacillus subtilis* NX-2 via Solid-State Fermentation. *Bioresour. Technol.* **2015,** *181,* 351–354.
9. Dhillon, G. S.; Brar, S. K.; Kaur, S.; Verma, M. Screening of Agro-Industrial Wastes for Citric Acid Bioproduction by *Aspergillus niger* NRRL 2001 through Solid State Fermentation. *J. Sci. Food. Agric.* **2013,** *93* (7), 1560–1567.
10. Chen, H.-z.; He, Q. Value-Added Bioconversion of Biomass by Solid-State Fermentation. *J. Chem. Technol. Biotechnol.* **2012,** *87* (12), 1619–1625.
11. Heerd, D.; Yegin, S.; Tari, C.; Fernandez-Lahore, M. Pectinase Enzyme-Complex Production by *Aspergillus* Spp. in Solid-State Fermentation: A Comparative study. *Food. Bioprod. Process.* **2012,** *90 (*2), 102–110.

12. Ruiz, H. A.; Rodríguez-Jasso, R. M.; Rodríguez, R.; Contreras-Esquivel, J. C.; Aguilar, C. N. Pectinase Production from Lemon Peel Pomace as Support and Carbon Source in Solid-state Fermentation Column-Tray Bioreactor. *Biochem. Eng. J.* **2012,** *65* (0), 90–95.

13. Bansal, N.; Tewari, R.; Soni, R.; Soni, S. K. Production of Cellulases from *Aspergillus niger* NS-2 in Solid- State Fermentation on Agricultural and Kitchen Waste Residues. *Waste Manage.* **2012,** *32* (7), 1341–1346.

14. Mazutti, M. A.; Zabot, G.; Boni, G.; Skovronski, A.; Oliveira, D. d.; Luccio, M. D.; Rodrigues, M. I.; Treichel, H.; Maugeri, F. Kinetics of Inulinase Production by Ssolid-State Fermentation in a Packed-Bed Bioreactor. *Food Chem.* **2010,** *120* (1), 163–173.

15. Rajendran, A.; Thangavelu, V. Utilizing Agricultural Wastes as Substrates for Lipase Production by *Candida Rugosa* NCIM 3462 in Solid-State Fermentation: Response Surface Optimization of Fermentation Parameters. *Waste Biomass Valorization.* **2012,** *4,* 347–357.

16. Waghmare, S.; Kulkarni, S.; Ghosh, J. Chitinase Production in Solid-state Fermentation from *Oerskovia xanthineolytica* NCIM 2839 and its Application in Fungal Protoplasts Formation. *Curr. Microbiol.* **2011,** *63* (3), 295–299.

17. Suryanarayan, S. Current Industrial Practice in Solid-State Fermentations for Secondary Metabolite Production: The Biocon India Experience. *Biochem. Eng. J.* **2003,** *13* (2–3), 189–195.

18. Durand, A. Bioreactor Designs for Solid-State Fermentation. *Biochem. Eng. J.* **2003,** *13* (2–3), 113–125.

19. Ruiz-Leza, H. A.; Rodríguez-Jasso, R. M.; Rodríguez-Herrera, R.; Contreras-Esquivel, J. C.; Aguilar, C. N. Diseño de Biorreactores Para Fermentación en Medio Sólido. *Rev. Mex. Ing. Quim.* **2007,** *6* (1), 33–40.

20. van Breukelen, F. R.; Haemers, S.; Wijffels, R. H.; Rinzema, A. Bioreactor and Substrate Selection for Solid-State Cultivation of the Malaria Mosquito Control Agent *Metarhizium Anisopliae. Process. Biochem.* **2011,** *46* (3), 751–757.

21. Rodríguez-Fernández, D. E.; Rodríguez-León, J. A.; de Carvalho, J. C.; Karp, S. G.; Sturm, W.; Parada, J. L.; Soccol, C. R. Influence of Airflow Intensity on Phytase Production by Solid-State Fermentation. *Bioresour. Technol.* **2012,** *118* (0), 603–606.

22. von Meien, O. F.; Luz Jr, L. F. L.; Mitchell, D. A.; Pérez-Correa, J. R.; Agosin, E.; Fernández-Fernández, M.; Arcas, J. A. Control Strategies for Intermittently Mixed, Forcefully Aerated Solid-State Fermentation Bioreactors Based on the Analysis of a Distributed Parameter Model. *Chem. Eng. Sci.* **2004,** *59* (21), 4493–4504.

23. Mitchell, D. A.; von Meien, O. F.; Krieger, N. Recent Developments in Modeling of Solid-State Fermentation: Heat and Mass Transfer in Bioreactors. *Biochem. Eng. J.* **2003,** *13* (2–3), 137–147.

24. Ashley, V. M.; Mitchell, D. A.; Howes, T. Evaluating Strategies for Overcoming Overheating Problems during Solid-State Fermentation in Packed Bed Bioreactors. *Biochem. Eng. J.* **1999,** *3* (2), 141–150.

25. Fanaei, M. A.; Vaziri, B. M. Modeling of Temperature Gradients in Packed-Bed Solid-State Bioreactors. *Chem. Eng. Process. Process. Intensif.* **2009,** *48* (1), 446–451.

26. Fernández-Fernández, M.; Pérez-Correa, J. R. Realistic Model of a Solid Substrate Fermentation Packed-Bed Pilot Bioreactor. *Process. Biochem.* **2007,** *42* (2), 224–234.

27. Singhania, R. R.; Patel, A. K.; Soccol, C. R.; Pandey, A. Recent Advances in Solid-State Fermentation. *Biochem. Eng. J.* **2009,** *44* (1), 13–18.

28. Jahromi, M. F.; Liang, J. B.; Ho, Y. W.; Mohamad, R.; Goh, Y. M.; Shokryazdan, P. Lovastatin Production by *Aspergillus terreus* Using Agro-Biomass as Substrate in Solid-State Fermentation. *BioMed. Res. Int.* **2012**, 196264.

29. Velmurugan, P.; Hur, H.; Balachandar, V.; Kamala-Kannan, S.; Lee, K.-J.; Lee, S.-M.; Chae, J.-C.; Shea, P. J.; Oh, B.-T. *Monascus* Pigment Production by Solid-State Fermentation with Corn Cob Substrate. *J. Biosci. Bioeng.* **2011**, *112* (6), 590–94.

30. Chávez-González, M. L.; Rodríguez-Durán, L. V.; Cruz-Hernández, M. A.; Hernández, R.; Prado-Barragán, L. A.; Aguilar, C. N. *Effect of Packing Density and Aeration Rate on Tannase Production by Aspergillus niger GH1 in Solid State Fermentation,* In *Chemistry and Biotechnology of Poliphenols,* Sabu, A., Roussos, S., Aguilar, C. N., Eds.; CiBET Publishers: India, 2011; pp 101–111.

31. Raghavarao, K. S. M. S.; Ranganathan, T. V.; Karanth, N. G. Some Engineering Aspects of Solid-State Fermentation. *Biochem. Eng. J.* **2003**, *13* (2â€"3), 127–135.

32. Chen, H.-Z.; Xu, J.; Li, Z.-H. Temperature Control at Different Bed Depths in a Novel Solid-State Fermentation System with Two Dynamic Changes of Air. *Biochem. Eng. J.* **2005**, *23* (2), 117–122.

33. Chen, H. *Modern Solid State Fermentation - Theory and Practice*; Springer: Londres, 2013.

34. Hamidi-Esfahani, Z.; Shojaosadati, S. A.; Rinzema, A. Modelling of Simultaneous Effect of Moisture and Temperature on *A. Niger* Growth in Solid-State Fermentation. *Biochem. Eng. J.* **2004**, *21* (3), 265–272.

35. Karimi, A.; Shojaosadati, S. A.; Hejazi, P.; Vasheghani-Farahani, E.; Hashemi, M. Porosity Changes during Packed Bed Solid-Sstate Fermentation. *J. Ind. Eng. Chem.* **2014**, *20* (6), 4022–4027.

36. Mitchell, D. A.; Krieger, N.; Stuart, D. M.; Pandey, A. New Developments in Olid-Dtate Fermentation II. Rational Approaches to the Design, Operation, and Scale-up of Bioreactors. *Process. Biochem.* **2000**, *35* (10), 1211–1225.

37. Raimbault, M.; Alazard, D. Culture Method to Study Fungal Growth in Solid Fermentation. *Eur. J. Appl. Microbiol. Biotechnol.* **1980**, *9* (3), 199–209.

38. Roussos, S.; Raimbault, M.; Prebois, J. P.; Lonsane, B. K. Zymotis, a Large Scale Solid State Fermenter Design and Evaluation. *Appl. Biochem. Biotechnol.* **1993**, *42* (1), 37–52.

39. Mitchell, D. A.; von Meien, O. F. Mathematical Modeling as a Tool to Investigate the Design and Operation of the Zymotis Packed-Bed Bioreactor for Solid-State Fermentation. *Biotechnol. Bioeng.* **2000**, *68* (2), 127–135.

40. Nandakumar, M. P.; Thakur, M. S.; Raghavarao, K. S. M. S.; Ghildyal, N. P. Substrate Particle Size Reduction by *Bacillus cagulans* in Solid-State Fermentation. *Enzyme. Microb. Technol.* **1996**, *18* (2), 121–125.

41. Gervais, P.; Molin, P. The Role of Water in Solid-State Fermentation. *Biochem. Eng. J.* **2003**, *13* (2–3), 85–101.

42. Barstow, L. M.; Dale, B. E.; Tengerdy, R. P. Evaporative Temperature and Moisture Control in Solid Substrate Fermentation. *Biotechnol. Tech.* **1988**, *2* (4), 237–242.

43. Castro, A. M.; Castilho, L. R.; Freire, D. M. G. Performance of a Fixed-Bed Solid-State Fermentation Bioreactor with Forced Aeration for the Production of Hydrolases by *Aspergillus awamori. Biochem. Eng. J.* **2015**, *93,* 303–308.

44. Ryoo, D.; Murphy, V. G.; Karim, M. N.; Tengerdy, R. P. Evaporative Temperature and Moisture Control in a Rocking Reactor for Solid Substrate Fermentation. *Biotech. Tech.* **1991**, *5* (1), 19–24.

45. Thibault, J.; Pouliot, K.; Agosin, E.; Pérez-Correa, R. Reassessment of the Estimation of Dissolved Oxygen Concentration Profile and $K_L a$ in Solid-State Fermentation. *Process. Biochem.* **2000,** *36* (1–2), 9–18.

46. Patel, N.; Thibault, J. Enhanced in Situ Dynamic Method for Measuring $K_L a$ in Fermentation Media. *Biochem. Eng. J.* **2009,** *47* (1–3), 48–54.

47. Sahir, A. H.; Kumar, S.; Kumar, S. Modelling of a Packed Bed Solid-State Fermentation Bioreactor Using the *N*-Tanks in Series Approach. *Biochem. Eng. J.* **2007,** *35* (1), 20–28.

48. Mitchell, D. A.; Cunha, L. E. N.; Machado, A. V. L.; de Lima Luz Jr, L. F.; Krieger, N. A Model-Based Investigation of the Potential Advantages of Multi-Layer Packed Beds in Solid-State Fermentation. *Biochem. Eng. J.* ***2010,*** *48* (2), 195–203.

49. Muller dos Santos, M.; Souza da Rosa, A.; Dal'Boit, S.; Mitchell, D. A.; Krieger, N. Thermal Denaturation: Is Solid-State Fermentation Really a Good Technology for the Production of Enzymes? *Bioresour. Technol.* **2004,** *93* (3), 261–268.

50. Tao, N-guo; Shi, W-qing; Liu, Y-jin; Huang, S-rong. Production of Feed Enzymes from Citrus Processing Waste by Solid-State Fermentation with *Eupenicillium javanicum.* *Int. J. Food Sci. Technol.* 2011, *46* (5), 1073–1079.

51. Han, W.; Lam, W. C.; Melikoglu, M.; Wong, M. T.; Leung, H. T.; Ng, C. L.; Yan, P.; Yeung, S. Y.; Lin, C. S. K. Kinetic Analysis of a Crude Enzyme Extract Produceddvia Solid State Fermentation of Bakery Waste. *ACS Sustainable. Chem. Eng.* 2015, *3* (9), 2043–2048.

52. Demir, H.; Tari, C. Bioconversion of Wheat Bran for Polygalacturonase Production by *Aspergillus sojae* in Tray Type Solid-State Fermentation. *Int. Bioteterior. Biodegrad.* **2016,** *106,* 60–66.

53. Rezaei, F.; Joh, L. D.; Kashima, H.; Reddy, A. P.; VanderGheynst, J. S. Selection of Conditions for Cellulase and Xylanase Extraction from Switchgrass Colonized by *Acidothermus cellulolyticus.* *Appl. Biochem. Biotechnol.* **2011,** *164* (6), 793–803.

CHAPTER 14

EFFECTS OF GLUTATHIONE, PHOSPHONATE, OR SULFONATED CHITOSANS AND THEIR COMBINATION ON SCAVENGING FREE RADICALS

TAMER M. TAMER[1,2], KATARÍNA VALACHOVÁ[1*],
AHMED M. OMER[2], MAYSA M. SABET[1,2], and LADISLAV ŠOLTÉS[1]

[1]*Laboratory of Bioorganic Chemistry of Drugs, Institute of Experimental Pharmacology and Toxicology, Bratislava 84104, Slovakia*

[2]*Polymer Research Department, Advanced Technologies and New Materials Research Institute (ATNMRI), City of Scientific Research and Technological Applications (SRTA-City), New Borg El-Arab, Alexandria 21934, Egypt*

*Corresponding author. E-mail: katarina.valachova@savba.sk

CONTENTS

ABSTRACT

To inhibit reactive oxygen species-induced hyaluronan degradation, protective effects of glutathione, chitosan derivatives (phosphonate chitosan and sulfonated chitosan) individually and in a combination with glutathione were analyzed. A source of reactive oxygen species was cupric ions and ascorbate—the so-called Weissberger's biogenic oxidative system (WBOS). Both phosphonate and sulfonated chitosan enhanced the degradation of hyaluronan by WBOS. On the other hand, glutathione itself or in the presence of the chitosan derivative reduced the rate of hyaluronan degradation.

The radical scavenging capacity of the analyzed substances was assessed also by the standard colorimetric ABTS assay; thereby their electron donor properties are determined. Both chitosan derivatives applied individually demonstrated a minimal scavenging capacity against ABTS$^{\cdot+}$. However, those capacities were significantly enhanced on combining each chitosan derivative with glutathione.

14.1 INTRODUCTION

Reactive oxygen species is an expression used to describe a number of reactive particles and free radicals originated from oxygen. The uncontrolled generation of oxygen-based radicals is adverse to all aerobic species. Reactive oxygen species, produced as byproducts through, for example, the mitochondrial reactions can create a number of harmful events. It was formerly thought that only phagocytic cells were responsible for the production of reactive oxygen species and their participation in number of host cell defense mechanisms. In the hosts, reactive oxygen species code defense through phagocytes that induce a burst of reactive oxygen species against the pathogens present in wounds, leading to their destruction. During the burst period, moreover, an excess of reactive oxygen species leakage into the neighboring environment has additional bacteriostatic effects. Reactive oxygen species perform a central role in the orchestration of the typical wound healing sensor. They are secondary messengers to many immunocytes and non-lymphoid cells, which are required in the repair process, and are important in coordinating the recruitment of lymphoid cells to the wound site and efficient tissue repair.

Reactive oxygen species also control the ability to regulate the generation of blood vessels (angiogenesis) at the injury site and the optimal perfusion of blood into the wound-healing area. According to benefits of reactive oxygen species in wound healing and the extended quest for therapeutic procedures to treat injuries in acute and chronic wounds, the manipulation of reactive oxygen species represents a promising avenue for improving wound-healing responses when they are delayed.

Antioxidants are a first defense line toward uncontrolled reactive oxygen species production and free radical damages of tissues. Glutathione is an endogenous biothiol: a tripeptide comprising L-glutamate, L-cysteine, and glycine and has various significant functions inside cells, such as signal transduction, catalysis, gene metabolism, apoptosis, and expression.[1] Glutathione is the principal intracellular thiol playing a notable role in the maintenance of the intracellular redox states, since this oligopeptide significantly protects tissues by quenching reactive oxygen species.[2-8]

Chitosan as a natural cationic polysaccharide has molecular weight in a range from few hundred to several million Daltons. It is synthesized by the deacetylation of chitin, which is obtained from the shells of crabs and shrimps. From a chemical point of view, chitosan is a copolymer of ß-(1→4)-2-acetamido-2-deoxy-D-glucopyranose and deacetylated unit ß-(1→4)-2-amino-2-deoxy-D-glucopyranose with the degree of deacetylation greater than 60%. Chitosan is characterized by its amine groups that are converted to a cationic form in acidic medium. The occupation of different types of functional groups distributed in repeating unit: $-NH_2$ at the C2 position in addition to $-OH$ at the C3 and C6 positions, respectively, simplify its chemical modifications.[9-12]

Chitosan possesses several useful properties such as biodegradability, biocompatibility, hydrophilicity as well as it has antioxidative and antimicrobial properties.[13-17] Cationic structure of chitosan simplified its use in several applications such as drug delivery, water treatment, biomedical engineering, fuel cell, and food packaging.[18-27] Chitosan as a polysaccharide is capable of forming membranes for healing skin wounds. Better physicochemical properties can be obtained by adding hyaluronan into chitosan to form a membrane; however, an adverse effect is the limited solubility of chitosan in neutral aqueous media. Therefore, when an acid has been used to enforce chitosan solubility, it is necessary to remove the adsorbed acid from the formed membranes. A proper solution could be to replace chitosan by its water soluble derivative such as phosphonate chitosan or sulfonated chitosan.

Hyaluronan is a linear high molecular weight, natural polysaccharide composed of alternating D-glucuronic and N-acetyl-D-glucosamine residues linked with 1,4-β and 1,3-β bonds. It is a common component of synovial fluid and extracellular matrix. Hyaluronan ranks among glycosaminoglycans and it is structurally the most simple among them. Hyaluronan is the only one not covalently associated with a core protein, not synthesized in Golgi apparatus, and the only non-sulfated glycosaminoglycan. Hyaluronan and its derivatives are applied in ophthalmology, orthopedic surgery, otolaryngology, dermatology, plastic surgery, wound healing, and drug delivery.[28-31]

In this study, radical scavenging activities of phosphonate chitosan, sulfonated chitosan, glutathione and their combination were assessed by the ABTS assay and rotational viscometry.

14.2   MATERIALS AND METHODS

14.2.1   MATERIALS

Chitosan was purchased from Acros Organics™, USA (mean molecular weight: 100–300 kDa). 2-Chloroethylphosphonic acid was purchased from Sigma-Aldrich, Germany. Acetone was supplied by El-Gomhouria Co., Egypt. High-molecular-weight hyaluronan was purchased from Lifecore Biomedical, Chaska, USA (M_w = 1.93 MDa). The analytical purity grade NaCl and $CuCl_2 \cdot 2H_2O$ were from Slavus, Bratislava, Slovakia. $K_2S_2O_8$ (p.a. purity, max 0.001% of nitrogen) and L-ascorbic acid were from Merck, Darmstadt, Germany. 2,2-Azinobis-[3-ethylbenzothiazoline-6-sulfonic acid] diammonium salt (ABTS; purum, > 99%) was from Fluka, Seelze, Germany.

14.2.2   METHODS

14.2.2.1   PREPARATION OF PHOSPHONATE CHITOSAN

Chitosan (0.1 M) was dispersed in 500 mL of 0.1 M sodium hydroxide solution. The solution was kept under stirring overnight. The temperature of the solution was raised to 70°C simultaneously with the addition of 0.1 M of 2-chloroethylphosphonic acid; the reaction performed for 3 h. The reaction was stopped by cooling, and the product was obtained *via* precipitation in acetone. The yield of phosphonate chitosan was 70%.

14.2.2.2   PREPARATION OF SULPHONATED CHITOSAN

Chitosan (10 mM) was dispersed in methanol (50 mL), different amounts of 1,3-propane sultone (10 and 20 mM) were added. The mixture was stirred and heated at 65°C for 4 h under reflux conditions. The reaction was stopped by cooling. The product was filtered and washed several times with methanol to remove unreacted 1,3-propane sultone. The resultant yellow powder of sulfonated chitosan was dried overnight. The yields for both chitosan derivatives were 85%.

14.2.3 ABTS ASSAY

The ABTS$^{\cdot+}$ cation radical was formed by dissolving $K_2S_2O_8$ (3.3 mg) in H_2O (5 mL), followed by addition of ABTS (17.2 mg) and stored overnight in the dark below 0°C. ABTS$^{\cdot+}$ solution (1 mL) was diluted with water to a final volume 60 mL. All stock solutions of phosphonate chitosan or sulfonated chitosans (20 and 40 mg/mL) and glutathione (40, 200, and 400 μM) were prepared in distilled water. A modified ABTS assay[32] was used to assess the radical-scavenging activity of phosphonate chitosan, sulfonated chitosans, glutathione, as well as of their combinations by using a UV-1800 spectrophotometer (SHIMADZU, Japan). The UV/VIS spectra were recorded in time intervals 2–25 min in 1 cm quartz UV cuvette after admixing of the substance solution (50 μL) to the ABTS$^{\cdot+}$ cation radical solution (2 mL).[33]

14.2.4 PREPARATION OF STOCK AND WORKING SOLUTIONS

The samples of hyaluronan (14 mg) were dissolved in 0.15 M of aqueous NaCl solution for 24 h in the dark. Hyaluronan solutions were prepared in two steps: first, 4.0 mL and after 6 h, 3.90, 3.85, or 3.80 mL of 0.15 M NaCl were added when working in the absence or presence of the samples. Solutions of ascorbate and glutathione (each one 16 mM), cupric chloride (160 μM) were prepared in 0.15 M NaCl.

14.2.5 UNINHIBITED HYALURONAN DEGRADATION

First, hyaluronan degradation was induced by the Weissberger biogenic oxidative system (WBOS), which is composed of $CuCl_2$ (1.0 μM) and ascorbic acid (100 μM). The procedure was as follows: a volume of 50 μL of 160 μM $CuCl_2$ solution was added to the hyaluronan solution (7.90 mL), and the mixture was left to stand for 7 min 30 s at room temperature after a 30 s stirring. Then, 50 μL of ascorbic acid solution (16 mM) was added to the solution and stirred for 30 s. The solution was then immediately transferred into the viscometer Teflon® cup reservoir.

14.2.6 INHIBITED HYALURONAN DEGRADATION

The procedures to investigate the effectiveness of glutathione itself, and phosphonate chitosan or sulfonated chitosan individually or in the presence of glutathione were as follows:

i.   A volume of 50 µL of 160 µM $CuCl_2$ solution was added to the hyaluronan solution (7.85 or 7.80 mL), and the mixture, after a 30 s stirring, was left to stand for 7 min 30 s at room temperature. Then, 50 µL glutathione (0.049–4.9 mg/mL), phosphonate chitosan or sulfonated chitosan (20 or 40 mg/mL) was added individually or in a combination to reach 0.125 or 0.25 mg/mL concentrations for phosphonate or sulfonated chitosan and a concentration range 1–100 µM for glutathione in the hyaluronan reaction mixture, followed by stirring again for 30 s. Finally, 50 µL of ascorbic acid solution (16 mM) was added to the reaction vessel, and the mixture was stirred for 30 s. The solution was then immediately transferred into the viscometer Teflon® cup reservoir.

ii. In the second experimental regime a procedure similar to that described in i) was applied, however, after standing for 7 min 30 s at room temperature 50 µL of ascorbic acid solution (16 mM) was added to the mixture and a 30 s stirring followed. One hour later, finally, 50 µL of glutathione (0.049–4.9 mg/mL), phosphonate or sulfonated chitosan (20 or 40 mg/mL) individually or in the combination with glutathione were added to the reaction vessel, followed by stirring for 30 s after addition of each component. The reaction mixture was then immediately transferred into the viscometer Teflon® cup reservoir.

14.2.7 VISCOSITY MEASUREMENTS

Dynamic viscosity of the reaction mixture (8 mL) containing hyaluronan (1.75 mg/mL), ascorbate (100 µM) plus Cu(II) ions (1 µM) in the absence and presence of samples started to be monitored by a Brookfield LVDV-II+PRO digital rotational viscometer 2 min after the addition of all reactants

(Brookfield Engineering Labs., Inc., Middleboro, MA, USA) at $25.0 \pm 0.1°C$ and at a shear rate of 237.6 s^{-1} for 5 h in the Teflon® cup reservoir.[34,35]

14.3 RESULTS AND DISCUSSION

Schemes 14.1 and 14.2 present the synthesis of phosphonate chitosan from chitosan and 2-chloroethylphosphonic acid or the synthesis of sulfonated chitosan from chitosan and 1,3-propane sultone.

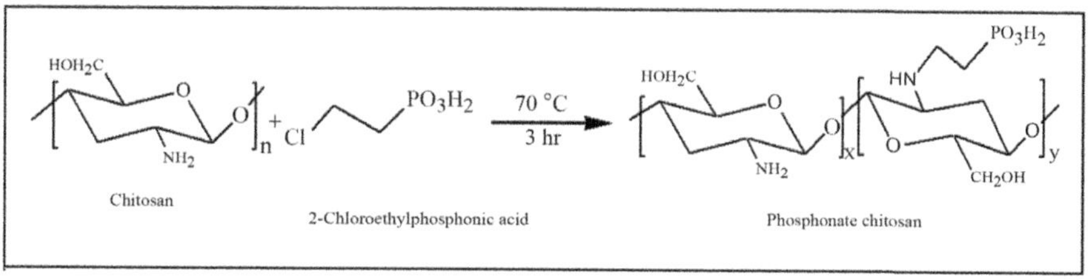

SCHEME 14.1 Preparation of phosphonate chitosan.

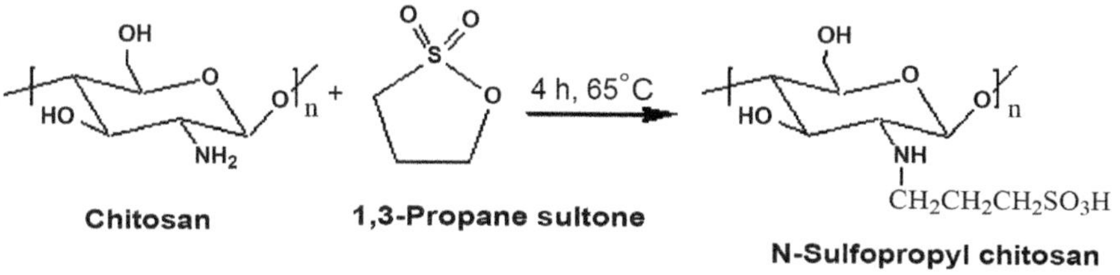

SCHEME 14.2 Preparation of sulphonated chitosan.

14.3.1 ROTATIONAL VISCOMETRY

Figure 14.1, left panel, curve 0 displays the result, where hyaluronan was exposed to degradation by WBOS. The decrease in the dynamic viscosity (η) of the hyaluronan solution within 5 h was 6.25 mPa·s. As seen glutathione dose-dependently inhibited the degradation of hyaluronan induced by ·OH radicals; at 100 µM concentration a complete inhibition of hyaluronan degradation can be stated (left panel, curve 3). Glutathione at 10-times lower concentration was shown to be protective in part (curve 2). Even at 1 µM concentration glutathione was slightly effective (left panel, curve 1). The dose-dependent glutathione protective effect can be claimed even at the situation when the tripeptide was applied into the reaction vessel 1 h after the hyaluronan degradation began (cf. Fig. 14.1, right panel). However, under such an experimental setting, at the lowest glutathione loading a slight pronouncing of hyaluronan degradation was observed (Fig. 14.1, right panel, curve 1).

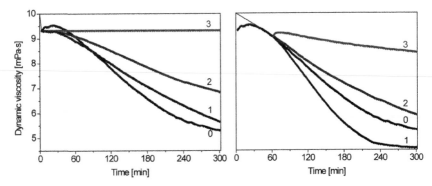

FIGURE 14.1 Time-dependent changes in dynamic viscosity of the hyaluronan solution induced by WBOS in the presence of glutathione at concentrations: 0 μM (curves 0), 1 μM (curves 1), 10 μM (curves 2) and 100 μM (curves 3). Glutathione was added to the reaction system before hyaluronan degradation begins (left panel) and 1 h later (right panel).

Phosphonate chitosan added into the hyaluronan solution at concentrations 20 or 40 mg/mL led to a pronounced degradation of hyaluronan, whereas the decreases in η were 7.1 or 7.25 mPa·s, respectively (Fig. 14.2, left panel, curves 1 or 2). When phosphonate chitosan was added to the hyaluronan reaction mixture 1 h after the reaction onset the decreases in η were 6.25 or 6.57 mPa·s, respectively (Fig. 14.2, right panel, curves 1 or 2).

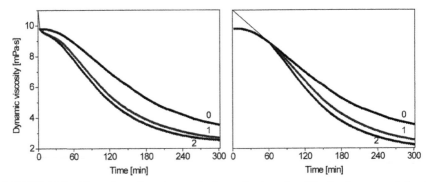

FIGURE 14.2 Time-dependent changes in dynamic viscosity of the hyaluronan solution in the presence of phosphonate chitosan added at concentrations 20 mg/mL (curves 1) or 40 mg/mL (curves 2). Hyaluronan degradation was induced by WBOS (curves 0). Phosphonate chitosan was added to the reaction system before hyaluronan degradation begins (left panel) and 1 h later (right panel).

Several mechanisms were used to explain the chitosan actions against hydroxyl radicals.[36–38] The presence of hydroxyl and amine groups along the polymer backbone may act as a cornerstone of this mechanism. Thus we expected that these groups by their donation of hydrogen atom(s) will decrease the harmful effect of hydroxyl radicals generated by WBOS.[27] Our opposite observations can be simply explained by the fact that the substitution of hydroxyl and amine groups, that is, their convertion to phosphonic groups most plausibly limit the protective—free radical quenching by phosphonate chitosan and, oppositely, the presence of acidic (phosphonic) groups stimulates the degradation of hyaluronan by WBOS (see Fig. 14.2).

As illustrated in Figure 14.3, left panel, hyaluronan, exposed to WBOS, was degraded, whereas the decrease in the η of the hyaluronan solution within 5 h was 4.02 mPa·s (curve 0). Sulfonated chitosan-1 (0.125 mg/mL) pronounced hyaluronan degradation, whereas the decrease in η of the hyaluronan solution was 6.53 mPa·s (curve 1). Addition of sulfonated chitosan-1 at a higher concentration (0.25 mg/mL) resulted in a less rapid degradation of hyaluronan and the η value of the solution was 5.41 mPa·s (curve 2). The addition of sulfonated chitosan-2 at the lower concentration (right panel, curve 1) led to a significant pronouncing of hyaluronan degradation. Increasing the concentration of sulfonated chitosan-2 (0.25 mg/mL) decreased the rate of hyaluronan degradation (curve 2) as seen in Figure 14.3, right panel.

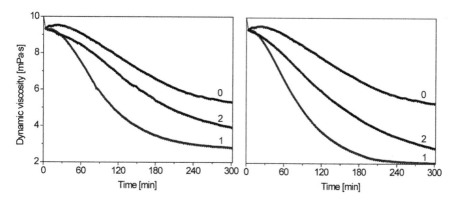

FIGURE 14.3   Time-dependent changes in dynamic viscosity of the hyaluronan solution in the presence of sulfonated chitosan-1 (left panel) or sulfonated chitosan-2 (right panel) added at concentrations 20 mg/mL (curves 1) or 40 mg/mL (curves 2). Hyaluronan degradation was induced by WBOS (curves 0). Chitosan derivatives were added to the reaction system before initiating hyaluronan degradation.

Further, the sulfonated chitosans were added to the reaction mixture 1 h later, that is, in the propagation period of the hyaluronan degradation preceding now mostly by alkoxy-/peroxy-type hyaluronan C-macroradicals (Fig. 14.4). A process similar to a pro-oxidative effect of sulfonated chitosan-1 at both concentrations can be stated. Decreases in the η of the hyaluronan solution within 5 h were 5.67 and 4.47 mPa·s for sulfonated chitosan-1 added at concentrations 20 and 40 mg/mL, respectively (left panel). As seen in right panel, for sulfonated chitosan-2 the values within 5 h were 6.32 and 5.22 mPa·s at concentrations 20 and 40 mg/mL, respectively, which means that sulfonated chitosan-2 promoted the degradation of hyaluronan more significantly than sulfonated chitosan-1.

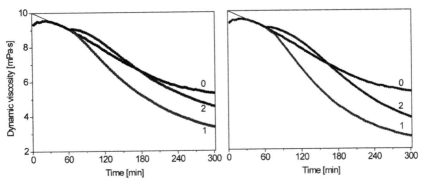

FIGURE 14.4   Time-dependent changes in dynamic viscosity of hyaluronan solution in the presence of sulfonated chitosan-1 (left panel) or sulfonated chitosan-2 (right panel) added at concentrations 20 mg/mL (curves 1) or 40 mg/mL (curves 2). Hyaluronan degradation was induced by WBOS (curves 0). Sulfonated chitosans were added to the reaction system 1 h later.

Glutathione (10, 50, and 100 μM) added to phosphonate chitosan (0.125 mg/mL) led to a dose-dependent decrease in the rate of hyaluronan degradation whereas η was lower by 3.3, 1.23, and 0.85 mPa·s, respectively (Fig. 14.5, left panel). A similar result was observed, when the phosphonate chitosan and glutathione mixture was added during hyaluronan degradation, that is, during the propagation phase (Fig. 14.5, right panel).

Addition of glutathione (10, 50, and 100 μM) into sulfonated chitosan at the higher concentration (0.25 mg/mL) enhanced hyaluronan degradation (Fig. 14.6, left panel). The value of η of the reaction mixture within 5 h was lower by 5.88, 2.66, and 2.23 mPa·s (curves 1, 2, and 3). The similar effect was reached, when the combination of sulfonated chitosan and glutathione (50 and 100 μM) was added 1 h later (right panel, curves 2 and 3). The addition of the combination of sulfonated chitosan with glutathione (10 μM) resulted in a mild pro-oxidative effect (right panel, curve 1).

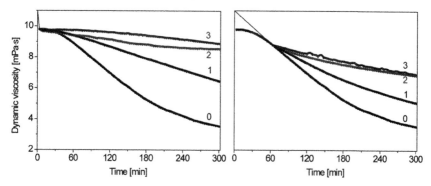

FIGURE 14.5 Time-dependent changes in dynamic viscosity of hyaluronan solution in the presence of phosphonate chitosan (0.125 mg/mL) with the addition of glutathione at concentrations: 10 μM (curve 1), 50 μM (curve 2) and 100 μM (curve 3). Hyaluronan degradation was induced by WBOS (curve 0). Phosphonate chitosan with glutathione were added to the reaction system before hyaluronan degradation begins (left panel) and 1 h later (right panel).

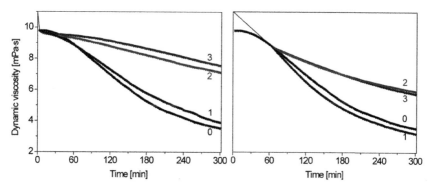

FIGURE 14.6 Time-dependent changes in dynamic viscosity of the hyaluronan solution in the presence of phosphonate chitosan (0.25 mg/mL) with the addition of glutathione at concentrations: 10 μM (curve 1), 50 μM (curve 2) and 100 μM (curve 3). Hyaluronan degradation was induced by WBOS (curve 0). Phosphonate chitosan with glutathione were added to the reaction system before hyaluronan degradation begins (left panel) and 1 h later (right panel).

In the measuring material, the hydrogen atoms coming from unsubstituted —NH₂ or substituted —NH— groups and those of —OH groups, which spread along the backbone of the sulfonated chitosan, along with the presence of sulfonic groups increased the acidity of the sulfonated chitosan, thus enhancing its self-gelation *via* intramolecular hydrogen bonds. Simultaneously, the acidic nature of sulfonic groups enhanced free radical degradation of hyaluronan by WBOS.

As seen in Figure 14.7, left panel, curve 2, the addition of glutathione (10 µM) into sulfonated chitosan-1 moderately inhibited hyaluronan degradation and only a negligible decrease in the η of the hyaluronan solution was observed in the combination with glutathione at 100 µM concentration (curve 3). A pro-oxidative effect was observed when glutathione (1 µM) was admixed to sulfonated chitosan-1 (left panel, curve 1).

On the other hand, when adding glutathione (100 µM) to sulfonated chitosan-2, no protection against hyaluronan degradation was monitored and sulfonated chitosan-2 with lower concentrations of glutathione served as a pro-oxidant (right panel, curves 1 and 2).

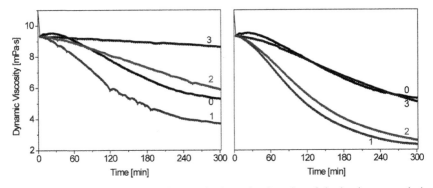

FIGURE 14.7 Time-dependent changes in dynamic viscosity of the hyaluronan solution in the presence of sulfonated chitosan-1 (left panel) or sulfonated chitosan-2 (right panel) (0.125 mg/mL) with the addition of glutathione at concentrations: 1 µM (curve 1), 10 µM (curve 2) and 100 µM (curve 3). Hyaluronan degradation was induced by WBOS (curve 0). The mixture was added to the reaction system before hyaluronan degradation begins.

As seen in Fig.14.8, left panel the combination of sulfonated chitosan-1 and the lowest concentration of glutathione added to the hyaluronan reaction mixture 1 h later was shown to have a pro-oxidative effect. The decrease in η of hyaluronan solution within 5 h was 5.67 mPa·s (curve 1). A preventive action against forming alkoxy- and peroxy-type hyaluronan *C*-macroradicals was observed in part only for a combination of glutathione (100 µM) and sulfonated chitosan-1. In this case of the hyaluronan solution within 5 h was lower by 1.63 mPa·s. On the other hand, the curve 2 was almost identical to the reference (curve 0).

The addition of glutathione into sulfonated chitosan-2 (right panel) caused a dose-dependent pro-oxidative effect, especially at glutathione concentrations 1 and 10 μM (curves 1 and 2).

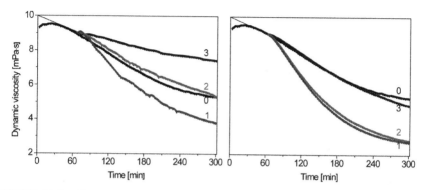

FIGURE 14.8 Time-dependent changes in dynamic viscosity of the hyaluronan solution in the presence of sulfonated chitosan-1 (left panel) and sulfonated chitosan-2 (right panel) at the concentration 0.125 mg/mL with the addition of glutathione at concentrations: 1 μM (curve 1), 10 μM (curve 2) and 100 μM (curve 3). Hyaluronan degradation was induced by WBOS (curve 0). The mixture was added to the reaction system 1 h later.

As evident in Scheme 3, at the initiation period of hyaluronan degradation just first 1 μM of ascorbate reacts from its total concentration 100 μM. The formed product is first 1 μM of H_2O_2, because in the reactive system in a given time just 1 μM of Cu(II) ions participates in the reaction.

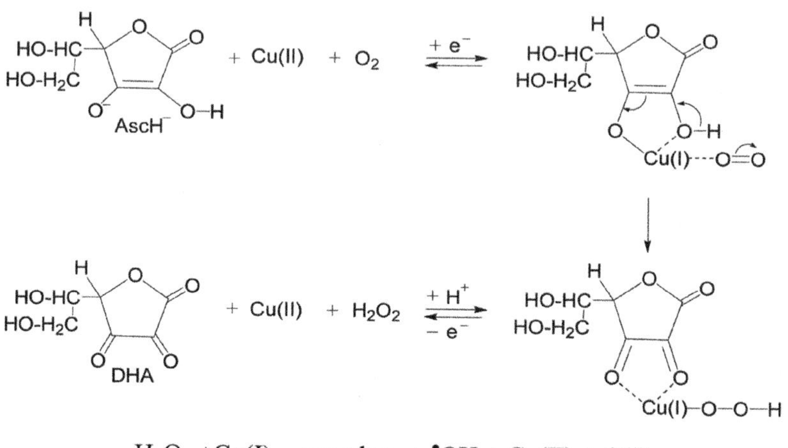

$$H_2O_2 + Cu(I)\text{---complex} \rightarrow {}^\bullet OH + Cu(II) + OH^-$$

SCHEME 3 The Weissberger biogenic oxidative system.

The formed amount of hydrogen peroxide is due to the presence of the intermediate complex Cu(I) immediately decomposed to form maximally 1 µM of ·OH. However, liberated Cu(II) ions with the rest of the unreacted ascorbate (99 µM) immediately form next 1 µM of hydrogen peroxide, followed again by formation of 1 µM ·OH. However, since at the beginning of the reaction within the system of reactants there is an ascorbate excess, the generated nascent ·OH radicals are continually scavenged as shown in findings based on spin-trapping EPR spectroscopy.[39] Simultaneously, a trace fraction of nascent (not scavenged) ·OH radicals can however react with the hyaluronan macromolecule by abstracting H· radical, resulting in the formation of a hyaluronan C-macroradical, further denoted as A·. Hyaluronan in the following reactions is denoted as HA.

$$HA + ·OH \rightarrow A· + H_2O \tag{14.1}$$

Under aerobic conditions, the alkyl-type macroradical A· reacts rapidly with the molecule of dioxygen to form a peroxy-type C-macroradical, hereafter denoted as AOO·. The formed intermediate AOO· may react with an adjacent hyaluronan macromolecule—and in that way the radical chain reaction propagates rapidly:

$$AOO· + HA \rightarrow AOOH + A·, \text{ propagation of the radical chain reaction} \tag{14.2}$$

After "collision" with hyaluronan macromolecule [cf. reaction (14.2)], the generated peroxy-type C-macroradical yields a high-molecular-weight hydroperoxide (AOOH), which subsequently, in the presence Cu(I) ions can yield an alkoxy-type C-macroradical (AO·)

$$\{AOOH + Cu(I) \rightarrow AO· + HO^- + Cu(II)\} \tag{14.3}$$

This macro-radical (AO·) is the assumed intermediate of the main chain-cleavage resulting in biopolymer fragments. The solution is characterized by a reduced viscosity.[40]

14.4.2 ABTS ASSAY

Data in Table 14.1 show the ability of phosphonate chitosans alone and in the presence of glutathione to reduce ABTS·⁺. The percentages of reduced ABTS·⁺ after 10 min were 47.6, 39.2, and 20% when phosphonate chitosan was examined at 0.5, 1.0, and 2.5 mg/mL concentrations. The addition of glutathione at different concentrations 40, 200, or 400 µM to phosphonate chitosan had a positive effect on reduction of ABTS·⁺. The least unreacted

ABTS˙, that is, 5.5 % was determined when using phosphonate chitosan (1.0 and 2.5 mg/mL) with glutathione (400 µM).

TABLE 14.1 Percentage of Unreacted ABTS˙⁺ After Addition of Phosphonate Chitosan in the Absence and Presence of Glutathione Determined After 10 min.

	Glutathione [µM]			
	0	40	200	400
Phosphonate chitosan (0.5 mg/mL)	47.6	40.4	9.9	8.2
Phosphonate chitosan (1.0 mg/mL)	39.2	29.9	7.7	5.5
Phosphonate chitosan (2.5 mg/mL)	20.0	20.4	6.1	5.5

Table 2 shows the percentage of reducing ABTS˙⁺ with sulfonated chitosan-1 or -2 in the absence and presence of glutathione. The amount of unreacted ABTS⁺ after 10 min-measurement was 85.4 and 84.9% for sulfonated chitosan-1 and sulfonated chitosan-2, respectively. The percentage of reduction of ABTS˙⁺ was triggered after addition of glutathione to both sulfonated chitosans. A mixture of sulfonated chitosan-2 with glutathione was shown to be less effective in reducing ABTS˙⁺ than the mixture sulfonated chitosan-1 with glutathione.

TABLE 14.2 Percentage of Unreacted ABTS˙⁺ After Addition of Sulfonated Chitosans the Absence and Presence of Glutathione Determined After 10 min.

	Glutathione [µM]			
	0	4.0	40	400
Sulfonated chitosan-1 (0.125 mg/mL)	85.4	75.7	33.6	12.5
Sulfonated chitosan-2 (0.125 mg/mL)	84.9	77.9	46.9	23.5

Regarding to the effect of glutathione itself, after a 10 min measurement 4.3% of ABTS˙⁺ remained unreacted after using glutathione (100 µM) as a scavenger (not published).

ABTS⁺, being a singly positive charged cationic radical, is soluble in both aqueous and organic solvents and is not affected by ionic strength, so it has been used in multiple media to determine both hydrophilic and lipophilic antioxidant capacity. ABTS˙⁺ exhibits a bluish-green color with maximum absorbance values at approximately 730 nm. The assay uses intensely colored cation radicals of ABTS˙⁺ to test the ability of antioxidants to quench radicals, the color rapidly decreased after addition of a tested antioxidant, which acts like a donor of electron.[41]

$$ABTS^{\cdot+} + e^- \quad \rightarrow \quad ABTS$$

Bluish-green　　　Colorless

14.5　CONCLUSION

This study proved that glutathione itself at a concentration range $1-100$ μM inhibited hyaluronan degradation induced predominantly by hydroxyl radicals. Glutathione at concentrations 100 and 10 μM retarded hyaluronan degradation when added 1 h after the onset of the degradation reaction. Both sulfonated chitosans and phosphonate chitosan promoted degradation of hyaluronan. The addition of glutathione to sulfonate or phosphonate chitosan resulted in a more significant protective effect against hyaluronan degradation compared to chitosan derivatives themselves. The results of the ABTS assay showed that all chitosan derivatives negligibly reduced $ABTS^+$. In contrast, the combination of glutathione with an individually added chitosan derivative led to its higher radical scavenging capacity.

ACKNOWLEDGMENTS

The study was supported by the grants VEGA 2/0065/15 and APVV-15-0308.

KEYWORDS

- ʟ-glutathione
- glycosaminoglycans
- water treatment
- hyaluronan solution
- free radical scavenging activity

REFERENCES

1. Jefferies, H.; Coster, J.; Alzan, K.; Bot, J.; Mccauley, D. R.; Hall, J. C. Glutathione. *ANZ. J. Surg.* **2003,** *73*(7), 517–522.

2. Milne, L.; Nicotera, P.; Orrenius, S.; Burkitt, M. J. Effects of Glutathione and Chelating Agents on Copper-mediated DNA Oxidation: Pro-oxidant and Antioxidant Properties of Glutathione. *Arch. Biochem. Biophys.* **1993**, *304* (1), 102–109.

3. Kidd, P. M. Glutathione: Systemic Protectant against Oxidative and Free Radical Damage. *Altern. Med. Rev.* **1997**, *2*, 155.

4. Jimenéz, I.; Speisky, H. Effects of Copper Ions on the Free-radical Properties of Reduced Glutathione: Implications of a Complex Formation. *J. Trace Elem. Med. Biol.* **2000**, *14* (3), 161–167.

5. Hultberg, M.; Hultberg, B. The Effect of Different Antioxidants on Glutathione Turnover in Human Cell Lines and Their Interaction with Hydrogen Peroxide. *Chem. Biol. Interact.* **2006**, *163*, 192–198.

6. Franco, R.; Schoneveld, O. J.; Pappa, A.; Panayiotidis, M. I. The Central Role of Glutathione in the Pathophysiology of Human Diseases. *Arch. Phys. Biochem.* 2007, *113*, 234.

7. Cuddihy, S. L.; Parker, A.; Harwood, D. T.; Vissers, M. S.; Winterbourn, C. C. Ascorbate Interacts with Reduced Glutathione to Scavenge Phenoxyl Radicals in HL60 Cells. *Free Radic. Biol. Med.* **2008**, *44* (8), 1637–1644.

8. Valachová, K.; Tamer M. T.; Šoltés, L. Comparision of Free-radical Scavenging Properties of Glutathione under Neutral and Acidic Conditions. *J. Nat. Sci. Sustain. Technol.* **2014**, *8*, 645–660.

9. Hejazi, R.; Mansoor, A. Chitosan-based Gastrointestinal Delivery Systems. *J. Control Release.* **2003**, *89*, 151–156.

10. Ahmadi, F.; Oveisi, Z.; Mohammadi Samani, S.; Amoozgar, Z. Chitosan Based Hydrogels: Characteristics and Pharmaceutical Applications. *Res. Pharm. Sci.* **2015**, *10*, 1–16.

11. Shaari, N.; Kamarudin, S. K. Chitosan and Alginate Types of Bio-membrane in Fuel Cell Application: An Overview. *J. Power Sources.* **2015**, *289*, 71–80.

12. Omer, A. M.; Tamer, T. M.; Hassan, M. A.; Rychter, P.; Mohy Eldin, M. S. Koseva, N. Development of Amphoteric Alginate/Aminated Chitosan Coated Microbeads for Oral Protein Delivery. *Int. J. Biol. Macromolec.* **2016**, *92*, 362–370.

13. Zhang, Y.; Cui, Z.; Liu, C.; Xing, W.; Zhang, J. Implantation of Nafion® Ionomer into Polyvinyl Alcohol/Chitosan Composites to Form Novel Proton-Conducting Membranes for Direct Methanol Fuel Cells. *J. Power Sources.* **2009**, *194*, 730–736.

14. Mohy Eldin, M. S.; Soliman, E. A.; Hashem, A. I.; Tamer, T. M. Antibacterial Activity of Chitosan Chemically Modified with New Technique. *Trends Biomater. Artif. Organs.* **2008**, *22*, 121–133.

15. Mohy Eldin, M. S.; Soliman, E. A.; Hashem, A. I.; Tamer, T. M. Antimicrobial Activity of Novel Aminated Chitosan Derivatives for Biomedical Applications. *Adv. Polytechn.* **2012**, *31*, 414–428.

16. Mohy Eldin, M. S.; Soliman, E. A.; Hashem, A. I.; Tamer, T. M.; Sabet, M. M. Antifungal Activity of Aminated Chitosan against Three Different Fungi Species. In *Key Engineering Materials-Current State-of-the-Art on Novel Materials;* Apple Academic Press: NJ, 2013; Vol 1, pp 515–431.

17. Khalifa, I.; Barakat, H.; El-Mansy, H. A.; Soliman, S. A. Improving the Shelf-life Stability of Apple and Strawberry Fruits Applying Chitosan-incorporated Olive Oil Processing Residues Coating. *Food Pack. Shelf Life.* **2016**, *9*, 10–19.

18. Tripathi, S.; Mehrohta, G. K.; Duta, P. K. Physicochemical and Bioactivity of Crosslinked Chitosan–PVA Film for Food Packaging Applications. *Int. J. Biol. Macromol.* **2009**, *45*, 372–376.

19. Mohy Eldin, M. S.; Hashem, A. I.; Omer, A. M.; Tamer, T. M. Preparation, Characterization and Antimicrobial Evaluation of Novel Cinnamyl Chitosan Schiff Base. *Int. J. Adv. Res.* **2015**, *3*, 741–755.

20. Mohy Eldin, M. S.; Hashem, A. I. Omer, A. M.; Tamer, T. M. Wound Dressing Membranes Based on Chitosan: Preparation, Characterization and Biomedical Evaluation. *Int. J. Adv. Res.* **2015**, *3* (8), 908–922.

21. Mohy Eldin, M. S.; Tamer, T. M.; Abu Saied, M. A.; Soliman, E. A.; Madi, N. K.; Ragab, I.; Fadel, I. Click Grafting of Chitosan onto PVC Surfaces for Biomedical Applications. *Adv. Polym. Tech.* 2015, 1–12.

22. Mohy Eldin, M. S.; Omer, A. M.; Wassel, M. A..; Tamer, T. M; Abd-Elmonem, M. S.; Ibrahim, S. A. Novel Smart pH Sensitive Chitosan Grafted Alginate Hydrogel Microcapsules for Oral Protein Delivery: I. Preparation and Characterisation. *J. Appl. Pharm. Sci.* **2015**, *7* (10), 320–326.

23. Mohy Eldin, M. S.; Omer, A. M.; Wassel, M. A.; Tamer, T. M. Abd-Elmonem, M. S; Ibrahim, S. A. Novel Smart pH Sensitive Chitosan Grafted Alginate Hydrogel Microcapsules for Oral Protein Delivery: II. Evaluation of the Swelling Behavior. *J. Appl. Pharm. Sci.* **2015**, *7* (10), 331–337.

24. Kenawy, E.; Abdel-Hay, F. I.; Mohy Eldin, M. S.; Tamer, T. M.; Ibrahim, E. M. A. Novel Aminated Chitosan-aromatic Aldehydes Schiff Bases: Synthesis, Characterization and Bio-evaluation. *Int. J. Adv. Res.* **2015**, *3* (2), 3563–572.

25. Tamer, T. M.; Valachová, K.; Mohy Eldin, M. S.; Šoltés, L. Free Radical Scavenger Activity of Chitosan and its Aminated Derivative. *J. Appl. Pharm. Sci.* **2016**, *6* (4), 195–201.

26. Tamer, T. M.; Valachová, K.; Mohy Eldin, M. S.; Šoltés, L. Scavenger Activity of Cinnamyl Chitosan Schiff Base. *J. Appl. Pharm. Sci.* **2016**, *6* (1), 130–136.

27. Valachová, K.; Tamer, T. M.; Mohy Eldin, M. S.; Šoltés, L. Radical-Scavenging Activity of Glutathione, Chitin Derivatives and Their Combination. *Chem. Pap.* **2016**, *70* (6), 820–827.

28. Adamia, S.; Maxwell, C. A.; Pilarski, L. M. Hyaluronan and Hyaluronan Synthase: Potential Therapeutic Targets in Cancer. **2005**, *5* (1), 3–14.

29. Stern, R.; Asari, A. A.; Sugahara, K. N. Hyaluronan Fragments: An Information-Rich System. *Eur. J. Cell Biol.* **2006**, *85* (8), 699–715.

30. Kogan, G.; Šoltés, L.; Stern, R.; Gemeiner, P. Hyaluronic Acid: A Natural Biopolymer with a Broad Range of Biomedical and Industrial Applications. *Biotechnol. Lett.* **2007**, *29*, 17–25.

31. Kalpakcioglu, B.; Senel, K. The Interrelation of Glutathione Reductase, Catalase, Glutathione Peroxidase, Superoxide Dismutase and Glucose-6-Phosphate Peroxidase in the Patogenesis of Rheumatoid Arthritis. *Clin. Rheumatol.* **2008**, *27* (2), 141–145.

32. Rapta, P.; Valachová, K.; Gemeiner, P.; Šoltés, L. High-molar-mass Hyaluronan Behaviour during Testing its Radical Scavenging Capacity in Organic and Aqueous Media: Effects of the Presence of Manganese(II) Ions. *Chem. Biodivers.* **2009**, *6*, 162–169.

33. Hrabárová, E.; Valachová, K.; Rapta P.; Šoltés, L. An Alternative Standard for Trolox-equivalent Antioxidant-capacity Estimation Based on Thiol Antioxidants. Comparative 2,2'-azinobis[3-ethylbenzothiazoline-6-sulfonic acid] Decolorization and Rotational Viscometry Study Regarding Hyaluronan Degradation. *Chem. Biodivers.* **2010**, *7*, 2191–2200.

34. Valachová, K.; Vargová, A.; Rapta, P.; Hrabárová, E.; Dráfi, F.; Bauerová, K.; Juránek, I.; Šoltés, L. Aurothiomalate as Preventive and Chain-breaking Antioxidant in Radical Degradation of High-molar-mass Hyaluronan. *Chem. Biodivers.* **2011**, *8*, 1274–1283.

35. Topoľská, D.; Valachová, K.; Rapta, P.; Šilhár, S.; Panghyová, E.; Horváth, A.; Šoltés, L. Antioxidative Properties of *Sambucus nigra* Extract. *Chem. Pap.* **2015**, *69*, 1202–1210.

36. Xie, W; Xu, P.; Liu, Q. Antioxidant Activity of a Water-soluble Chitosan Derivates. *Bioorg. Med. Chem. Lett.* **2001**, *11* (13), 1699–1701.

37. Yang, S.; Guo, Z.; Miao, F.; Qin, S. The Hydroxyl Radical Scavenging Activity of Chitosan, Hyaluronan, Starch and Their O-carboxymethylated Derivatives. *Carbohydr. Polym.* 2010, *82* (4), 1043−1045.

38. Rajalakshmi, A.; Krithiga, N.; Jayachitra, A. Antioxidant Activity of the Chitosan Extracted from Shrimp Exoskeleton. *Middle-East J. Sci. Res.* **2013**, *16* (10), 1446−1451.

39. Šoltés, L.; Stankovská, M.; Brezová, V.; Schiller, J.; Arnhold, J.; Kogan, G.; Gemeiner, P. Hyaluronan Degradation by Cupric Chloride and Ascorbate: Rotational Viscometric, EPR Spin Trapping, and MALDI-TOF Mass Spectrometric Investigations. *Carbohydr. Res.* **2006**, *341*, 2826–2834.

40. Rychlý, J.; Šoltés, L.; Stankovská, M.; Janigová, I.; Csomorová, K.; Sasinková, V.; Kogan, G.; Gemeiner, P. Unexplored Capabilities of Chemiluminescence and Thermoanalytical Methods in Characterization of Intact and Degraded Hyaluronans. *Polym. Degrad. Stab.* **2006**, *91*, 3174–3184.

41. Re, R.; Pellegrini, N.; Proteggente, A.; Pannala, A.; Yang, M.; Rice-Evans, C. Antioxidant Activity Applying an Improved ABTS Radical Cation Decolorization Assay. *Free Radical Biol. Med.* **1999**, *26* (9–10), 1231–1237.

INDEX

Printed and bound by CPI Group (UK) Ltd, Croydon, CR0 4YY

23/10/2024

01777701-0013